Progress in Nonhistone Protein Research

Volume I

Editor

Isaac Bekhor, Ph.D.
Professor
Laboratory for Molecular Genetics
University of Southern California
Los Angeles, California

Associate Editors

Carol J. Mirell, Ph.D.
Laboratory for Molecular Genetics
University of Southern California
Los Angeles, California

C. C. Liew, Ph.D.
Professor
Department of Clinical Biochemistry
The Banting Institute
University of Toronto
Toronto, Ontario, Canada

CRC Press
Taylor & Francis Group
Boca Raton London New York

CRC Press is an imprint of the
Taylor & Francis Group, an informa business

First published 1985 by CRC Press
Taylor & Francis Group
6000 Broken Sound Parkway NW, Suite 300
Boca Raton, FL 33487-2742

Reissued 2018 by CRC Press

Library of Congress Cataloging in Publication Data
Main entry under title:

Progress in nonhistone protein research.

Bibliography.
Includes indexes.
1. Nonhistone chromosomal proteins. I. Bekhor, Isaac. II. Mirell. Carol J. III. Liew. C. C.
[DNLM: 1. Chromosomal Proteins. Non-Histone.
QU 58 P964]
QP552.N62P76 1985 574.19'245 84-9308
ISBN 0-8493-5528-1 (v. 1)
ISBN 0-8393-5579-X (v. 2)

A Library of Congress record exists under LC control number: 84009408

Publisher's Note
The publisher has gone to great lengths to ensure the quality of this reprint but points out that some imperfections in the original copies may be apparent.

Disclaimer
The publisher has made every effort to trace copyright holders and welcomes correspondence from those they have been unable to contact.

ISBN 13: 978-1-315-89696-0 (hbk)
ISBN 13: 978-1-351-07606-7 (ebk)

Visit the Taylor & Francis Web site at http://www.taylorandfrancis.com and the
CRC Press Web site at http://www.crcpress.com

FOREWORD

Nonhistone chromosomal (NHC) proteins are assumed to include regulatory proteins which modulate genetic expression by activating or repressing specific genes. This assumption has lured many investigators to search for sequence-specific DNA-binding proteins.

At the present time, however, the field of nonhistone protein research has been sharply curtailed. Many investigators have shifted their efforts to DNA cloning and sequencing. Molecular biologists have cloned hundreds of genes in the past several years, and the data for nucleotide sequences of genes and gene fragments have grown at a dashing pace. Sequencing data continue to provide valuable insight as to how genes are structured, but have given little information as to how gene expression is manipulated. In particular, the nucleotide base sequences of gene control regions, from higher organisms, have provided little insight as to how gene control is accomplished. It is possible that DNA sequencing has been overemphasized, and that the secondary structure of DNA, which may be due to the DNA sequence itself or possibly due to the intimate association of DNA with special proteins, is responsible for the ultimate mechanism of control.

Why have so many well qualified and highly creative investigators abandoned their search for the function of nonhistones? These proteins constitute one of the least understood groups of proteins whose functions are extremely difficult to define. At one time the genome was not understood. Now many investigators may claim that by close examination of the DNA we will learn how to control the gene. How realistic is this belief? From studies on prokaryotes, it is clear that DNA-binding proteins play a vital role in determining the various functions necessary for genetic survival of that organism. Bacteria provide an excellent system for the determination of a specific genetic function of a defined protein; the genome can be modified by the production of mutants, and the functional changes can be monitored. Chromosomal proteins of higher organisms are not amenable to such a study. Clearly the struggle to find defined functions of nonhistones requires the skills of the creative investigator, who is willing to expend his efforts even though the field does not give way to easy answers and breakthroughs. Many investigators have thus been discouraged. We believe that more encouragement should be given to those who are willing to continue in this difficult area of research, and that more rapid progress would be made as a reflection of this positive attitude.

In this spirit we have asked various investigators to contribute information recently obtained in their laboratories; we have also encouraged them to express their sentiments as to how these proteins functionally interact with DNA. This undertaking has not been easy since many investigators, discouraged by the inherent difficulties in studying these proteins, have left the field. These highly heterogeneous proteins are not readily extractable with physiological buffers, nor are they easy to isolate or purify. Then, even if our efforts at purification are successful, how are we to ascertain their biological activity? Clearly, these proteins must do more than merely fill in a nuclear space. It is scientifically correct to assume that each protein must possess a specific function. What is that function? These functions must be investigated and tabulated. The task is obviously frustrating and formidable, yet exciting and promising. Resolving the questions of gene control is the reward offered to those scientists whose persistence and pioneering attitudes allow them to continue in this complex study; this is the reward offered to the scientists contributing to this volume.

These volumes are rich with information on HMG proteins (Bhorjee, Bustin, Reeck, Teller, Isackson, Cox, Manning, Teng, Teng), on specific nuclear antigens (Chiu, Epstein, Clevenger, Pettijohn, Sculley, Spelsberg), on a specific phosphoprotein (Halikowski, Zhao), on salt-soluble nonhistones (Horst, Rackowicz-Szulczyńska, Wang, Chamberlin), on salt-insoluble nonhistones (Duerre, Traynor, Leatherbury, Horst, Rackowicz-Szulczyńska, Halikowski, Zhao, Pettijohn, Wang, Chamberlin, Hentzen, Sculley, Spelsberg), on hormone interactions with DNA (Wang, Chamberlin), on nonhistones with enzyme functions (Miyazaki, Yamashita, Kihira, Hagiwara, Nagao, Yamashita, Horio, Ruchirawat, Becker, Andoh, Ishii, Nagasawa, Seno), on nonhistones binding carcinogens (Trafzer, Rho), on Epstein-Barr nuclear antigen (Sculley, Spelsberg), on nuclear matrix salt-insoluble DNA (Vogelstein, Small, Robinson, Nelkin, Pogo, Procyk, Razin, Chernokhvostov, Yarovaya, Georgiev), on nuclear matrix salt-insoluble nonhistones (Pogo, Procyk, Razin, Chernokhvostov, Yarovaya, Georgiev), on methods for analysis of nonhistones in both normal and malignant cells (Rodriguez, Ramagli, Johnston), and on the use of monoclonal antibodies to study functions of nonhistones (Epstein, Clevenger, Pettijohn). Clearly, these two volumes encompass many of the difficult areas encountered in research on nonhistones. The scope is rather broad, yet it is focused on one theme: How do these nonhistones participate in the control of gene function?

Our purpose is to encourage research on nonhistones and to stimulate the imagination of other investigators whose future efforts might result in new discoveries as to the significance of these proteins. We hope that future volumes in this series will contain data from many more laboratories, which have been "turned-on" to NHC protein research.

I. Bekhor
C. J. Mirell
C. C. Liew

THE EDITOR

Isaac Bekhor, Ph.D., is Chief, Laboratory for Molecular Genetics, University of Southern California, and Professor of Biochemistry and Molecular Biology, School of Dentistry, University of Southern California, Los Angeles, California.

Dr. Bekhor graduated from the University of California, Los Angeles, with a B.S. in Chemistry, and obtained his Ph.D. in 1966 in Biochemistry, University of Southern California, Los Angeles, where the Graduate School presented him with the Harry J. Deuel, Jr. Award in Biochemistry in recognition of outstanding research. Between 1966 and 1969 Dr. Bekhor was a postdoctoral Fellow at the California Institute of Technology, Pasadena. In 1970 he joined the University of Southern California School of Dentistry Faculty. In 1971 he won a 5-year Career Development Award from the National Institutes of Health.

Dr. Bekhor is a member of the American Association for The Advancement of Science, American Chemical Society, American Society for Cell Biology, International Biochemical Society, International Society for Developmental Biologists, Southern California Craniofacial Group and West Coast Chromatin Group. He has been a consultant to *Industry on Enzyme Technology* for over 15 years.

His major research interest is the participation of nonhistone chromosomal proteins in gene regulation in eukaryotes. He has published more than 50 papers in *Molecular Biology*, and is currently cloning specific genes to facilitate his studies on the possible regulatory roles of specific nonhistones.

CONTRIBUTORS

Toshiwo Andoh
Professor
Department of Hygienic Chemistry
Meiji College of Pharmacy
Tokyo, Japan

Frederick F. Becker, M.D.
Professor of Pathology
The University of Texas System Cancer Center
M. D. Anderson Hospital and Tumor Institute
Houston, Texas

Isaac Bekhor, Ph.D.
Professor
Laboratory for Molecular Genetics
University of Southern California
Los Angeles, California

Jen-Fu Chiu, Ph.D.
Associate Professor
Department of Biochemistry
College of Medicine
University of Vermont
Burlington, Vermont

Charles V. Clevenger
M.D.-Ph.D. Program
Northwestern University
Chicago, Illinois

John A. Duerre, Ph.D.
Professor
Department of Microbiology
School of Medicine
University of North Dakota
Grand Forks, North Dakota

Alan L. Epstein, M.D., Ph.D.
Assistant Professor
Department of Pathology
University of Southern California
Los Angeles, California

Hideaki Hagiwara, Ph.D.
Postdoctoral Fellow
Department of Biology
University of California, San Diego
La Jolla, California

Marvin J. Halikowski, M.Sc.
Graduate Student
Department of Clinical Biochemistry
The Banting Institute
University of Toronto
Toronto, Ontario, Canada

Peter C. Hentzen, Ph.D.
Postdoctoral Fellow
Laboratory for Molecular Genetics
University of Southern California
Los Angeles, California

Takekazu Horio, Ph.D.
Professor
Institute for Protein Research
Osaka University
Osaka, Japan

Antoni Horst, M.D., Ph.D.
Professor
Institute of Human Genetics of the Polish Academy of Sciences
Poznan, Poland

Yasunori Kihira, Ph.D.
Research Associate
School of Dentistry
Tokushima University
Tokushima, Japan

Kazuyuki Kshii
Research Associate
Department of Hygienic Chemistry
Meiji College of Pharmacy
Tokyo, Japan

Jean-Numa Lapeyre, Ph.D.
Assistant Professor
The University of Texas System Cancer Center
M. D. Anderson Hospital and Tumor Institute
Houston, Texas

Keith C. Leatherbury
Medical Student
School of Medicine
University of North Dakota
Grand Forks, North Dakota

C. C. Liew, Ph.D.
Professor
Department of Clinical Biochemistry
The Banting Institute
University of Toronto
Toronto, Ontario, Canada

Kaoru Miyazaki, Ph.D.
Research Associate
Institute for Protein Research
Osaka University
Osaka, Japan

Yoshinobu Nagao, Ph.D.
Research Associate
Department of Biochemistry
Kanazawa Medical University
Ishikawa, Japan

Kinzo Nagasawa
Professor
Department of Hygienic Chemistry
Faculty of Pharmaceutical Sciences
Kitasato University
Tokyo, Japan

David E. Pettijohn, Ph.D.
Professor
Department of Biochemistry/Biophysics/Genetics
University of Colorado Health Sciences Center
Denver, Colorado

Ewa Rakowicz-Szulczńska, Dr. Rer. Nat., Ph.D.
Institute of Human Genetics of the Polish Academy of Sciences
Poznan, Poland

Mathuros Ruchirawat
Project Investigator
The University of Texas System Cancer Center
M.D. Anderson Hospital and Tumor Institute
Houston, Texas

Nobuko Seno
Professor of Biochemistry
Department of Chemistry
Ochanomizu University
Tokyo, Japan

Michael D. Traynor
Medical Student
School of Medicine
University of North Dakota
Grand Forks, North Dakota

Jinpei Yamashita, Ph.D.
Associate Professor
Institute for Protein Research
Osaka University
Osaka, Japan

Nobuhiko Yamashita
Institute for Protein Research
Osaka University
Osaka, Japan

Ming-Shum Zhao, M.D.
Assistant Professor
Department of Biochemistry
Chinese Academy of Medical Sciences
Beijing, People's Republic of China

TABLE OF CONTENTS

Volume I

Chapter 1

THE STRUCTURAL AND FUNCTIONAL ROLES OF SPECIFIC NONHISTONE PROTEIN FRACTIONS IN CHROMATIN

Jen-Fu Chiu

TABLE OF CONTENTS

I. INTRODUCTION

All somatic cells within a given organism are endowed with the same genetic information. During cytodifferentiation, however, considerable segments of cellular DNA are transcriptionally inactivated and only a small portion of the genome remains active in the process of RNA transcription. The mechanism of gene selection during differentiation is still not clear. Accumulated experimental evidence suggests that chromosomal proteins play an important role in the process of cellular differentiation, oncogenesis, and genetic regulation.[1-5]

In the cell nucleus, DNA interacts with histones and nonhistone proteins to form chromatin, which is the principal site of the genetic memory and its expression. The basic chromatin fiber is now believed to consist of a chain of repeating subunits, nucleosomes, each made up of about 145 to 200 base pairs (bp) of DNA wrapped around a core of eight of the small histones (two each of H2A, H2B, H3 and H4).[6-10] Histone H1 and nonhistone chromosomal proteins apparently interact with the outer surface of the DNA-histone bead.[7,10] Obviously, the basic chromatin fiber must be folded or organized into a higher-order structure.[7,10] The detailed architecture is not yet understood.

The organization of chromatin in the nucleus is such that biological processes (that is, replication or transcription) are unlikely to occur without some degree of disruption of the overall structure, or even nucleosome structure. Recent comparisons between the structure of active and inactive genes have shown that active genes probably have a more open structure than those that are inactive.[11-13] Weintraub and Groudine[11] first reported that the active globin gene in chick reticulocyte nuclei was preferentially susceptible to DNase I digestion. This differential sensitivity of active loci has been confirmed using several other systems.[12-16] Although the type of structural alteration leading to DNase I sensitivity has not yet been clearly identified, this differential susceptibility of active genes must reflect a change in conformation or structure at some level. What kind of structure could confer such sensitivity? Protein-free DNA might be digested preferentially. However, the same sequences that are digested by DNase I are resistant to staphylococcal nuclease in a way that indicates that proteins are present. The former observations also show that the DNase I sensitive regions are also packaged with proteins in a nucleosome-like structure.[17-19] To reconcile this conclusion with the observed sensitivity to pancreatic DNase, it is logical to suggest that some of the nonhistone proteins may play important determinative and structural roles in this process.

The sensitivity of pancreatic DNase digestion in active chromatin could arise not only from an alteration within each core particle through the chemical modification of histones,[20,21] but could also reflect a perturbation of the higher-order packing of the nucleosomes, perhaps connected with loss or modification of histone H1 molecules and nonhistone proteins. In fact there is some evidence that indicates that the higher-order structure may be altered in active genes.[22] Certain nonhistone proteins have been shown to associate with activation of gene expression. Weintraub and his associates[23,24] have demonstrated that when the two nuclear proteins HMG-14 and HMG-17 are selectively eluted from chicken embryo erythrocytes, preferential DNase I sensitivity of the globin gene is lost, and when HMG-14 and -17 are reconstituted to HMG-depleted chromatin, the DNase I sensitivity of the globin is restored. Moreover, when avian erythrocyte chromatin is subjected to a limited digestion with DNase I, the HMG proteins are selectively released.[25] This does not occur when erythrocyte chromatin is treated with staphylococcal nuclease. No alteration in the DNase I sensitivity of the ovalbumin gene, which is not expressed in these cells, is noted with removal or readdition of the HMG proteins. This appears to be a general phenomenon in that the DNase I sensitivity of most actively transcribed genes appears to be associated with binding of HMG-14 and -17 to chromatins.[24] This specificity is not maintained, however, when the HMG proteins are reassociated with pure DNA, suggesting that their binding requires the conformation provided by chromatins or nucleosomes.

Recently, Weisbrod and Weintraub[26] used an affinity column containing agarose cross-linked with HMG-14 and HMG-17 to isolate actively transcribed chromatin subunits (nucleosomes) and to confirm that HMG-14 and -17 bind specifically to active chromatin. Although the HMG proteins have been linked with active gene expression of cell-specific protein products, the ubiquitous presence of HMG-14 and -17 in all somatic cells and the requirement of special conformation of chromatins for their binding implies that additional nonhistone proteins may serve to modulate the gene activities during cellular maturation and differentiation.

Our laboratory has been interested in the identification of cell-specific nonhistone proteins as markers for studying the structural and functional roles of nonhistone proteins in chromatin during cell differentiation and transformation. Nevertheless, attempts to identify and characterize the specific nonhistone protein fractions have been hampered due to their extreme insolubility and heterogeneity. To circumvent these problems, we have produced antibodies against nonhistone proteins and their specific conformation in chromatin. Using these antibodies we have probed the cell- or tissue-specific organization of chromatins and the composition of nonhistone proteins in various cell and tissue systems. Immunological techniques are useful as analytical tools for comparative purposes in that they enable comparisons of large numbers of proteins in a relatively convenient way. Specific antibodies can be produced to obtain information on the *in situ* organization of specific chromosomal components in native chromatin without markedly altering the structure of chromatin and chromosomes;[27-29] to estimate sequence homologies between related chromosomal components derived from various sources,[30-33] to detect structural alterations associated with functional changes in the genome;[34-36] and to visualize the location of a particular modified base or protein in metaphase or polytene chromosomes.[34,35,37]

We have been using immunological techniques to analyze the changes of chromosomal nonhistone proteins and their complexes with DNA in chromatins during cellular differentiation and neoplastic transformation. Our selection of immunological specificity as the principal assay procedure is based on evidence that there are immunospecific nonhistone protein species in chromatin which interact with homologous DNA in a highly specific manner. The immunospecificity of chromosomal nonhistone protein-DNA complexes changes during cellular differentiation.[38-44] In this chapter, some of the recent work on the changes of nonhistone chromosomal proteins and chromatin structures in differentiating erythroid cells and in rat colon and liver during chemical carcinogenesis are reviewed.

II. NONHISTONE PROTEINS AND CHROMATIN STRUCTURE IN CHICKEN ERYTHROID CELLS

Programmed, spatiotemporal expression of cellular genes forms the basis for differentiation and organogenesis in higher eukaryotes. Since chromosomal nonhistone proteins are believed to play essential roles in genetic expression and regulation, considerable effort has been expended to determine the role of various chromosomal proteins in maintenance of chromatin structure and to relate this structure to the expression of specific genes.

The maturation of the avian erythrocyte involves a series of nuclear and cytoplasmic events which lead to an almost complete cessation of metabolic function.[45-51] During the course of differentiation, the nucleus of the reticulocyte becomes pycnotic with condensed chromatin.[45,49,50] DNA and RNA synthesis in these nuclei is severely diminished.[45-48] This suggests that the basic transcriptional conformation of the DNA in chromatin has been changed. An examination of the changes in nuclear proteins and chromatin structure might give insight into the nature of the mechanisms involved.

We are exploring this further by examining the extent of species specificity of nonhistone proteins, as well as specificity for different stages of erythroid cell development.[52,53] We

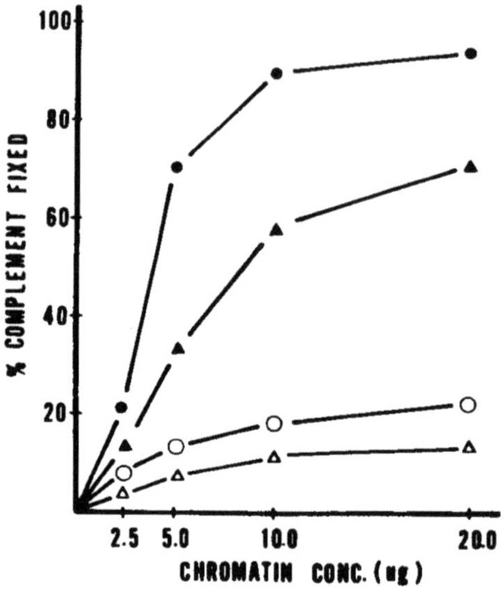

FIGURE 1. Extent of species specificity of the chicken reticulocyte antigen. Chicken reticulocytes (●); goose erythrocytes (▲); frog erythrocytes (○); and DMSO-treated Murine Friend erythroleukemia cells (△).

have produced antibodies against chicken reticulocyte dehistonized chromatin. Examination of immunological reactivity and accessibility of differentiating erythroid chromatin revealed information about the alteration of chromatin configuration during maturation of chick erythroid cells.

A. Specificity of Antigenic Nonhistone Proteins

Reticulocyte chromatin was dehistonized by the method of Chytil and Spelsberg[38] and used to induce specific antibodies in rabbits. The chromatin is dehistonized because the presence of large quantities of histone could induce preferential antibodies to histones. This is especially important for us to avoid since histone H5 is avian erythrocyte-specific.

By using the microcomplement fixation test the antibodies react strongly with reticulocyte chromatin, react very slightly with chicken erythrocyte chromatin, and have essentially no reaction when tested against chicken liver chromatin, chromatin from large bowel transplantable tumors in rats, or chromatin isolated from transplantable Morris hepatoma 7777 in rats.[52] The reticulocyte chromatin always has three times as much antibody binding activity as the erythrocyte chromatin.

The specificity of chicken erythroid cell chromatin was compared with goose and frog chromatins and mouse erythroleukemia cell chromatin to determine species specificity. As illustrated in Figure 1 only the goose, which is closely related to the chicken, shows cross-reactivity. The nuclear antigens were not detected immunologically in frog nucleated erythrocyte chromatin and dimethylsulfoxide (DMSO).

B. Changes in Immunoreactivity of Nonhistone Proteins During Erythroid Cell Differentiation

In normal adult chickens, mature, genetically inactive erythrocytes represent over 99% of the total circulating red-cell population.[46] After chemically induced anemia, the value drops to about 3 to 5% as the mature erythrocyte is replaced by a population of reticulocytes. These cell types vary in their capacity for hemoglobin synthesis from essentially no synthesis in small erythroblasts to a peak of synthetic activity in the reticulocyte and then to a decline

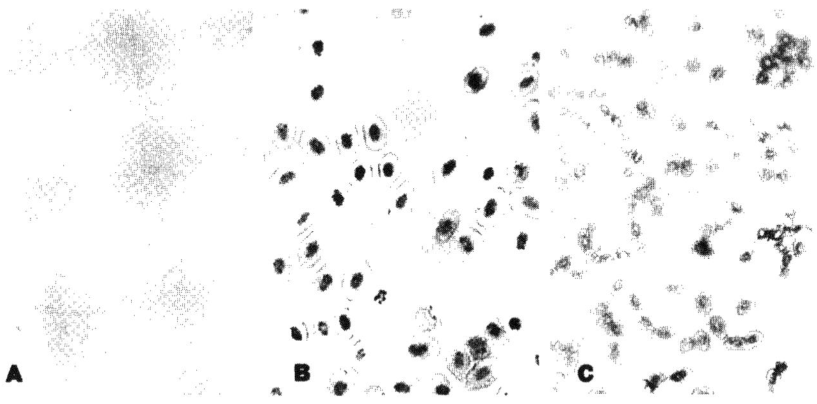

FIGURE 2. Horseradish peroxidase localization of chicken reticulocyte nuclear antigen. All localizations were performed on blood smears which were pretreated to remove endogenous peroxidases before being exposed to antibody made against chicken reticulocyte nuclear antigen. (A) Intact chicken erythrocytes; (B) intact chicken reticulocytes; (C) dextran sulfate-treated chicken erythrocytes. (Magnification × 960.)

in synthesis in the late erythrocyte.[54] Gene inactivation is relatively rapid. We found that the immunological activity and reticulocyte population of chicken erythroid cells increased as phenylhydrazine was applied to the chicken. The withdrawal of phenylhydrazine from experimental animals resulted in the gradual loss of the reticulocyte-specific antigen from the erythroid cell chromatin.[53]

Another approach used to identify the presence of reticulocyte-specific antigens was the horseradish peroxidase bridge technique. The results of the antigen localization procedure are shown in Figure 2. This figure clearly shows that the antigens are accessible to antibodies in the nuclei of reticulocytes (Figure 2B), but not in their cytoplasm nor in the nuclei or cytoplasm of intact erythrocytes (Figure 2A). The chromatin of nucleated erythroid cells becomes increasingly condensed as maturation progresses and transcriptional activity of the cell declines to a low level. This condensation probably causes a relatively nonspecific repression of all genetic activity in these metabolically inert cells. The apparent loss of reticulocyte-specific antigens from erythroid cell chromatin during maturation may be due to structural hindrance caused by condensation of the chromatin. We reasoned that if our ideas concerning a structural masking of the nuclear antigen were correct, it should be possible to expose the antigens by treating otherwise intact erythrocytes with dextran sulfate. The dextran sulfate should induce a decondensation of the chromatin within the erythrocyte,[55] allowing the antigens to become available for binding with antibodies which would be visible with the horseradish peroxidase bridge technique. Erythrocyte blood smears were incubated in a moist chamber with a solution of dextran sulfate (400 $\mu g/m\ell$). After the incubation the treated erythrocytes were spread in the same manner as the reticulocytes and intact erythrocytes. Figure 2C illustrates that the dextran sulfate treatment allowed the antibodies to bind nuclear antigens. Although the level of binding is still less than the amount found in chicken reticulocyte, it is considerably greater than that of the intact erythrocytes in Figure 2A.

C. Changes in Chromatin Structure During Erythroid Cell Differentiation

As mentioned in the Introduction, the exceptional sensitivity and selectivity of antigen-antibody interactions has the potential of becoming a major tool in studies of chromatin function and structure. We therefore used the antibodies against chicken reticulocyte dehistonized chromatin to further explore the changes in chromatin structure during erythroid cell differentiation.

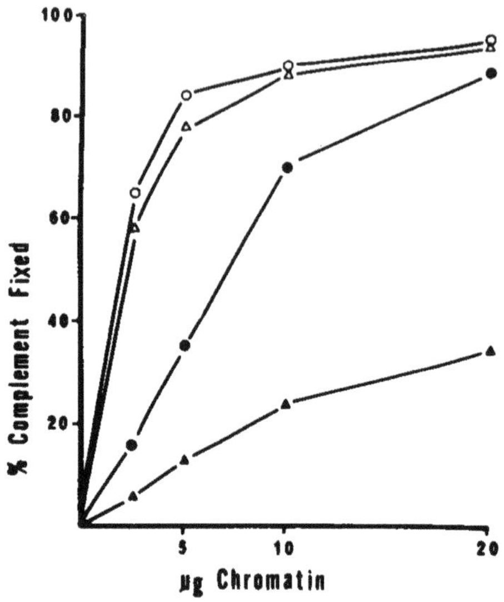

FIGURE 3. Effect of sonication on the accessibility of erythroid nuclear antigens in the complement-fixation test with antibodies to dehistonized chicken reticulocyte chromatin. Sonicated reticulocyte chromatin (○); sonicated erythrocyte chromatin (△); intact reticulocyte chromatin (●); intact erythrocyte chromatin (▲).

The data presented in Figure 2 indicate that the antigen complexes could actually be present in both reticulocytes and erythrocytes, but that chromatin condensation causes a structural masking of the complexes which coincides with the known decline in transcriptive activity of the erythrocyte. However, when the chromatin is no longer intact (that is, after it has been exposed to dehistonization of sonication) it is not possible to distinguish between reticulocyte and erythrocyte chromatins. Our earlier experimental data demonstrated that following dehistonization, both reticulocyte and erythrocyte chromatins have a higher complement-fixing capacity than intact reticulocyte chromatin.[52] It was found that 2.5 μg of dehistonized chromatin from either erythrocyte or reticulocyte (measured as DNA) fixes four times as much complement as 2.5 μg of intact reticulocyte chromatin.

Almost identical results are achieved when the chromatins from both types of blood cells are mechanically sheared by sonication. The results are shown in Figure 3. Again, the intact chromatins show an increase in complement-fixing activity over the standard level of reticulocytes. These data are similar to that of chromatin dehistonization; the immunoactivity of the disrupted chromatin from either reticulocytes or erythrocytes is almost fourfold higher than that of intact reticulocyte chromatin.

To further confirm the effect of chromatin condensation during erythroid maturation on its immunological activity, erythrocyte chromatin was pretreated with several polyanions which should induce a decondensation of the chromatin within the erythrocyte.[55] As expected, all polyanions (i.e., dextran sulfate, heparin, and phenylhydrazine) enhanced the immunological activity of erythrocyte chromatin making it similar to that of reticulocyte chromatin.[53] These data strongly suggest that the reticulocyte nuclear antigens are present in the erythrocyte but are masked in some way by the structural complexity of the intact erythrocyte. The erythrocyte chromatin is very condensed relative to the reticulocyte chromatin. When the native structure of the chromatin is drastically altered by removing the histones, sonication, or decondensation by polyanions, the nuclear antigenic protein-DNA complexes present cannot be structurally hidden, hence the increase in measurable immunoactivity.

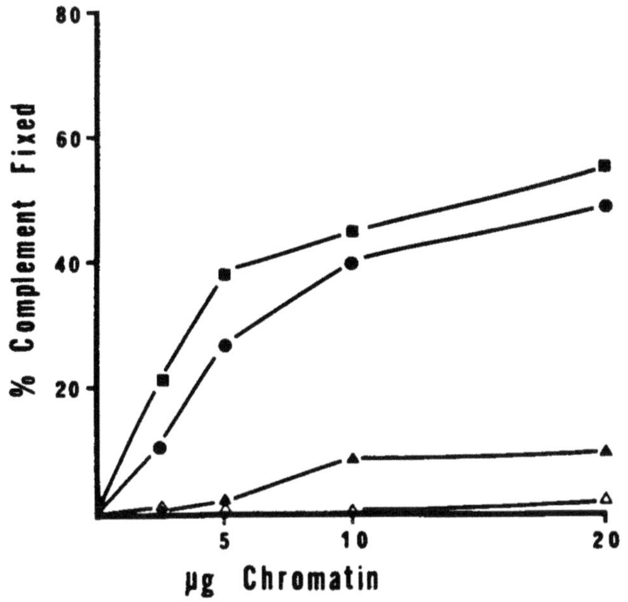

FIGURE 4. Nuclear antigenic activity in sodium dodecyl sulfate extractable (tightly bound to DNA) nonhistone chromosomal proteins from chicken reticulocytes. Proteins were reconstituted to chicken DNA before being assayed. Tightly bound nonhistone chromosomal proteins (■); whole chicken reticulocyte chromatin (●); whole erythrocyte chromatin (▲); chicken liver chromatin (△).

D. Erythroid-Specific Nuclear Antigen and Globin Gene Binding Protein

We have found that the nonhistone proteins required for the specificity of the antigenic nonhistone protein-DNA complexes are very tightly bound to DNA. We extracted reticulocyte chromatin with 2.5 *M* NaCl-5 *M* urea to remove histone proteins and the majority of nonhistone proteins. We extracted the remainder of the tightly bound nonhistones with 2% $NaDodSO_4$.[56] Very little evidence of antigenic activity was found in the high salt-urea extractable fractions; however, as shown in Figure 4, the fraction extracted with $NaDodSO_4$ binds slightly more complement than reticulocyte chromatin. In order to obtain immunoactivity, we removed $NaDodSO_4$ from the proteins by column chromatography through Dowex® AG1-X2 resin[57] since the presence of $NaDodSO_4$ can interfere with the complement-fixation test. We reconstituted the proteins to chicken DNA before testing in the microcomplement fixation test because protein or DNA alone does not have immunoactivity; protein-DNA complexes are required.[53]

Recently Bekhor and Mirell[58] have reported a class of nonhistone proteins from chicken erythroid cells which remain bound to DNA in 2 *M* NaCl. They postulated that these proteins include globin regulators. These proteins are only present in erythrocyte and reticulocyte cells, but not in other tissues. This protein fraction is specially localized on the globin gene DNA sequence.[58] These findings led to speculation that the erythroid-specific nuclease antigens may be the same class of nonhistone proteins reported by Bekhor and associates. Pumo et al.[59] therefore compared the immunoactivity and gel patterns of these two fractions of nonhistone proteins.

As illustrated in Figure 5, the results of microcomplement-fixation tests comparing the reactivity of whole chicken reticulocyte chromatin with the nonhistone protein tightly bound to DNA isolated by the procedure of Bekhor and Mirell.[58] The level of activity per microgram of chromatin (measured as DNA in 5 *M* urea) is similar in both the standard reticulocyte

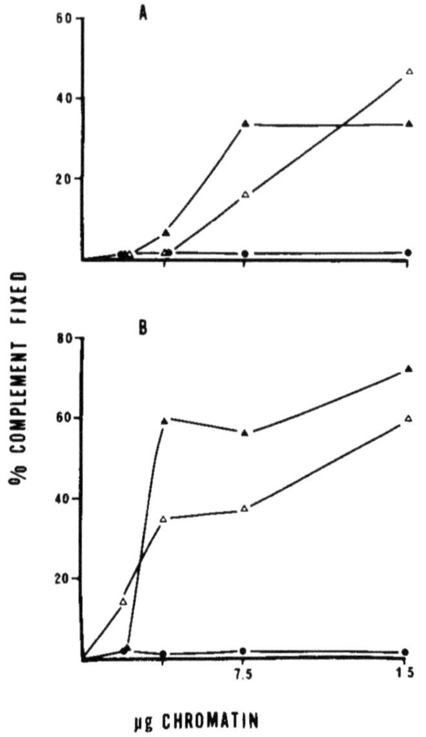

FIGURE 5. Complement fixation tests to determine the amount of chicken reticulocyte nuclear antigen present in nonhistone tightly bound to DNA isolated by the procedure of Bekhor and Mirell.[58] (A) Complement fixation test with antiserum prepared according to Chytil and Spelsberg[38] used at a 1:150 dilution; (B) complement-fixation test with antisera from rabbit injected with nonhistone protein-DNA complex after treatment with 50 m*M* Tris (pH 8.0), 2.5 *M* NaCl, 5 *M* urea. The dilution of antiserum is 1/200. Reticulocyte chromatin (▲); tightly binding nonhistone chromatin proteins (△); erythrocyte chromatin (●).

chromatin and the Bekhor and Mirell fraction. The antisera for the experiment in Figure 5A were raised against dehistonized chicken reticulocyte chromatin. The antisera for the experiment in Figure 5B were raised against tightly bound nuclear protein-DNA complexes. The tightly bound nuclear protein-DNA complexes were prepared by extracting chromatin with 50 m*M* Tris-HCl (pH 8.0) — 2.5 *M* NaCl — 5 *M* urea to remove most histones and nonhistone proteins.[60] The residue was used as tightly bound nuclear protein-DNA complexes. As shown in Figure 5, both fractions of whole chromatin and tightly bound nuclear protein-DNA complexes have similar antigenic characteristics.

Figure 6 shows the gel electrophoresis profile comparing the tissue-specific immunogenic nuclear proteins and the protein fractions obtained following the procedure of Bekhor and Mirell.[58] The immunogenic nuclear proteins were isolated from chromatin previously treated with 2.5 *M* NaCl — 5 *M* urea-50 m*M* Tris (pH 8.0) to remove most of the nuclear proteins which are not immunogenic. The chromatin residue was extracted with 2% sodium dodecyl sulfate (SDS) to isolate nuclear antigen by following the procedure of Elgin and Bonner.[56] As shown in Figure 6, there are some differences in relative quantities of some of the protein bands but all major species are present in each sample.

In order to further compare the similarity between erythroid immunogenic nuclear nonhistone proteins, the immunoblotting technique[61,62] was applied. Chromosomal nonhistone

FIGURE 6. Polyacrylamide gel electrophoretic profiles of chicken reticulocyte nonhistone proteins which remained tightly bound to DNA during two different isolation procedures. All major bands are present in both samples. (A) Proteins removed from DNA by extraction with 2% sodium dodecyl sulfate (12); (B) proteins removed from DNA by DNase I (3.1.4.5) digestion of the nonhistone protein-DNA complex described by Bekhor and Mirell.[58]

proteins of Bekhor's fractions, reticulocyte chromatin, and erythroid chromatin were resolved by polyacrylamide gel electrophoresis, transferred to nitrocellulose sheets, reconstituted to DNA, and antigens were localized immunochemically. We found that some immunogenic proteins were present in Bekhor's fraction of nonhistone proteins.

III. IMMUNOSPECIFIC NUCLEAR ANTIGENS IN COLON ADENOCARCINOMA

There is a growing amount of evidence indicating that misprogramming of genetic expression occurs in cancer.[63-67] Transcriptional modifications could be accompanied by parallel changes in macromolecules that comprise and regulate the genetic apparatus. Because of the extensive physical and probably functional heterogeneity of chromosomal components, it is extremely difficult to identify the biologically active components simply by the process of elimination and analysis. We have been using immunological techniques to analyze the change of chromosomal components during large bowel carcinogenesis.[68-71]

A. Preparation and Specific Antisera Against Rat Colon Tumor Antigens

We have produced specific antisera against large bowel carcinoma dehistonized chromatin as described in the last section. These antisera were tissue- and tumor-specific. They only

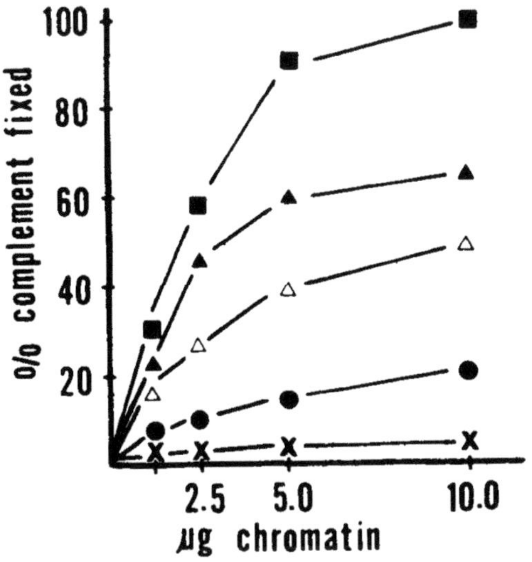

FIGURE 7. Complement fixation of various tissue chromatins isolated from different animals in the presence of antiserum against dehistonized chromatin of transplantable colon adenocarcinoma of the rat (■); human colon cancer (Vanderbilt Hospital; patient No. 76-5313) (▲); transplantable colon adenocarcinoma #51 of the mouse (△); normal rat colon (●); and rat Novikoff hepatoma, normal human colon, human breast cancer, or HeLa cells (×).

reacted with chromatin isolated from rat large bowel adenocarcinoma. The chromatins isolated from Novikoff hepatoma, Morris hepatoma 7777, Ehrlich ascites tumor, human breast cancer, human HeLa cells, human normal colon, and rat normal colon epithelium showed no affinity for the antisera against rat colon adenocarcinoma (Figure 7). With the use of immunoabsorption procedures, we have verified the specificity of antiserum. Chromatin preparations from normal rat colon or transplantable colon adenocarcinoma were incubated with antisera and centrifuged, and the supernatants were assayed by the complement-fixation technique. Antiserum absorbed with normal rat colon chromatin still fixed the complement in the presence of the rat or human colon adenocarcinoma chromatin. However, antiserum that was absorbed with chromatin from rat transplantable colon adenocarcinoma did not fix the complement in the presence of either rat or human colon tumor chromatin.[69]

The most sensitive, convenient, and accurate way to detect and quantitate the interaction between chromatins and antidehistonized chromatin sera is the solid-phase radioimmunoassay which we adopted from the method recently described by Romani et al.[72] with some modification. Figure 8 presents the dependence of the ^{125}I-protein A bound on the sera concentration added to the reaction mixture containing 10 µg chromatin. The amount of ^{125}I-protein A bound is also dependent on the concentration of chromatins used (Table 1). The tumor specificity of antisera investigated previously by the microcomplement-fixation technique have been reinvestigated by the solid-phase radioimmunoassay technique. The data, presented in Table 1, confirmed the earlier data that antisera against large bowel adenocarcinoma dehistonized chromatin is tumor-specific.

B. Appearance of Tumor-Specific Nuclear Antigens in Colon Chromatin During Large Bowel Carcinogenesis

In another series of experiments, the changes in the immunospecificities of the nonhistone protein-DNA complexes from the large bowel mucosae of rats treated with 1,2-dimethylhydrazine (MDH) were studied. DMH is a powerful carcinogen with a high propensity for

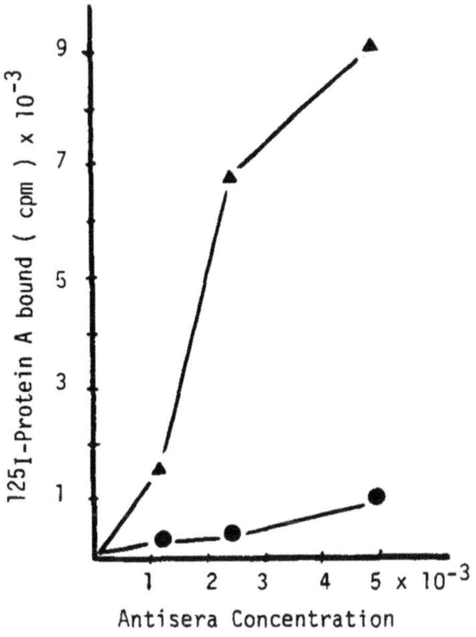

FIGURE 8. Dependence of ^{125}I-protein A binding on the concentration of antisera. Antigens were chromatins (10 μg) from rat large bowel adenocarcinoma (▲) and normal rat colon (●).

Table 1
CHARACTERIZATION OF ANTISERA AGAINST COLON DEHISTONIZED CHROMATIN BY SOLID RIA

Chromatin conc. (μg)	Normal colon chromatin	Colon tumor chromatin
2.5	250	2705[a]
5.0	676	4250
10.0	1635	9553

[a] Total ^{125}I-protin A bound (cpm). A 1/200 dilution of antisera was used in this experiment. Experimental procedure is described in the Methods section.

producing cancer of the large intestine in various rodent species.[73] Chromatins prepared from the colon of rats treated with DMH for 2, 4, 6, 11, 20, and 31 weeks were assayed for complement fixation in the presence of transplantable colon adenocarcinoma dehistonized chromatin antiserum. As shown in Figure 9, a significant change in the immunospecificity of chromatin occurred earlier than any histologic evidence of dysplasia or neoplasia.[68] There is no further evidence of nuclear or cytoplasmic alteration in the colon mucosae of rats treated with MDH until after 5 weeks of treatment when compared to control rats.

Accumulated evidence demonstrated that certain chemical agents can either promote or inhibit tumor formation in animals during chemical carcinogenesis.[74-76] The effect of tumor promoters (e.g., sodium barbiturate) and carcinogen inhibitors (e.g., disulfiram and butylated hydroxytoluene) on colon tumor-specific nuclear antigen activity was studied in our laboratory. We demonstrated the promotional effect of sodium barbiturate on tumor-specific

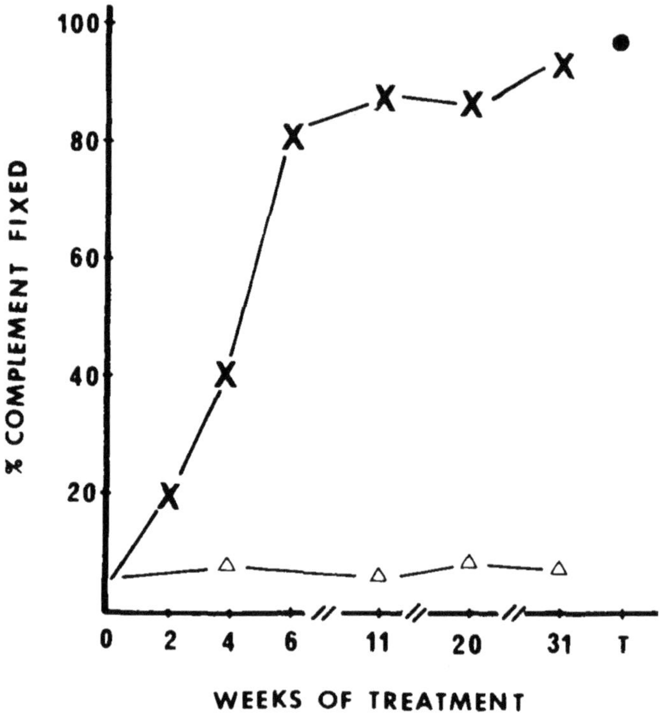

FIGURE 9. Time course of the increasing concentration of nuclear antigens in chromatin preparations (10 μg of DNA) isolated from colon of rats treated with DMH (X). The complement fixation assays were performed in the presence of antisera against rat transplantable colon adenocarcinoma dehistonized chromatin. T = chromatin isolated from rat transplantable colon adenocarcinoma (●); control chromatins isolated from rat treated with normal saline (△).

nuclear antigen activity in rats treated with DMH.[71] We also demonstrated that the rise of immunoactivity in carcinogen-treated colon chromatin can be abolished by simultaneous treatment with disulfirant or butylated hydroxytoluene.

Administration of DMH to Sprague-Dawley rats resulted in the development of intestinal and rectal adenocarcinomas.[73] In contrast, application of the same experimental protocols to Lubond Wistar rats failed to induce tumors.[77] It has been demonstrated for many chemical carcinogens that the microsomal mixed-function oxidase system converts these compounds to a proximate carcinogenic form.[78] The difference in susceptibility to DMH between the Sprague-Dawley and Lubond Wistar rats was partially related to metabolic activation of DMH. The ineffective metabolic activation of DMH in Lubond Wistar rats may be due to absent or insufficient amounts of specific enzyme in their microsomal mixed-function oxidase system.

Recently we demonstrated that DMH induced colon tumors in all Sprague-Dawley rats tested. Sodium barbiturate increased by threefold the colon tumor number in DMH-treated Sprague-Dawley rats. In contrast, the incidence of colon tumor in DMH-treated Wistar rats was very low (16.6% incidence). Sodium barbiturate did not enhance colon tumor formation in Wistar rats. We also found that the colon chromatin isolated from Wistar rats injected with DMH did not exhibit any tumor-specific immunoactivity with antibody against rat colon adenocarcinoma dehistonized chromatin. Sodium barbiturate also did not enhance the immunoactivity of chromatin (Figure 10). On the other hand, Sprague-Dawley rats that were given DMH for only 5 weeks developed tumor-specific nuclear antigenic activity in colon

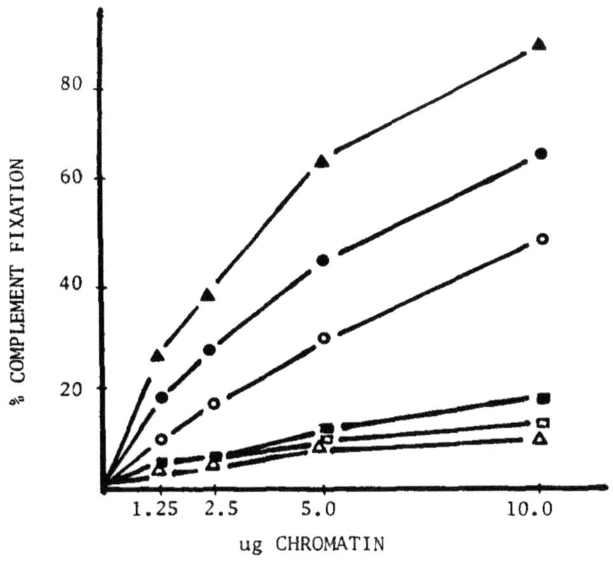

FIGURE 10. The complement-fixation activities of colon chromatins isolated from Wistar and Sprague-Dawley rats treated with DMH and sodium barbiturate. Chromatin from rat colon adenocarcinoma (▲); chromatin from normal rat colon (△); colon chromatin isolated from Wistar (□) and Sprague-Dawley (○) rats treated with DMH for 5 weeks; colon chromatins isolated from Wistar (■) and Sprague-Dawley (●) rats treated with DMH plus sodium barbiturate for 5 weeks.

chromatin. DMH-treated Sprague-Dawley rats given sodium barbiturate in the water developed higher immunoactivity than did those given water alone. These data indicate that the specific nuclear antigen is closely associated with colonic carcinogenesis.

To further confirm that tumor-specific antiserum activity is not merely against DMH or its metabolite which contaminate with chromatin during the isolation of nuclei, we have also studied the immunoactivity of liver chromatin isolated from Sprague-Dawley rats treated with DMH for 5 weeks. Interestingly, liver chromatin did not show any immunoactivity with antibody against colon tumor dehistonized chromatin.

C. Characterization of Colon Tumor-Specific Nuclear Antigens

Is colon tumor-specific antigen a CEA? From our earlier experimental results on immunolocalization studies[79] in which the antibodies localized with nuclear material but not with the cytoplasm, it appears that the possibility of specific colon antigen being a CEA is very slim. To obtain the direct evidence that colon tumor-specific nuclear antigens are not CEA, we isolated CEA-like substances from rat transplantable colon adenocarcinoma as described by Krupey et al.[80] These substances were then assayed for their immunoactivity with antisera against colon adenocarcinoma dehistonized chromatin. We found that neither free CEA-like substances nor their complexes with DNA exhibited affinity to this antiserum. These results imply that these antisera are not CEA.

Tumor-specific nuclear antigens are chromosomal nonhistone protein-DNA complexes. The treatment with DNase resulted in a complete loss of immunoactivity. However, DNA is not the antigen; chromosomal nonhistone proteins are principal components responsible for immunoactivity. That the antigenic activity of chromatins depends on the protein components of the chromatins and not the DNA is shown in Figure 11, which compares chromatins reconstituted from nonhistone proteins isolated either from normal rat colon or transplantable large bowel adenocarcinoma.

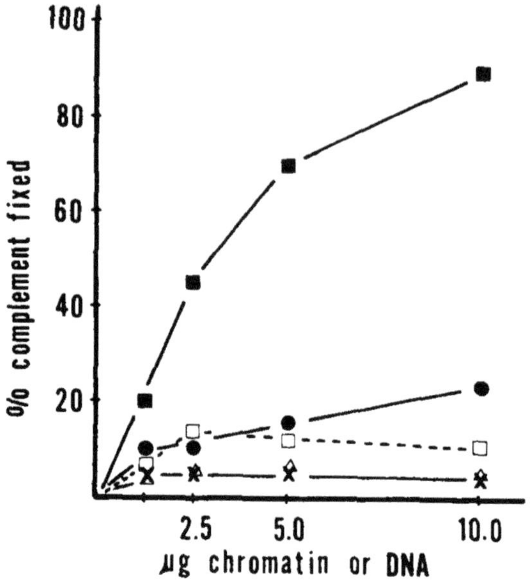

FIGURE 11. Effects of enzymatic digestion on the complement fixation of transplantable colon adenocarcinoma chromatin in the presence of antiserum against dehistonized chromatin of transplantable colon adenocarcinoma of the rat. Adenocarcinoma was incubated at 37°C for 1 hr with either H_2O (■), 100 μg DNase I per milliliter + 100 μg DNase II per milliliter (□), or 100 μg pronase per milliliter (x). Normal colon chromatin (●) and DNA (△) were also used for control experiment.

Busch and his associates[81,82] have identified tumor antigen associated with nucleoli. Their antigen is a protein, not a DNA-protein complex. We were, nonetheless, curious whether the nuclear antigen recognized by our tumor-specific antibody was actually with the nucleoli since we had made no attempt to specifically remove nor enrich for nucleoli in our antigen preparations. However, we could not detect any difference between the chromatin fractions highly enriched in nucleoli or whole chromatin. The results were identical in three separate nucleolar preparations. We preabsorbed our antibody with a nucleolar preparation and it still retained a high level of activity with rat transplantable colon adenocarcinoma chromatin compared with normal chromatin.

The chromosomal nonhistone proteins which are responsible for immunoactivity are the tightly DNA-binding proteins. Chromatin was extracted with 5 *M* urea — 2.5 *M* NaCl — Tris-HCl (pH 8.0) to remove most of the nuclear proteins. The DNA containing a small fraction of tightly DNA-binding proteins sedimented in the pellet after centrifugation at 100,000 × *g* for 48 hr at 4°C. The soluble protein fraction (composing over 95% of total nuclear proteins) is reconstituted with DNA. The immunoactivity of reconstituted soluble protein-DNA complexes and insoluble chromatin fraction is shown in Figure 12. Most of the tumor-associated colon nuclear antigens are located in the tightly DNA-binding protein fraction which still forms a complex with pelleted DNA.

In order to know how many nonhistone proteins are actually responsible for the immunoactivity, we identified the specific complex(es) by combining the techniques of protein blotting and immunolocalization. Chromosomal nonhistone proteins isolated from large bowel tumor were resolved by polyacrylamide gel electrophoresis (PAGE), transferred to nitrocellulose sheets, reconstituted to DNA, and antigens were localized immunochemically. As shown in Figure 13, there are more than 30 protein bands resolved on the gel and most of these protein bands were effectively transferred to nitrocellulose sheets. Surprisingly,

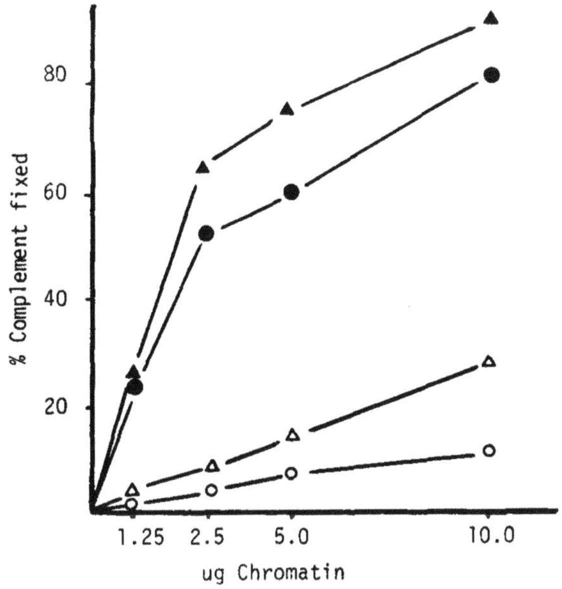

FIGURE 12. Immunoreactivity of native large bowel adenocarcinoma (●) and normal rat colon (○) chromatins, fractionated chromatin (▲), and reconstituted chromatin (△). Over 95% of chromosomal proteins were extracted with urea-salt buffer and reconstituted to rat DNA (△). The residual chromatin fraction (▲) composed tightly DNA-binding proteins and DNA complexes.

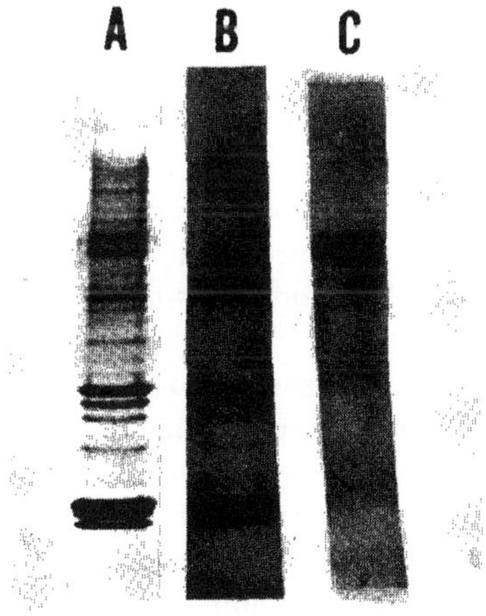

FIGURE 13. (A) SDS-gel pattern of colon tumor nonhistone proteins; (B) the pattern of proteins transferred to nitrocellulose sheet; (C) immunolocalization of antigenic protein on nitrocellulose as in (B).

there is only one major protein band immunochemically localized by one of our three antisera preparations. The molecular weight of this major antigenic protein is about 67,000. Very recently Duhl et al.[83] have also found that two antigenic nonhistone proteins with molecular weights of 67,000 and 92,000 were identified only with the human colon adenocarcinoma chromatins. Boffa et al.[84,85] have also found that the synthesis of two nuclear proteins with molecular weights of 44,000 and 62,000 is selectively accelerated within 4 weeks after administration of the carcinogen. These nuclear proteins are not similar to our antigenic nonhistone proteins. Antigenic nonhistone proteins are tightly DNA-binding proteins; however, the nuclear proteins reported by Boffa et al. can be easily extracted from chromatin with 5 *M* urea — 2.0 *M* NaCl — 50 m*M* Tris-HCl (pH 8.0).

IV. HEPATOMA-ASSOCIATED NONHISTONE PROTEINS

Using antisera against dehistonized chromatin from adult rat liver, Chytil et al.[42] showed gradual changes in the antigenicity of chromosomal nonhistone proteins in developing embryonic and post partum livers of rats. Hnilica and associates,[39,43,86,87] using antibodies to Novikoff hepatoma dehistonized chromatin have also found that hepatocarcinogenesis results in the formation of new histone protein-DNA complexes specific for the cancerous growth. However, these antibodies (Novikoff hepatoma antisera) could not differentiate immunochemically in nonhistone protein-DNA complexes derived from various tumors, regardless whether the tumors were of rat, mouse, or human origins. Chiu et al.[43] showed that the change in immunochemical specificity of the nonhistone protein-DNA complexes can be detected relatively early during experimental azo-dye hepatocarcinogenesis in rats and that this change is different from changes produced by cellular response following partial hepatectomy.[88] Since Novikoff hepatoma is originated from bile duct cell and is not true hepatoma (originated from hepatocyte), recently we have produced antisera against Morris hepatoma 7777 dehistonized chromatin. Using these antisera we detected that hepatoma-associated nonhistone proteins appeared in liver at early stages during hepatocarcinogenesis. Hepatoma-associated nonhistone proteins were not present in rat fetal liver chromatin.

A. Enzyme-Linked Immunosorbent Assay for Chromatin Antigens

Several immunological techniques such as complement fixation,[38-43,89] radioimmunoassay,[72,90] immunolocalization,[52,82,89,91-93] and double diffusion[44] have been applied to study the specificity of nonhistone proteins and chromatin structure with the antibodies to chromosomal proteins or dehistonized chromatins. Recently, we have developed an enzyme-linked immunosorbent assay (ELISA) suitable for use with nuclear antigens.[94] This assay is considerably more sensitive than the microcomplement-fixation test. It has the further advantage that one avoids recurrent problems of anticomplimentarity with different antigen or antibody preparations. The ELISA does not necessitate the use of the short-lived radioactive reagents required for radioimmunoassay.

The ELISA was optimized in our laboratory for use with chromatins.[94] Briefly, chromatin (0.1 to 10 μg/100 μℓ) suspended in PBS (phosphate buffered saline, 0.01 *M* sodium phosphate pH 7.2, 0.14 *M* sodium chloride) was pipetted into a 96-well polyvinyl chloride (PVC) microtiter plate (Costar) and incubated overnight at 0 to 4°C in a moisture atmosphere. Plates were washed with 1% (w/v) bovine serum albumin (Fraction V, Sigma) in PBS as the plate was incubated at 37°C for 45 min in a moist atmosphere. Following these washes with PBS/Tween® 20, antidehistonized 7777 chromatin antibody was added at the appropriate dilution in 1% BSA/PBS and the plate was incubated for 4 hr at 22°C in a moist atmosphere. The plate was washed, then goat anti-rabbit immunoglobulin conjugated to horseradish peroxidase was added and the plate was incubated at 37°C for 45 min. The plate was washed and a stain solution of 0.1% (w/v) *o*-phenylene diamine, 0.006% (v/v) H_2O_2 in 0.1 *M* sodium citrate pH 5.0 was added. The staining reaction was stopped by the addition of H_2SO_4 to 0.5 *M*. The absorbance at 490 nm was read on a microelisa reader.

Figure 14A shows that this ELISA procedure detects specific antibody reaction with nuclear antigen prepared from two species. Total chromatin (0.05 mg/mℓ) from either chicken reticulocyte or normal rat liver are easily distinguished by antisera prepared against dehistonized chicken reticulocyte chromatin even at 1/64,000 dilution. Dilutions of antisera less

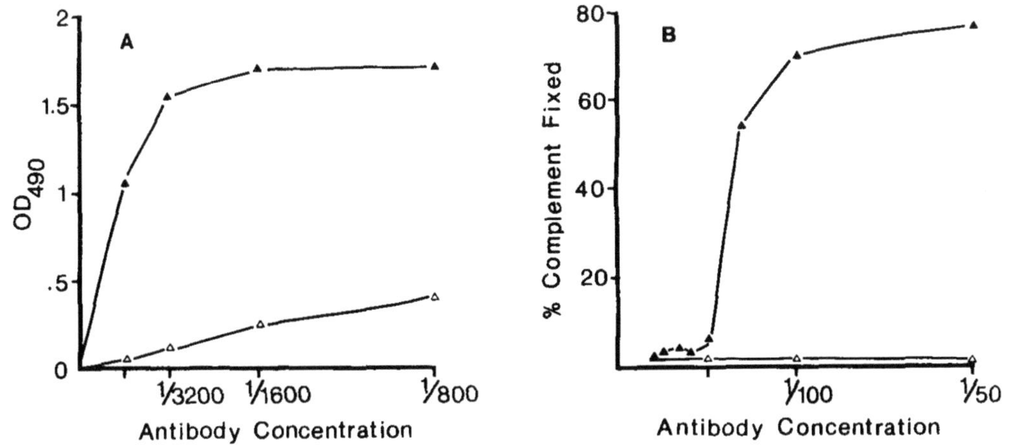

FIGURE 14. Titration of antisera prepared against chicken reticulocyte dehistonized chromatin by (A) ELISA and (B) microcomplement fixation technique. The antisera was tested with chicken reticulocyte whole chromatin (▲) and normal rat liver chromatin (△) as a nonreactive control. In ELISA, 5 μg of chromatins were used and 10 μg were used in the mirocomplement-fixation assay.

than 1/800 have large amounts of nonspecific binding. To allow comparison between the ELISA and the currently more commonly used microcomplement-fixation test, Figure 14B illustrates the results of an antibody titration with the same antigen and antibody used in Figure 14A by complement fixation. It is immediately obvious that the complement-fixation assay is much less sensitive than the ELISA assay. Significantly more antisera is necessary to detect immunoreactivity with the complement-fixation assay.

B. Production and Characterization of Specific Antibodies Against Hepatoma Dehistonized Chromatin

The nuclei and chromatins were isolated from transplantable Morris hepatoma 7777 as described by Chiu et al.[45] Dehistonized chromatins were prepared and used as antigen to immunized rabbits according to the method of Chiu et al.[43] Antisera were then subjected to immunoabsorption to normal rat liver chromatin as described.[43] Two cycles of immunoabsorption were determined to be optimum.

The antibodies raised against Morris hepatoma 7777 dehistonized chromatin in rabbit react strongly with hepatoma 7777 chromatin by ELISA assay. A certain degree of cross-reactivity with normal rat liver chromatin for two cycles react specifically with Morris hepatoma chromatin but not with normal liver chromatin as shown in Figure 15A. The microelisa procedure provides a sensitive method to detect immunoreactivity, antisera diluted to 1/3200 was able to detect reactivity with 0.1 μg of Morris hepatoma 7777 chromatin. When these antisera were used in microcomplement-fixation assays, a 1/200 dilution detected reactivity with 2.5 μg of hepatoma chromatin. When these antisera were absorbed to Morris hepatoma 7777 chromatin for one cycle, the immunoreactivity was essentially abolished (Figure 15B).

In an attempt to correlate the immunoreactivity to the growth rate of various tumors and normal rat liver, several Morris hepatomas and fetal liver hepatomas were compared. As can be seen in Figure 16 all the chromatin preparations isolated from Morris hepatomas immunoreacted to the antibodies against Morris hepatoma 7777 dehistonized chromatin. It suggested that all hepatomas assayed share some of the hepatoma-associated nuclear antigens. However, there is no positive correlation between immunoreactivity and the growth rate of hepatomas. Further data to support this notion is immunoreactivity of fetal liver chromatin. Fetal liver is fast-growing tissue but its chromatin did not have affinity to the antisera (data not shown).

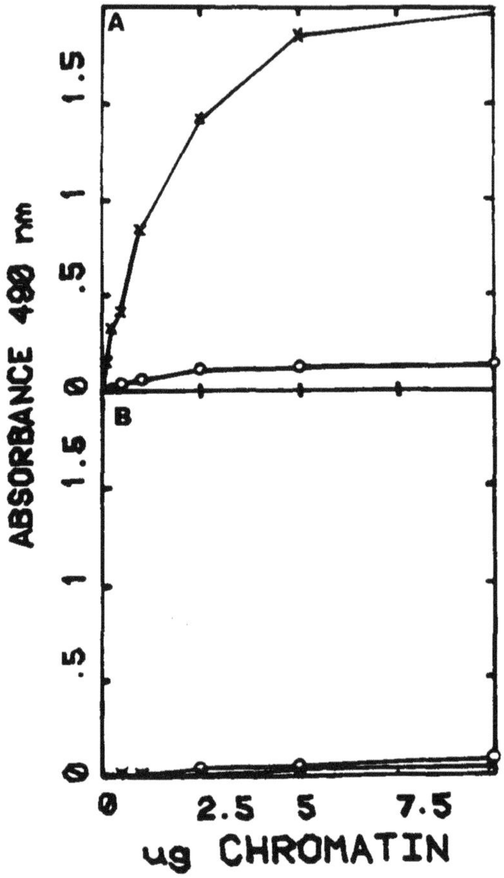

FIGURE 15. (A) Immunoreactivity of antisera absorbed with rat liver chromatin for two cycles. Titration of chromatin against antisera at a dilution of 1/3200 was measured by ELISA, using normal rat liver chromatin (O) and Morris hepatoma 7777 (X). (B) Immunoreactivity of antisera absorbed with Morris hepatoma 7777 chromatin for 1 cycle. Titration of chromatin against antisera at a dilution of 1/3200 was measured by ELISA. Normal rat liver chromatin (O) and Morris hepatoma 7777 chromatin (x) were assayed.

To better define the fraction of nonhistone responsible for the antigenicity of Morris hepatoma 7777 chromatin, the immunoblot technique was used as described earlier. Figure 17A is a polyacrylamide gel stained with Coomassie brilliant blue G that demonstrates the diversity of chromosomal proteins from normal rat liver and Morris hepatoma 7777. The other three sets of these proteins on identical gels were transferred to nitrocellulose. One set of these proteins on the nitrocellulose paper was stained with Amido black. As shown in Figure 17B it provides evidence that the transferred protein profiles were representative of those in the original polyacrylamide gel.

When the antigen profiles were visualized using antisera to hepatoma 7777 that was not preabsorbed to normal rat liver chromatin, the heterogeneity of patterns observed were as shown in Figure 17D. Comparison of the antigen profiles for normal rat liver and hepatoma 7777 revealed that the nonabsorbed antisera reacts predominantly with proteins present in the hepatoma 7777 chromatin and with only a few proteins in normal rat liver chromatin. The strongest immunoreactivity appeared in the high molecular weight region prominent

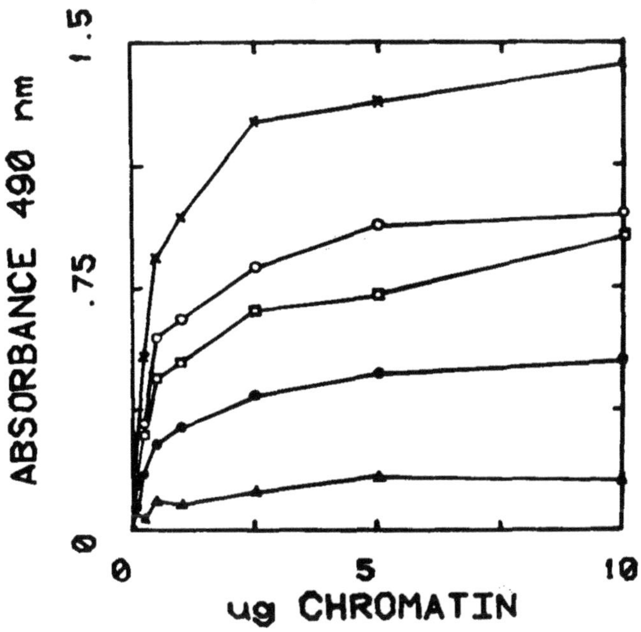

FIGURE 16. Cross-reactivity of antisera to chromatin from various Morris hepatomas. ELISA was used to measure immunoreactivity of antisera diluted to 1/3200 against chromatin from Morris hepatomas 7777 (x), 7288 (○), 5123tc (□), 7800 (●), and normal rat liver chromatin.

band at 200 K. In normal rat liver, a strong band was visible at 74 K and other bands appeared at the low molecular weight region. When two-times preabsorbed antisera was used, the immunoreactivity of normal rat liver chromosomal proteins disappeared (Figure 17C).

C. The Appearance of Hepatoma-Associated Nonhistone Proteins in Rat Liver Under Hepatocarcinogenesis

Chromatins from livers of Fischer rats maintained for various time periods on the diet containing hepatocarcinogen 3′-methyl-1,4-diamethylaminoazabenzene (3′-MDAB) were assayed for immunoreactivity by the ELISA technique. The results depicted in Figure 18 indicate an increase in immunoreactivity in the chromatin with length of time on diet. Chromatin isolated from the 30-day time point livers demonstrated a substantial degree of immunoreactivity. The immunoreactivity of liver chromatin was gradually increased as the length of carcinogen treatment increased. Its activity increased dramatically in the liver chromatins of rats treated with carcinogen for 165 days or longer. The profound difference in antigenicity of the liver chromatin of rats in the earlier stage (30 to 145 days) of carcinogenesis in comparison to later stages (160 days and 185 days) may be due to the heterogeneity in cellular composition of the livers at these time points. At the later time points, a much higher proportion of the liver is neoplastic.[96,97]

As shown in Figure 18, we found that hepatoma-associated immunoreactivity of liver chromatins increased in rats treated with the carcinogen 3′-MDAB. In order to see if the increased immunoreactivity resulted from the appearance of hepatoma-specific antigenic chromosomal proteins observed in the immunoblots described above chromatins from livers of rats fed 3′-MDAB for varying lengths of time were assayed by the immunoblot technique. As shown in Figure 19, the process of hepatocarcinogenesis induced by 3′-MDAB was accompanied by the appearance of hepatoma-associated antigenic nonhistone proteins, especially in the molecular-weight region ranging from 169 to 220 K daltons.

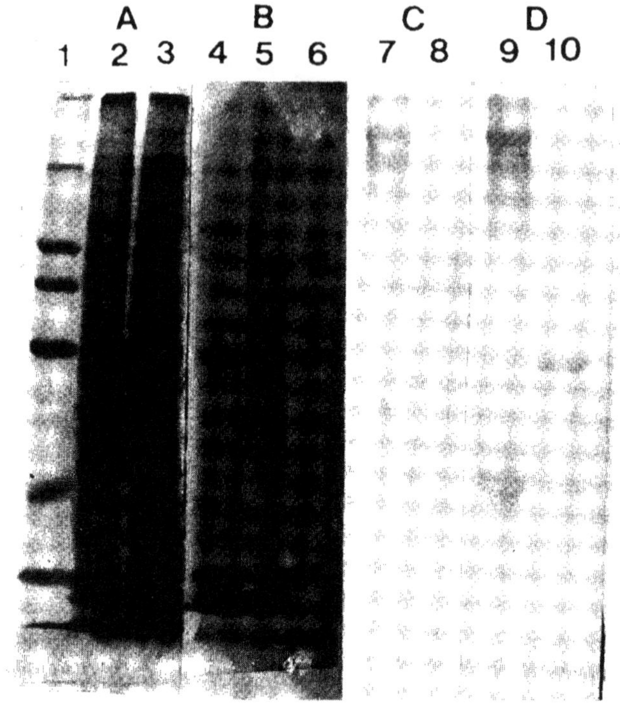

FIGURE 17. Immunoblot on normal and Morris hepatoma 7777 chromatin. (A) SDS-PAGE profiles of molecular weight standards (lane #1), Morris hepatoma 7777 chromatin (lane #2), and normal rat liver chromatin (lane #3); (B) Amido black stain of electrophoretic transfer to nitrocellulose paper of an SDS-PAGE identical to that in A, with molecular weight standards (lane #6), Morris hepatoma 7777 chromatin (lane #4), and normal rat liver chromatin (lane #5); (C) immunoblot with antisera absorbed to normal rat liver chromatin for two cycles on Morris hepatoma 7777 chromatin (lane #7) and normal rat liver chromatin (lane #8); (D) immunoblot with nonabsorbed antisera on Morris hepatoma 7777 (lane #9) and normal rat liver chromatin (lane #10).

D. The Presence of Hepatoma-Specific Chromosomal Nonhistone Proteins in Transcriptionally Active Chromatin

It has been shown that the chromosomal region which actively transcribed RNAs is sensitive to DNase I.[11,17] It will be very interesting to know if the hepatoma-associated immunogenic nonhistone proteins are present in active chromatin. We have digested the Morris hepatoma 7777 at normal rat liver chromatins with DNase I up to 15% digestion.

Proteins released by DNase I digestion and proteins present in undigested chromatin and total chromatin were subjected to the immunoblot analysis. Some antigenic protein bands were detected in the supernatant portion of digested hepatoma chromatin (transcriptionally active chromatin) as shown in Figure 20. Immunoreactive bands also appeared in the undigested hepatoma chromatin pellet (presumably inactive chromatin). However, the immunoreactive protein bands which were present in transcriptionally active hepatoma chromatin are different from the immunoreactive protein bands present in transcriptionally inactive chromatin. No immunoreactive protein bands were detected in both fractions of normal rat liver chromatin.

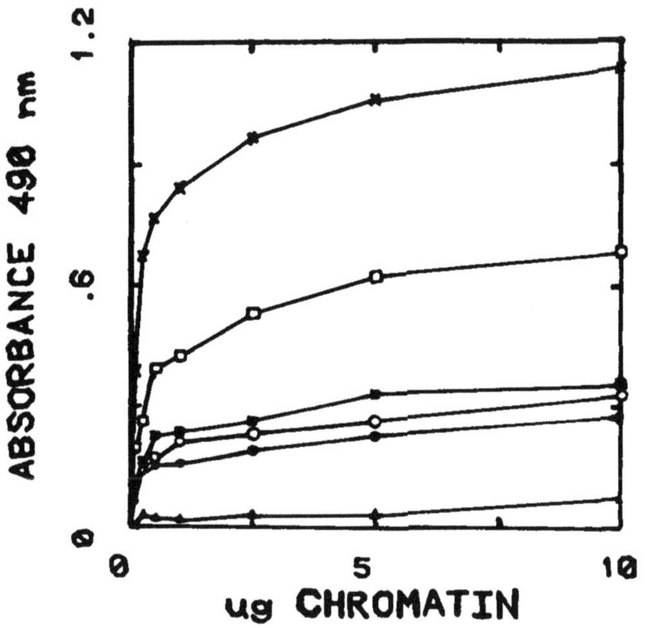

FIGURE 18. Immunoreactivity of liver chromatin from rats fed 3′-MDAB. Chromatin from control rats (▲), rats fed carcinogen for 30 days (●), 125 days (○), 140 days (■), 160 days (□) and 185 days (X) were assayed by ELISA.

V. CONCLUSIONS

Chromatin which contains complexes of DNA, histone, nonhistone proteins, and trace amounts of RNA is the principal site of genetic endowment and its expression in eukaryotic organisms. A detailed knowledge of the chemical identity and function of the components of chromatin is essential for understanding and manipulation of cell differentiation and genetic expression. This realization intensified the research on DNA, histones, and especially chromosomal nonhistone proteins. Development of techniques for the isolation of chromatin opened the way for studies on the biological and structural aspects of this important macromolecular complex. Systematic analysis of each individual nonhistone protein present in chromatin is a formidable task. However, elucidating specific properties which can be selectively associated with the individual protein species is essential for our understanding of chromosomal protein biochemistry and biology. We have been using an exceptionally sensitive and selective immunological technique as a major tool to study the chromatin structure and its components in differentiating cells and neoplastic tissues.

Using antisera against dehistonized chromatin from chicken reticulocyte, we showed the gradual changes in the antigenicity of chromatin in maturing chicken reticulocytes.[52,53,59] The nuclear nonhistone protein components of the nuclear antigens are very tightly bound to DNA. These proteins are a heterogeneous group of nonhistone proteins.[59] This fraction of DNA-binding proteins contains macromolecules which can influence the in vitro transcription of chromatin.[98] Bekhor and Mirell[58] reported a class of nonhistone proteins which remain bound to DNA in 2 *M* NaCl and postulated that these proteins include globin gene regulators. The proteins are only present in erythroid cells, not in oviduct cells. Recently, using the procedure of Bekhor and Mirell, we isolated a nonhistone protein fraction tightly bound to DNA which putatively has a role in globin gene regulation in chicken reticulocyte. This fraction was tested by gel electrophoresis and immunological activity, and appears by

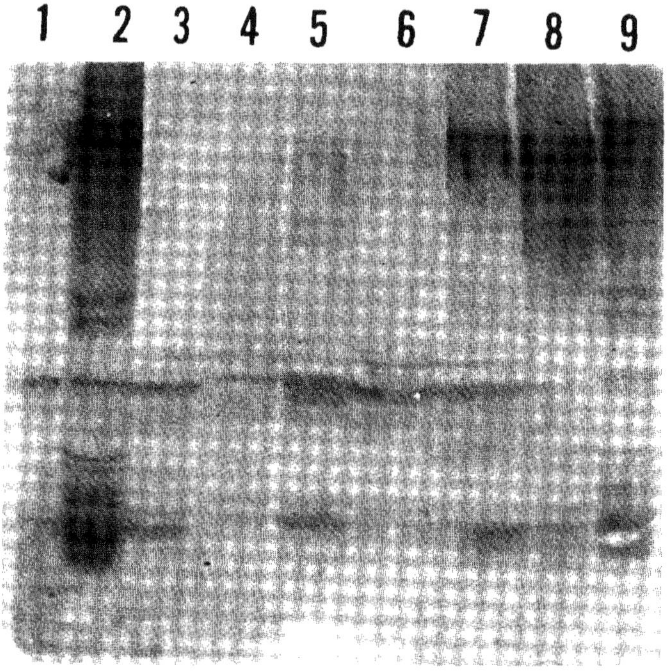

FIGURE 19. Immunoblot on chromatin from livers of rats fed 3′-MDAB. Immunolocalization of antigenic bands with nonabsorbed antisera on chromatin from control rats (lane #1), Morris hepatoma 7777 (lane #2), rats fed 3′-MDAB for 30 days (lane #3), 100 days (lane #4), 125 days (lane #5), 145 days (lane #6), 160 days (lane #7), 185 days (lane #8), and Morris hepatoma 7777 (lane #9).

these criteria very similar to the chicken nuclear antigen previously identified in reticulocyte chromatin and structurally altered erythrocyte chromatin.[52,53]

Accumulated evidence suggests that neoplastic transformation may be a manifestation of aberrant cellular differentiation.[99-101] A likely candidate for modulator of differential gene expression is the group of nonhistone proteins.[1-5] Using antibodies specific to tumor dehistonized chromatins, we demonstrated the presence of a fraction of nonhistone proteins in rat colon adenocarcinoma and hepatoma that is not detected in normal corresponding tissues. Hepatoma-associated nonhistone proteins were detected in Morris hepatomas 7777, 7288C, 7800 and 5123tc, and observed to gradually appear in liver during azo-dye-induced hepatocarcinogeneis. We suggest that this fraction of nonhistone proteins may be related to the malignant state and this could be a factor that initiates and propagates neoplastic transformation.

As mentioned previously, differential gene expression appears to be correlated with alterations in chromatin structure as revealed by preferential sensitivity to DNase I of active transcribed tissue-specific genes.[11,17,102-104] Detailed analysis of DNase I digestion of transcribed genes has revealed discrete sites that are extremely sensitive to DNase I clustered at or near the 5′-ends of the genes, suggesting an especially exposed chromatin structure at the 5′-terminus.[102,103,105-110] There is also an overall DNase I sensitivity that extends throughout the transcription unit. Investigation of the components necessary to establish and maintain the DNase I hypersensitive site, which appears to be a chromatin structure with a regulatory function, may provide an understanding of the mechanism of chromatin transcription and consequently the mechanism of neoplastic transformation. Hyperacetylation of histones appear to be correlated with increased transcriptional activity.[111,112] The nonhistone proteins

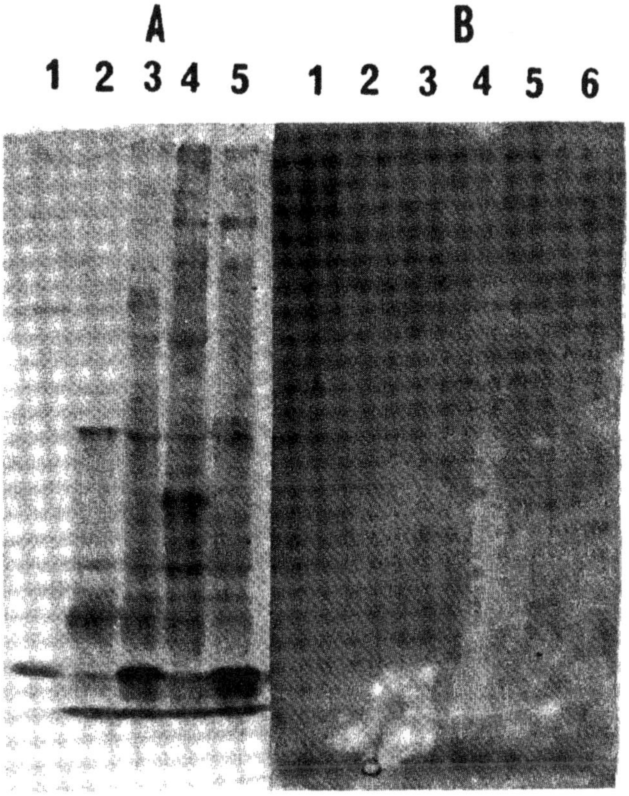

FIGURE 20. Immunoblot on chromatin proteins released by limited DNase I digestion. (A) Amido black stain on the electrophoretic transfer to nitrocellulose paper of molecular weight standards (lane #1), proteins released by DNase I digestion of Morris hepatoma 7777 chromatin (lane #2), proteins of undigested Morris hepatoma 7777 chromatin (lane #3), proteins released by DNase I digestion of normal rat liver chromatin (lane #4), and proteins of undigested normal rat liver chromatin (lane #5); (B) immunoblot on proteins of total Morris hepatoma 7777 chromatin (lane #1), normal rat liver chromatin (lane #2), proteins released by DNase I digestion of Morris hepatoma 7777 chromatin (lane #3), proteins released by digestion of normal rat liver chromatin (lane #4), proteins associated with undigested Morris hepatoma 7777 (lane #5), and proteins associated with undigested normal rat liver chromatin (lane #6).

HMG-14 and -17 have also been shown to play roles in the maintenance of DNase I sensitivity of transcribed sequences in chromatin.[23,24,113,114] As neither of these chromatin components exhibit tissue specificity, it has been surmised that other components exist that are responsible for conferring selective active chromatin structural alterations.[9,24,115]

In an effort to determine possible roles of nonhistone proteins in the alteration of chromatin structure and function in neoplastic vs. normal cells, we have studied the distribution of hepatoma-associated nonhistone proteins in transcriptionally active and inactive hepatoma chromatin fractions. Limited DNase I digestion of Morris hepatoma 7777 chromatin released immunological-specific hepatoma-associated nonhistone proteins which are not found in normal rat liver chromatin and transcriptionally inactive hepatoma chromatin, implying that these proteins are associated with active chromatin of the neoplastic state. These proteins may interact with the cellular genome in a way that induces the active conformation and result in the altered genetic expression that is a primary component of the etiology of neoplasia. Although a more direct means of demonstrating the roles of these nonhistone proteins is required, one may speculate that these specific nonhistone proteins may be involved in the modulation of chromatin structure and gene expression in the malignant state.

ACKNOWLEDGMENTS

The author would like to thank Mr. Robert Shaw for his devoted and patient secretarial efforts in preparing this manuscript. The work presented in this chapter was supported in part by Grants CA 25098 and PHS 5429-21-2 from the National Cancer Institute.

REFERENCES

1. **Allfrey, V. G.,** DNA-binding proteins and transcriptional control in prokaryotic and eukaryotic systems, in *Acidic Proteins of the Nucleus*, Cameron, I. L. and Jeter, J. R., Eds., Academic Press, New York, 1974, 1057.
2. **Baserga, R.,** Nonhistone chromosomal proteins in normal and abnormal growth, *Life Sci.*, 15, 1057, 1974.
3. **Elgin, S. C. R. and Weintraub, H.,** Chromosomal proteins and chromatin structure, *Annu. Rev. Biochem.*, 44, 725, 1975.
4. **Chiu, J.-F. and Hnilica, L. S.,** Nuclear nonhistone proteins: chemistry and function, in *Chromatin and Chromosome Structure*, Li, H. J. and Eckhard, R. A., Eds., Academic Press, New York, 1977, 193.
5. **Sevall, J. S., Jagodzinski, L. L., Tsai, S., Castro, C. E. and Lee, D.,** Nonhistone proteins and gene organization, in *The Cell Nucleus*, Vol. 4, Academic Press, New York, 1978, 319.
6. **Kornberg, R.,** Structure of chromatin, *Annu. Rev. Biochem.*, 46, 931, 1977.
7. **McGhee, J. D. and Felsenfeld, G.,** Nucleosome structure, *Annu. Rev. Biochem.*, 49, 1115, 1980.
8. **Wang, J. C.,** The path of DNA in the nucleosome, *Cell*, 29, 724, 1982.
9. **Mathis, D., Oudet, P., and Chambon, P.,** Structure of transcribing chromatin, *Prog. Nucleic Aci Res. Mol. Biol.*, 24, 1, 1982.
10. **Igo-Kemenes, T., Horz, W., and Zachau, H. G.,** Chromatin, *Annu. Rev. Biochem.*, 51, 89, 1982.
11. **Weintraub, H. and Groudine, M.,** Chromosomal subunits in active genes have an altered conformation, *Science*, 193, 848, 1976.
12. **Garel, A. and Axel, R.,** Selective digestion of transcriptionally active ovalbumin genes from oviduct nuclei, *Proc. Natl. Acad. Sci. U.S.A.*, 73, 3966, 1977.
13. **Bellard, M., Gannon, F., and Chambon, P.,** Nucleosome structure III: the structure and transcriptional activity of the chromatin containing the ovalbumin and globin genes in chick oviduct nuclei, *Cold Spring Harbor Symp. Quant. Biol.*, 42, 779, 1977.
14. **Panet, A. and Cedar, H.,** Selective degradation of integrated murine leukemia proviral DNA by deoxyribonucleases, *Cell*, 11, 933, 1977.
15. **Flint, S. J. and Weintraub, H. M.,** An altered subunit configuration associated with the actively transcribed DNA of integrated adenovirus genes, *Cell*, 12, 783, 1977.
16. **Levy, W. B. and Dixon, G. H.,** Renaturation kinetics of cDNA complementary to cytoplasmic polyadenylated RNA from ranbow trout testis. Accessibility of transcribed genes to pancreatic DNase, *Nucleic Acids Res.*, 4, 883, 1977.
17. **Lacy, E. and Axel, R.,** Analysis of DNA of isolated chromatin subunits, *Proc Natl. Acad. Sci. U.S.A.*, 72, 3978, 1975.
18. **Kuo, M. T., Sahasrabuddhe, C. G., and Saunders, G. F.,** Presence of messenger specifying sequences in the DNA of chromatin subunits, *Proc. Natl. Acad. Sci. U.S.A.*, 73, 1572, 1976.
19. **Foe, V. E., Wiekinson, A., and Laird, C. D.,** Comparative organization of active transcription units in *Oncopeltus fasciatus, Cell.*, 9, 131, 1976.
20. **Wallace, R. B., Sargent, T. D., Murphy, R. F., and Bonner, J.,** Physical properties of chemically acetylated rat liver chromatin, *Proc. Natl. Acad. Sci. U.S.A.*, 74, 3244, 1977.
21. **Boffa, L. C., Vidali, G., Mann, R. S., and Allfrey, V. G.,** Suppression of histone deacetylation *in vivo* and *in vitro* by sodium butyrate, *J. Biol. Chem.*, 253, 3364, 1976.

22. **Wu, C., Wong, Y. C., and Elgin, S. C. R.,** The chromatin structure of specific genes. II. Disruption of chromatin structure during gene activity, *Cell*, 16, 907, 1979.
23. **Weisbrod, S. and Weintraub, H.,** Isilation of a subclass of nuclear proteins responsible for conferring a DNase I-sensitive structure on globin chromatin, *Proc. Natl. Acad. Sci. U.S.A.*, 76, 630, 1979.
24. **Weisbrod, S., Groudine, M., and Weintraub, H.,** Interaction of HMG 14 and 17 with actively transcribed genes, *Cell*, 19, 289, 1980.
25. **Vidali, G., Boffa, L. C., and Allfrey, V. G.,** Selective release of chromosomal proteins during limited DNase I digestion of avian erythrocyte chromatin, *Cell*, 12, 409, 1977.
26. **Weisbrod, S. and Weintraub, H.,** Isolation of actively transcribed nucleosomes using immobilized HMG 14 and 17 and an analysis of α-globulin chromatin, *Cell*, 23, 391, 1981.
27. **McKnight, S. M. and Miller, O. L., Jr.,** Electron microscopic analysis of chromatin replication in the cellular blastoderm *Drosophila melanogaster* embryo, *Cell*, 19, 289, 1977.
28. **Bustin, M., Goldblatt, D., and Sperling, R.,** Chromatin structure visualization by immunoelectron microscopy, *Cell*, 7, 297, 1976.
29. **Bustin, M., Simpson, R. J., Sperling, R., and Goldblatt, D.,** Molecular homogeneity of the histone content of HeLa chromatin subunits, *Biochemistry*, 16, 5381, 1977.
30. **Bustin, M. and Stoller, B. D.,** Immunological relatedness of thymus and liver F_1 histone subfractions, *J. Biol. Chem.*, 248, 3506, 1978.
31. **Blankenstein, L. A., Stoller, B. D., Franklin, S. G., Zweidler, A., and Lay, S. B.,** Biochemical and immunological characterization of two distinct variants of histone H2A in Friend leukemia, *Biochemistry*, 16, 4557, 1977.
32. **Comings, D. E. and Harris, D. C.,** Nuclear proteins. II. Similarity of nonhistone proteins in nuclear sap and chromatin, and essential absence of contractile protein from mouse liver nuclei, *J. Cell Biol.*, 70, 440, 1976.
33. **Chiu, J. F., Chytil, F., and Hnilica, L. S.,** Onco-fetal antigens in chromatin of malignant cells, in *Onco-Developmental Gene Expression*, Fishman, W. H. and Sell, S., Eds., Academic Press, New York, 1976, 271.
34. **Silver, L. M. and Elgin, S. C. R.,** Immunofluorescent analysis of chromatin structure in relation to gene activity: a speculative essay, *Curr. Top. Dev. Biol.*, 13, 71, 1979.
35. **Silver, L. M. and Elgin, S. C. R.,** Immunological analysis of protein distributions in *Drosophila polytene* chromosomes, in *The Cell Nucleus*, Vol. 5, Busch, H., Ed., Academic Press, New York, 1978, 216.
36. **Bustin, M.,** Histone antibodies-structural probes for chromatin and chromosomes, in *The Cell Nucleus*, Vol. 4, Busch, H., Ed., Academic Press, New York, 1978, 196.
37. **Stoller, B. D.,** Serological analyses of histones, *Methods Cell Biol.*, 18, 105, 1977.
38. **Chytil, F. and Spelsberg, T. C.,** Tissue differences in antigenic properties of nonhistone protein-DNA complexes, *Nature (London) New Biol.*, 233, 215, 1971.
39. **Wakabayashi, K. and Hnilica, L. S.,** The immunospecificity to nonhistone protein complexes with DNA, *Nature (London) New Biol.*, 242, 153, 1973.
40. **Zardi, L., Lin, J. C., and Baserga, R.,** Immunospecificity to nonhistone chromosomal proteins of anti-chromatin antibodies, *Nature (London) New Biol.*, 245, 211, 1973.
41. **Spelsberg, T. C., Mitchell, W. M., Chytil, F., Wilson, E. M., and O'Malley, B. W.,** Chromatin of the developing chick oviduct: changes in the acidic proteins, *Biochim. Biophys. Acta*, 312, 765, 1973.
42. **Chytil, F., Glasser, S. R., and Spelsberg, T. C.,** Alterations in liver chromatin during perinatal development of the rat, *Dev. Biol.*, 73, 295, 1974.
43. **Chiu, J.-F., Hunt, M., and Hnilica, L. S.,** Tissue-specific DNA-protein complexes during azo dye hepatocarcinogenesis, *Cancer Res.*, 35, 913, 1975.
44. **Yeoman, L. C., Jordan, J. J., Busch, R. K., Taylor, C. W., Savage, H. E., and Busch, H.,** A fetal protein chromatin of Novikoff hepatoma and Walker 256 carcinosarcoma tumors that is absent from normal and regenerating rat liver, *Proc. Natl. Acad. Sci. U.S.A.*, 73, 3258, 1976.
45. **Lucas, A. M. and Jamroz, C.,** *Atlas of Avian Hematology*, U.S. Department of Agriculture, Washington, D.C., 1961.
46. **Seligy, V. and Miyagi, M.,** Studies of template activity of chromatin induced from metabolically active and inactive cells, *Exp. Cell Res.*, 58, 27, 1969.
47. **Attardi, G., Parnas, H., and Attardi, B.,** Pattern of RNA synthesis in duck erythroyctes in relationship to the stage of cell differentiation, *Exp. Cell Res.*, 62, 11, 1970.
48. **Williams, A. F.,** Deoxythymidine metabolism in avian erythroid cells, *J. Cell Sci.*, 11, 777, 1972.
49. **Davies, H. G.,** Electron-microscope observations on the organzation of heterochromatin in certain cells, *J. Cell Sci.*, 3, 129, 1968.
50. **Brasch, K., Setterfield, G., and Neelin, J. M.,** Effects of sequential extraction of histone proteins on structural organization of avian erythrocyte and liver nuclei, *Exp. Cell Res.*, 74, 27, 1972.

51. **Grasso, J. A., Woodward, J. W., and Swift, H.,** Cytochemical studies of nucleic acids and proteins in erythrocyte development, *Proc. Natl. Acad. Sci. U.S.A.*, 50, 134, 1963.
52. **Pumo, D. E., Wierzbicki, R., and Chiu, J.-F.,** Chicken reticulocyte nuclear antigen: its identification and relation to transcriptive activity in erythropoietic cells, *Biochemistry*, 19, 2362, 1980.
53. **Pumo, D. E., Wierzbicki, R., Sainten, A., and Chiu, J.-F.,** Immunospecificity of nuclear antigens in chicken erythroid cells, *Mol. Cell. Biochem.*, 32, 49, 1980.
54. **Williams, A. F.,** DNA synthesis in purified populations of avian erythroid cells, *J. Cell Sci.*, 10, 27, 1972.
55. **Ansevin, A. T., MacDonald, K. K., Smith, C. E., and Hnilica, L. S.,** Mechanism of chromatin template activation: physical evidence for destabilization of nucleoproteins by polyanions, *J. Biol. Chem.*, 250, 281, 1975.
56. **Elgin, S. C. R. and Bonner, J.,** Partial fractionation and chemical characterization of the major nonhistone chromosomal proteins, *Biochemistry*, 11, 772, 1972.
57. **Weber, K. and Kuter, D. J.,** Reversible denaturation of enzymes by sodium dodecyl sulfate, *J. Biol. Chem.*, 246, 4504, 1971.
58. **Bekhor, I. and Mirell, C. J.,** Simple isolation of DNA hydrophobically complexed with presumed gene regulatory proteins (M_3), *Biochemistry*, 18, 609, 1979.
59. **Pumo, D. E., Seward, E. A., Childs, S. J., and Chiu, J.-F.,** Erythroid specific nuclear antigen and globin gene binding protein, *Biochem. Biophys. Res. Commun.*, 97, 487, 1980.
60. **Chiu, J.-F., Wang, S., Fujitani, H., and Hnilica, L. S.,** DNA-binding chromosomal nonhistone proteins. Isolation, characterization, and tissue specificity, *Biochemistry*, 14, 4552, 1975.
61. **Towbin, H., Staehelin, T., and Gordon, J.,** Electrophoretic transfer of proteins from polyacrylamide gels to nitrocellulose sheets: procedure and some applications, *Proc. Natl. Acad. Sci. U.S.A.*, 76, 4350, 1979.
62. **Bowen, B., Steinberg, J., Laemmli, U. K., and Weintraub, H.,** The detection of DNA-binding proteins by protein blotting, *Nucleic Acids Res.*, 8, 1, 1980.
63. **Criss, W. E.,** A review of isozymes in cancer, *Cancer Res.*, 31, 1523, 1971.
64. **Roof, B. S., Carpenter, B., Fink, K. J., and Gordon, G. S.,** Some thoughts on nature of ectopic parathyroid hormones, *Am. J. Med.*, 50, 686, 1971.
65. **Ono, T. and Weinhouse, S.,** *Isozyme and Enzyme Regulation in Cancer*, Gann Monograph 13, University of Tokyo Press, Tokyo, 1972.
66. **Weinhouse, S., Stratton, J. B., and Morris, H. P.,** in *Cancer Enzymology*, Schultz, J. and Ahmed, D., Eds., Academic Press, New York, 1976, 41.
67. **Weber, G.,** in *Cancer Enzymology*, Schultz, J. and Ahmed, D., Eds., Academic Press, New York, 1976, 63.
68. **Chiu, J.-F., Decha-Umphai, W., Markert, C., and Little, B. W.,** Immunospecificity of nuclear nonhistone protein-DNA complexes in colon adenocarcinoma, *J. Natl. Cancer Inst.*, 63, 313, 1979.
69. **Chiu, J.-F., Pump, D. E., and Gootnick, D.,** Antigenic changes in nuclear chromatin in 1,2-dimethylhydrazine-induced colon carcinogenesis, *Cancer*, 45, 1193, 1980.
70. **Pumo, D. E., Gootnick, D., Little, B. W., and Chiu, J.-F.,** Colon tumor specific nuclear antigen with potential as a pre-tumor diagnostic probe, *Cancer Lett.*, 9, 285, 1980.
71. **Gabryelak, T., Pumo, D. E., and Chiu, J.-F.,** Changes in tumor-specific nuclear antigen activity in carcinogen-treated colon by tumor promoter and carcinogen inhibitors, *Cancer Res.*, 41, 3392, 1980.
72. **Romani, M., Vidali, G., Tahourdin, C. S. M., and Bustin, M.,** Solid phase radioimmunoassay for chromosomal components, *J. Biol. Chem.*, 255, 468, 1980.
73. **Druckrey, H.,** Production of colonic carcinoma by 1,2-dialkylhydrazines and asoxyalkanes, in *Carcinoma of the Colon and Antecedent Epithelium*, Burdette, W. J., Ed., Charles C Thomas, Springfield, Ill., 1970, 267.
74. **Boutwell, P. K.,** The functions and mechanism of promoters of carcinogenesis, *CRC Crit. Rev. Toxicol.*, 2, 419, 1969.
75. **Slaga, T. J., Sivak, R. K., and Boutwell, P. K., Eds.,** *Mechanisms of Tumor Promotion and Cocarcinogenesis*, Raven Press, New York, 1978.
76. **Wattenberg, L. W.,** Inhibitors of chemical carcinogenesis, *Adv. Cancer Res.*, 26, 197, 1978.
77. **Asao, T. and Pollard, M.,** Strain susceptibility and resistance to 1,2-dimethylhydrazine-induced enteric tumors germfree rats, *Proc. Soc. Exp. Biol. Med.*, 158, 89, 1978.
78. **Miller, J. A.,** Carcinogenesis by chemicals: an overview — G. H. A. Clowes Memorial lecture, *Cancer Res.*, 30, 559, 1970.
79. **Chiu, J.-F., Markert, C., and Decha-Umphai, W.,** Immunospecific nuclear antigens in colon adenocarcinomas, in *Carcino-Embryonic Proteins*, Vol. 2, Lehmann, F. G., Ed., Elsevier, Amsterdam, 1979, 581.
80. **Krupey, T., Wilson, T., Freedman, S. D., and Gold, P.,** The preparation of purified carcinoembryonic antigen of human digestive system from large quantities of tumor tissues, *Immunochemistry*, 9, 617, 1972.

81. **Davis, F. M., Gyorkey, F., Busch, R. K., and Busch, H.**, Nucleolar antigen found in several human tumors but not in the nontumor tissues studied, *Proc. Natl. Acad. Sci. U.S.A.*, 76, 892, 1979.
82. **Marashi, F., Davis, F. M., Busch, R. K., Savage, H. E., and Busch, H.**, Purification and partial characterization of nucleolar antigen-I of the Novikoff hepatoma, *Cancer Res.*, 39, 59, 1979.
83. **Duhl, D. M., Banjar, Z., Briggs, R. C., Page, D. L., and Hnilica, L. S.**, Tumor-associated chromatin antigens of human colon adenocarcinoma cell lines HT-20 and LoVo, *Cancer Res.*, 42, 594, 1982.
84. **Boffa, L. C., Vidali, G., and Allfrey, V. G.**, Selective synthesis and accumulation of nuclear nonhistone proteins during carcinogenesis of the colon induced by 1,2-dimethylhydrazine, *Cancer*, 36, 2356, 1975.
85. **Boffa, L. C. and Allfrey, V. G.**, Characteristic complements of nuclear nonhistone proteins in colonic epithelial tumors, *Cancer Res.*, 36, 2678, 1976.
86. **Wakabayachi, K., Wang, S., and Hnilica, L. S.**, Immunospecificity of nonhistone proteins in chromatin, *Biochemistry*, 13, 1027, 1974.
87. **Chiu, J.-F., Craddock, C., Morris, H. P., and Hnilica, L. S.**, Immunospecificity of chromatin nonhistone protein-DNA complexes in normal and neoplastic growth, *FEBS Lett.*, 42, 94, 1974.
88. **Chiu, J.-F., Wakabayshi, K., Craddock, C., Morris, H. P., and Hnilica, L. S.**, Changes in nuclear macromolecules during normal and neoplastic growth, in *Cell Cycle Controls*, Padilla, G. M., Zimmerman, A. M., and Cameron, I. L., Eds., Academic Press, New York, 1974, 308.
89. **Tahourdin, C. S. M. and Bustin, M.**, Chromatin subunits elicit species-specific antibioties against nucleoprotein antigenic determinants, *Biochemistry*, 19, 4387, 1980.
90. **Saumweber, H., Symmons, P., Kabisch, R., Will, H., and Bonhoffer, F.**, Monoclonal antibodies against chromosomal proteins of *Drosophila melanogaster*, *Chromosoma*, 80, 253, 1980.
91. **Bustin, M. and Neihardt, N. K.**, Antibodies against chromosomal MHMG proteins stain the cytoplasm of mammalian cells, *Cell*, 16, 181, 1979.
92. **Silver, L. M. and Elgin, S. C. R.**, Distribution patterns of three subfractions *Drosophila* nonhistone chromosomal proteins: possible correlations with gene activity, *Cell*, 11, 971, 1977.
93. **Chytil, F.**, Immunochemical analysis of nonhistone proteins, *Meth. Cell Biol.*, 18, 123, 1978.
94. **Pumo, D. E., Viceps-Madore, D., and Chiu, J.-F.**, Enzyme-linked immunosorbent assay for use with nuclear protein-DNA complex antigens, *Mol. Cell. Biochem.*, 51, 133, 1983.
95. **Chiu, J. F., Fujitani, H., and Hnilica, L. S.**, Methods for selective extraction of chromosomal nonhistone proteins, *Meth. Cell Biol.*, 14, 283, 1977.
96. **Farber, E.**, Similarities in the sequence of early histological changes induced in the liver of the rat by ethionine, 2-acetyl aminofluorene and 3-'methyl-4-dimethylaminoazobenzene, *Cancer Res.*, 16, 142, 1956.
97. **Price, J. M., Harman, J. W., Miller, E. C., and Miller, J. A.**, Progressive microscopic alterations in the livers of rats fed the hepatic carcinogens, 3'-methyl-4-dimethylaminoazobenzene and 4'-fluoro-4-dimethylaminoazobenzene, *Cancer Res.*, 12, 192, 1952.
98. **Chiu, J.-F., Tsai, Y. H., Sakuma, K., and Hnilica, L. S.**, Regulation of *in vitro* mRNA transcription by a fraction of chromosomal proteins, *J. Biol. Chem.*, 250, 9431, 1975.
99. **Mintz, B. and Illmensee, K.**, Normal genetically mosaic mice produced from malignant teratocarcinoma cells, *Proc. Natl. Acad. Sci. U.S.A.*, 72, 3585, 1975.
100. **Fahmy, M. J. and Fahmy, O. G.**, Intervening DNA insertions and the alteration of gene expression by carcinogen, *Cancer Res.*, 40, 3374, 1980.
101. **Weinhouse, S.**, Glycolysis, respiration and anomalous gene expression in experimental hepatomas: G.H.A. Clowes memorial lecture, *Cancer Res.*, 32, 2007, 1972.
102. **Stalder, J., Larsen, A., Engel, J. D., Dolan, M., Groudine, M., and Weintraub, H.**, Tissue-specific DNA cleavages in the globin chromatin domain introduced by DNase I, *Cell*, 20, 451, 1980.
103. **Wu, C. and Gilbert, W.**, Tissue-specific exposure of chromatin structure at the 5'-terminus of the rat preproinsulin II gene, *Proc. Natl. Acad. Sci. U.S.A.*, 78, 1577, 1981.
104. **Perle, M. A., Leonard, C. M., and Newman, S. A.**, Developmentally regulated nonhistone proteins: evidence for deoxyribonucleic acid binding role and localization near deoxyribonuclease-I sensitive domains of precartilage cell chromatin, *Biochemistry*, 21, 2379, 1982.
105. **Wu, C., Bingham, P. M., Livak, K. J., Holmgren, R., and Elgin, S. C. R.**, The chromatin structure of specific genes. I. Evidence for higher order domains of defined DNA sequences, *Cell*, 16, 979, 1979.
106. **Wu, C.**, The 5'-ends of Drosophila heat shock genes in chromatin are hypersensitive to DNase I, *Nature (London)*, 286, 854, 1980.
107. **Weintraub, H, Larsen, A., and Groudine, M.**, α-Globin gene switching during the development of chicken embryos: expression and chromosome structure, *Cell*, 24, 333, 1981.
108. **Elgin, S. C. R.**, DNase I-hypersensitive sites of chromatin, *Cell*, 27, 413, 1981.
109. **Weintraub, H., Beng, H., Groudine, M., and Graf, T.**, Temperature-sensitive changes in the structure of globin chromatin in lines of red cell precursors transformed by ts-AEV, *Cell*, 28, 931, 1982.
110. **Elgin, S. C. R.**, Chromatin structure, DNA structure, *Nature (London)*, 300, 402, 1982.

111. **Allfrey, V. G., Boffa, L. C., and Vidali, G.,** A DNA-binding protein class appearing in DMH-induced carcinogenesis, in *Biological Markers of Neoplasia: Basic and Applied Aspects,* Ruddon, R. W., Ed., Elsevier, New York, 1978, 351.
112. **Nelson, D., Covault, J., and Chalkley, R.,** Segregation of rapidly acetylated histones into a chromatin fraction released from intact nuclei by the action of micrococcaly nuclease, *Nucleic Acids Res.,* 8, 1745, 1980.
113. **Gazit, B., Panet, A., and Cedar, H.,** Reconstitution of a deoxyribonuclease I-sensitive structure on active genes, *Proc. Natl. Acad. Sci. U.S.A.,* 77, 1787, 1980.
114. **Sandeen, G., Wood, W. I., and Felsenfeld, G.,** The interaction of high mobility proteins HMG 14 and 17 with nucleosomes, *Nucleic Acids Res.,* 8, 3757, 1980.
115. **Howard, G., Abmeyer, S. M., Shinefeld, L. A., Sato, V. L., and Elgin, S. R.,** Monoclonal antibodies against a specific nonhistone chromosomal protein of Drosophila associated with active genes, *J. Cell Biol.,* 88, 219, 1981.

Chapter 2

TWO SPECIFIC GROUPS OF NHC PROTEINS INVOLVED IN GENE EXPRESSION*

C. C. Liew, M. J. Halikowski, and M. S. Zhao

TABLE OF CONTENTS

* This article is dedicated to Professor A. G. Gornall on his retirement from the Department of Clinical Biochemistry, University of Toronto.

It has been suggested that either the DNA structure by itself or the interaction between chromatin proteins and nucleic acids (both DNA and RNA) controls specific gene expression. In the following report, we will present evidence which has accumulated in our laboratory in recent years together with results from other investigators to show that nonhistone chromatin (NHC) proteins are involved in both gene structure and expression. However, we do not exclude the possibility that DNA itself is involved in the regulation of gene expression. Specifically we would like to address three major questions, each related to our ongoing research.

I. ARE NHC PROTEINS ASSOCIATED WITH NUCLEOSOME STRUCTURE?

The association of NHC proteins with nucleosome structure has been clearly established by several investigators.[1-4] We provided the initial evidence that chromatin subunits (i.e., 11S particles) derived from micrococcal nuclease digestion, followed by subsequent sucrose density gradient centrifugation revealed the presence of several groups of NHC proteins (Figure 1). Although a group of NHC proteins associated with the heterogeneous ribonucleoprotein particles (HnRNP) also co-sediments with the chromatin subunit, there remain certain specific NHC proteins in the nucleosomes. In particular, a highly phosphorylated NHC protein (B_2), with a molecular weight of 68,000 and pI 6.5 to 8.2, has been identified.[3,5-7] This protein was further characterized and found to be a major phosphorylated protein[3] in the nucleosomes of rat liver nuclei.

Subsequent to our findings, several NHC proteins associated with nucleosomes have been identified by other investigators (Table 1). Questions have arisen whether the NHC proteins are associated with linker DNA in the nucleosome, or indeed attached to the nucleosome, and also whether there is a stoichiometric relationship between nucleosomes and the presence of NHC proteins. So far there are no definitive answers to these questions, though the high mobility group (HMG) proteins have been examined extensively in relation to nucleosome structure and their interactions with nucleosome DNA.[2,8-10]

We have undertaken to establish the location of the phosphorylated protein B_2 in nucleosomes by both biochemical and immunological characterization.[6,22] Using histone H_1, which is believed to be associated with the linker DNA region,[23] two types of monomers could be resolved following micrococcal nuclease digestion and polyacrylamide gel analysis. It was found that the phosphorylated protein B_2 is associated with both monomers (MN_1 and MN_2). These observations were confirmed through the use of specific anti-B_2 antibody, which was produced by injection of highly purified B_2 phosphorylated protein into rabbits. The antiserum reacted specifically with both monomers (MN_1 and MN_2), enabling us to conclude that phosphorylated B_2 is definitely associated with mononucleosome structure. We could not, however, rule out its association with the linker DNA. In view of the dynamic nature of chromatin structure and function, we believe that most NHC proteins are associated with the nucleosome, not in a static condition, but rather in a state of dynamic flux. It is possible that the highly phosphorylated state of this protein enables it to slide along from one nucleosome to another in chromatin.

It has been established that nucleosomes are deoxyribonucleoprotein (DNP) particles consisting of 145 base pairs (bp) of DNA wrapped around the histone core octamer (2 H_{2a}:H_{2b} and 2 H_3:H_4) in $1^3/_4$ left-handed superhelical turns with 80 bp per turn.[24,25] The current view is that the locus of the NHC protein interaction is with DNA and the linker or spacer DNA.[1-4] We found that a brief digestion with micrococcal nuclease at low salt concentrations yields mononucleosomal particles which contain mostly 200 to 240 bp of DNA with histone H_1 and NHC proteins.[6,7] Using a longer period of digestion, a relatively resistant particle was obtained with 160 bp of DNA, without histone H_1; however, some NHC proteins remained in the particle.[28] If digestion was carried out in relatively high salt (e.g., 0.3 *M*)

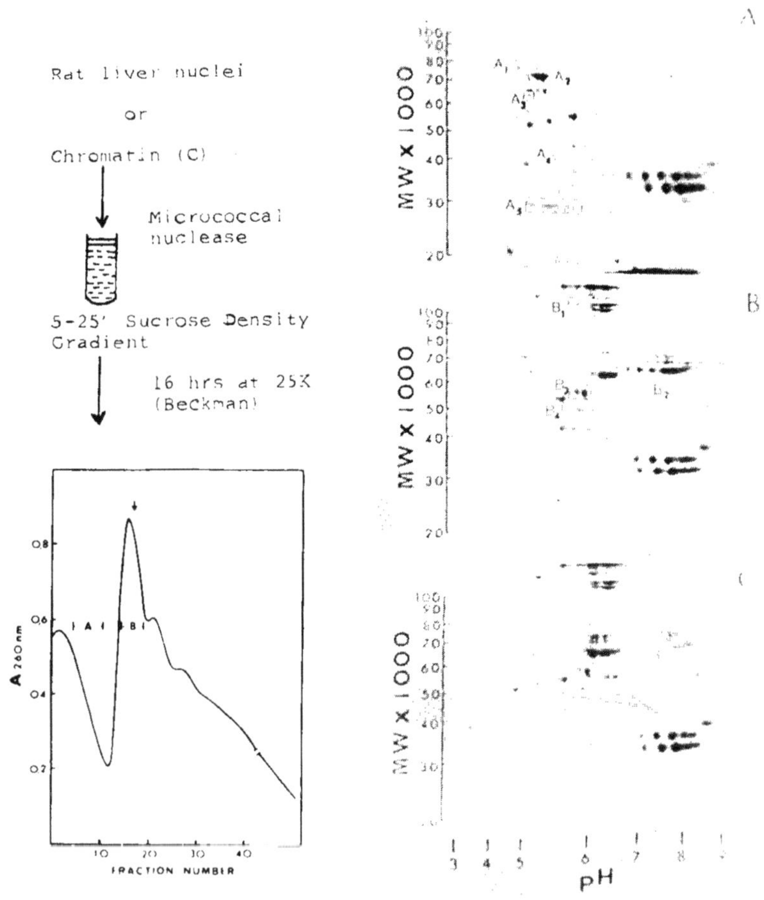

FIGURE 1. Liver nuclei were isolated from Wistar albino rats (150 g). Chromatin was prepared by the method of Reeder.[67] Nuclei or chromatin, (C), was digested with micrococcal nuclease as described previously.[5] The supernatant was layered onto 5 to 25% sucrose density gradient. The top fraction, A, and the 11S particles, B, were precipitated with 10% TCA, and NHC proteins were extracted by phenol-TEM buffer after removal of acid-soluble protein with 0.25 *N* HCl. Equal amounts (250 μg per gel) of each fraction of NHC protein were analyzed by one-dimensional isoelectricfocusing gel and followed by two-dimensional SDS-PAGE.[5] (A) Top fraction of the gradient which contained most of the nucleoplasmic proteins and proteins associated with the linker DNA; (B) the 11S particles which contained mainly the mononucleosomes and the heterogeneous ribonucleoprotein particles; (C) the whole spectrum of NHC proteins.

most of the NHC proteins were easily dissociated from the nucleosomes.[29] These observations indicate that the association of NHC proteins with core nucleosomes depends on both the ionic strength and the duration of digestion. In support, it has been shown that the ultrastructure of chromatin is absolutely dependent on the ionic strength (either mono- or divalent) used in the preparation of chromatin.[30,31] Given these facts, we have directed our research to a comparison of active and inactive chromatins, rather than products obtained by inducing in vitro conformational changes.

II. ARE THERE ANY SPECIFIC NHC PROTEINS ASSOCIATED WITH NUCLEOSOMES OF ACTIVE CHROMATIN?

The transcribed region of chromatin can be defined as the active polynucleosome fraction, since any given message actively transcribed must encompass a gene having more than the

Table 1
NHC PROTEINS IDENTIFIED IN THE NUCLEOSOME STRUCTURE

Proteins	Remarks	Ref.
High-mobility group proteins (HMG-14 and -17)	HMG-14 and -17 bind to two specific sites of core nucleosome of 160-bp DNA	8—10
Protein kinase	Associated with salt-washed nucleosome core	11, 12
Protein A_{24}	Ubiquitin moiety linked to residue 119 of histone H_{2a}	13—15
(ADP)-polymerase	Preferentially located in internucleosomal regions	16—18
B_2 phosphorylated proteins	Associated with mononucleosomes of actively transcribed chromatin	5, 6
DNA topoisomerase I	Nicking-closing enzyme required partially for nucleosome assembly	19
Nucleoplasmin	Required for nucleosome assembly	20, 21
RNA polymerase II	Associated with core particles from transcribed genes	21a

240 bp of a nucleosome. Since we have established that the phosphorylated protein B_2 is in the nucleosome structure, we have examined the question of whether or not the phosphorylated protein B_2 is associated with nucleosomes of active chromatin.

Two approaches have been taken. First, we used the conventional and classical fractionation of active and inactive chromatin by high salt, an approach carried out during the last decade by other investigators.[32-34] The method developed by Bekhor and co-workers was chosen. Their method has been modified (simplified in Figure 2) by several investigators in order to examine the following aspects of gene expression in eukaryotic cells: (1) characterization of tissue-specific NHC proteins,[35-36] (2) isolation of specific hormonal receptors or acceptors in chromatin,[37-39] and (3) production of specific antibody from the dehistonized chromatin proteins, in particular those responsible for carcinogenic induction.[40,41] This method is similar to the method currently used to prepare nuclear matrix, which has been demonstrated by several investigators to exhibit the involvement of cell-type specific DNA fragments associated with mRNA processing, DNA replication sites, receptor sites, etc.[29,33,42]

The second approach to fractionation uses nucleases as probes to differentiate the actively transcribed chromatin from the inactive fractions. Due to the conformation of chromatin, the actively transcribed regions are more sensitive to nucleases (such as micrococcal nuclease, DNase I and II) than nontranscribed regions. It has been found that most of the actively transcribed genes are rapidly degraded by treatment with nucleases. By using a molecular probe (e.g., the heat-shock gene or the globin gene), the DNase I hypersensitive site(s) of a specific gene can be identified in a specific tissue. For example, the DNase I hypersensitive site(s) of the globin gene was identified in the chick reticuloycte but not in brain or liver tissue.[43] Levy and Dixon[26] and Berkowitz and Riggs[27] have demonstrated that actively transcribed chromatin is degraded following mild treatment with micrococcal nuclease.

Using these two approaches, we first prepared actively transcribed and inactive chromatin by the methods of Bekhor et al.,[43-45] as outlined in Figure 2. We also used the regeneration induced by partial hepatectomy to achieve a further enhancement of gene expression in rat liver. As shown in Table 2, 16 hr after partial hepatectomy, when most of the genes in liver are highly expressed to prepare for cell growth and mitosis, the proteins of the DNA-P fraction, which contain most of the actively transcribed genes[43] were also highly phosphorylated.[44] Analyzed by the two-dimensional polyacrylamide gel electrophoretic (PAGE)

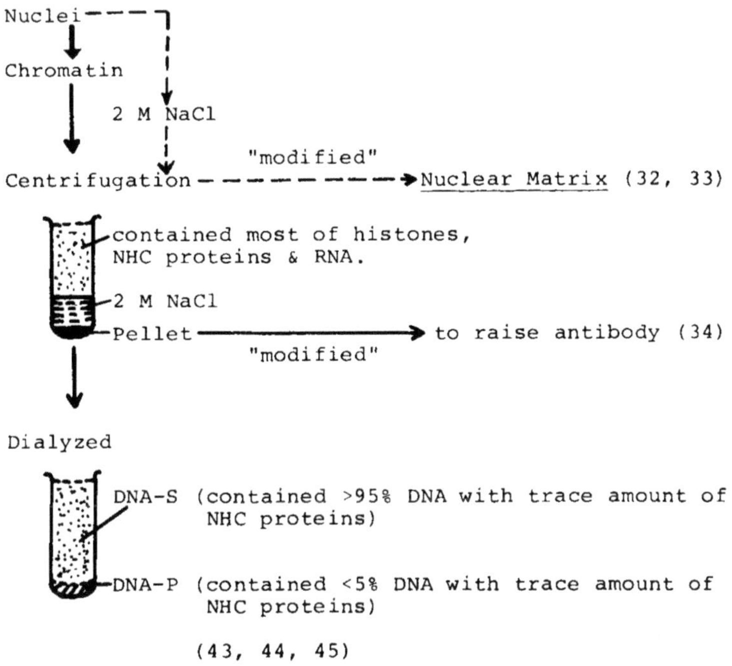

FIGURE 2. Chromatin was prepared from highly purified rat liver nuclei.[5] Chromatin was extracted by 2 *M* NaCl-1 m*M* phenylmethylsulfonyl fluoride (PMSF) to remove most of the histones, NHC proteins, and RNA. The resultant pellet, following ultracentrifugation for 20 hr, contained >99% DNA and trace amounts of tightly bound NHC proteins. This pellet was re-suspended in 10 m*M* Tris-HCl buffer and dialyzed extensively to solubilize the protein-free DNA (>95%). Following centrifugation a pellet (DNA-P) was obtained, which contained <5% DNA with tightly bound NHC proteins. Nuclear matrix was initially prepared by 2 *M* NaCl extraction to remove most of the nucleic acids and nucleoproteins, as depicted by the broken line (- - -). An alternative method used to prepare the tightly bound chromatin protein complex can be carried out as described by Chiu et al.[34]

Table 2
INCORPORATION OF ^{32}P INTO NHC PROTEINS

	DNA-S	DNA-P
Control	88	247
Partially hepatectomized	129	403

Note: NHC proteins of DNA-S and DNA-P were extracted by the phenol/TEM method.[66] This procedure ensures the removal of nucleic acids and phospholipids from NHC proteins.

system, a major phosphorylated protein with 68,000 mol wt and pI 6.5 to 8.2 was identified (Figure 3).[6] This protein, which has been identified previously in the nucleosome structure, is clearly associated with the DNA-P fraction containing actively transcribed gene sequences.

In an alternative approach, the isolated liver nuclei were subjected to mild nuclease digestion to release nucleosomes which are associated with actively transcribed genes. As shown in Figure 4, a much higher specific activity of ^{32}P incorporation into monomers,

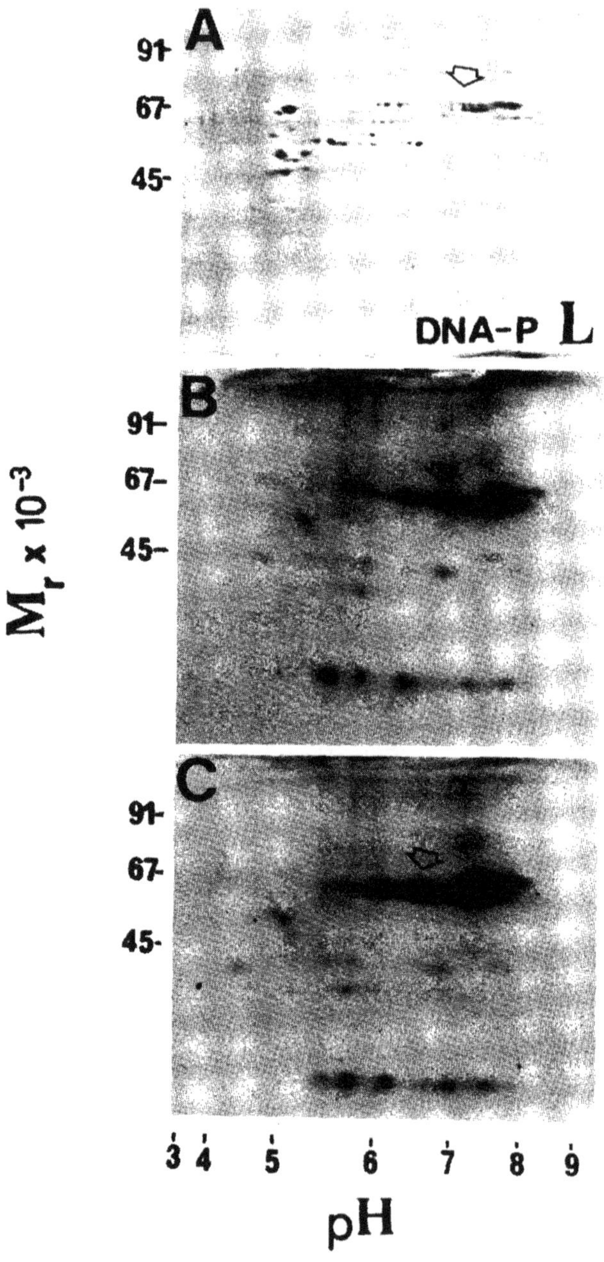

FIGURE 3. (^{32}P)-inorganic phosphate was injected i.p. into 150-g rats (5 mCi per rat) of both sham-operated (i.e., control) and 14 hr-postoperative partially hepatectomized rats for 2-hr pulse labeling. The DNA-P fraction was prepared to extract the NHC proteins. The NHC proteins were then analyzed by two-dimensional PAGE as described previously.[5] The electrophoretic pattern of the control as shown in panel A was identical to the partially hepatectomized rat (result not shown). However, autoradiographs of the control and the partially hepatectomized rats, showed distinctive differences in the intensity of the phosphorylated protein B_2 (arrow).

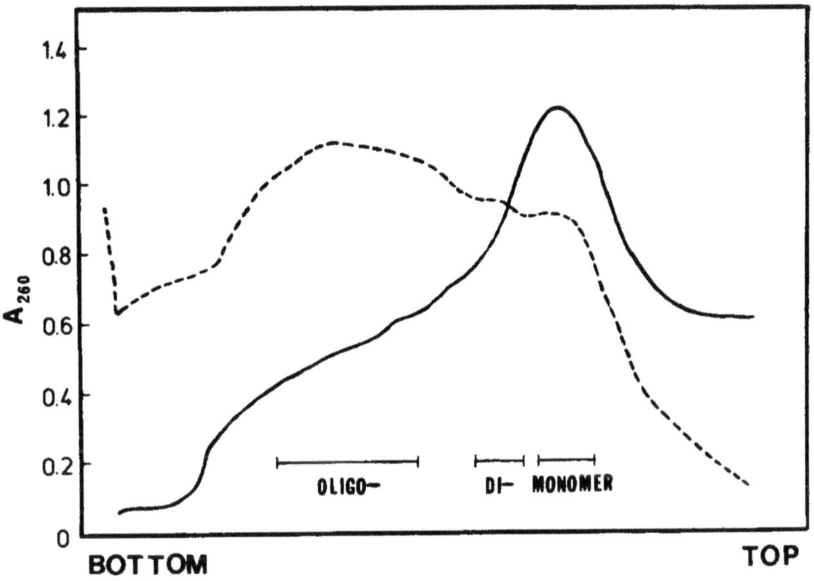

FIGURE 4. (^{32}P)-inorganic phosphate was given to a rat (150 g) i.p. for 2-hr pulse labeling. Isolated liver nuclei were digested with micrococcal nuclease (1.4 unit/A_{260}) for 5 and 15 min, respectively. The resultant supernatants were fractionated by 5 to 25% sucrose density gradient centrifugation. The peak of the monomers (equal amount) was then reacted with excess antiserum raised against phosphorylated protein B_2. Since the antibody of B_2 did not cross-react with proteins associated with heterogeneous ribonucleoprotein particles or other NHC proteins, the immunological precipitates of the monomers are specific to B_2.

which was immunoprecipitated following the 5-min treatment with micrococcal nuclease, was found. Again using the regenerating liver (i.e., highly activated gene transcription) vs its control relatively low level of transcription), we found an increase of protein phosphorylation to 144 and 177% at 5 and 15 min, respectively, in monomers released following treatment with micrococcal nuclease (Table 3). These highly phosphorylated fractions were specifically immunoprecipitated by an antibody to phosphorylated protein B_2.

In view of our current findings, the phosphorylated protein B_2 is definitely associated with nucleosomes possibly originating from actively transcribed chromatin.

III. WHAT IS THE ROLE OF NHC PROTEINS IN GENE EXPRESSION?

In view of recent findings in immunoglobulin genes,[47] oncogenes,[48,49] and calcitonin gene structure,[50] it has become evident that the DNA sequence by itself can mediate gene expression. However, the question remains as to why all cell types containing the same DNA information exhibit different cell-specific gene expression. For example, neuroendocrine genes are expressed only in nerve cells, not in other cell types. We believe that the interaction of chromatin proteins with nucleic acids must play an active role in gene expression. At least two groups of NHC proteins, which have been examined in our laboratory, provide evidence for this contention. One group of proteins is that known to be involved in the processing of mRNA maturation, i.e., the proteins associated with heterogeneous nuclear RNA (hnRNA).[51-54] The second is the group associated with active chromatin (i.e., the highly phosphorylated proteins) already discussed in this report.

Developmental studies of neonatal rat heart, as well as in the genesis of heart failure in cardiomyopathy, have shown that proteins associated with hnRNA, when analyzed by one-

Table 3
INCORPORATION OF ^{32}P INTO A PHOSPHOPROTEIN OF THE MONONUCLEOSOMES AS IDENTIFIED BY IMMUNOPRECIPITATION

	Radioactivity (cpm/A_{260})	
	5 min	15 min
Control sham-operated	439	316
Partially hepatectomized	632	560
% Change	144	177

Note: Rat liver nuclei labeled with ^{32}P in vivo were subjected to mild micrococcal nuclease digestion for 5 and 15 min, respectively. The peaks of mononucleosomal fractions were reacted with antiserum which was raised against phosphorylated protein B_2. The results of the precipitated complex of antigen-antibody using ^{125}I-protein A are presented. These precipitates were also dissolved in SDS sample buffer and the phosphorylated proteins were analyzed (results not shown). The protein B_2 was the major phosphorylated protein in the gel and was also found to contain mainly phosphorylated serine residues.[6]

and two-dimensional PAGE were strikingly altered.[55-57] In particular, we have been able to produce antibody to one of these hnRNA associated proteins (HrP_4) and found that there was a variation of this protein per nucleus in neonatal, adult, and aging hearts.[56]

Organization of a chromatin fiber to constitute a genetic unit which can be expressed requires at least 10 to 200 kb of DNA to be arranged into a higher-order structure. Evidence provided by Finch and Klug[58] suggests that the basic chromatin structure is constituted of approximately six nucleosomes per turn, to generate a 300-Å fiber. If this structure is taken to represent a genetic unit, at least 15 turns are required to constitute a transcription unit. It has also been shown[41-43,59] that active chromatin (euchromatin) is generally more open in configuration than the densely packaged inactive chromatin (heterochromatin). Active chromatin is also more sensitive to nuclease digestion than inactive chromatin.[2,41,43] A DNase I hypersensitive site has been detected in a specific gene probe, and these DNase I hypersensitive sites were not detected in any other cell types.[3-5,42-44,59-61]

These findings strongly indicate that in addition to the precise and conserved interaction between DNA and histones, the NHC proteins must play a crucial role in perturbing certain domain(s) affecting the structure and function of chromatin. The NHC proteins must be involved in designating a special configuration (e.g., domains) or packaging of the chromatin fiber for differential gene expression. In general, it is important to note that the organization of chromatin in the nucleus of eukaryotes defines the specific gene expression at differential stages of an organism. Since the amount of NHC proteins varies with cell development, it is reasonable to offer the model in Figure 5, which suggests that NHC proteins participate in gene regulation in two major fashions — structural and functional (detailed information regarding the involvement of NHC proteins in chromatin structure and function has been reviewed by Cartwright et al.).[62]

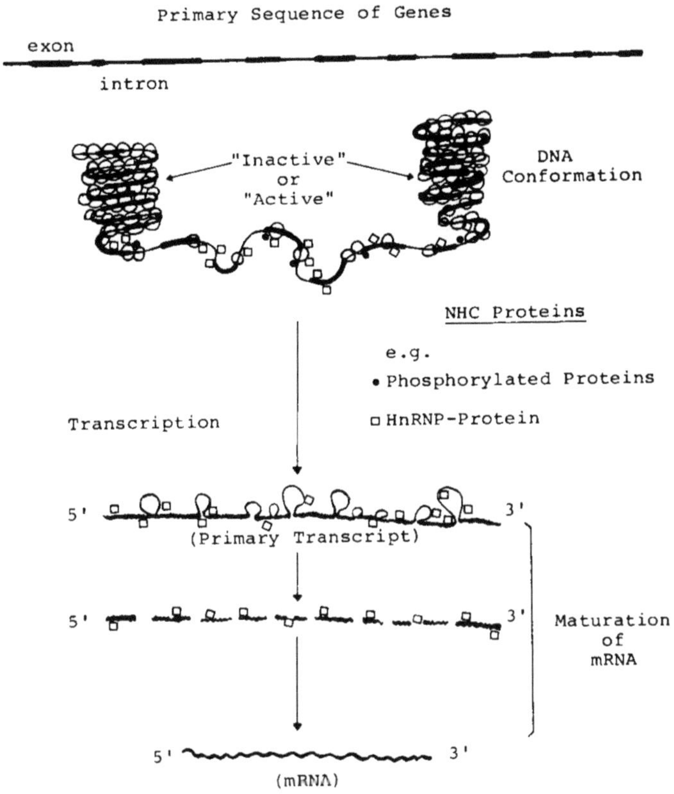

FIGURE 5. Model of conformational changes of DNA structure due to the presence of NHC proteins. Two specific groups of NHC proteins were examined. One group of proteins (□) identified as being involved in hnRNA processing to mature mRNA, as characterized by several investigators[51-54] were represented to be an integral part of chromatin protein. The second group of NHC proteins (●) is a highly phosphorylated protein B_2 (68,000 mol wt, pI 6.5 to 8.2) and was found to be associated with mononucleosomes of active chromatin.[5,6]

With regard to the functional aspect, it has been known for years[11] that NHC proteins are associated with many enzyme activities, e.g., protein kinase, DNA, RNA, (ADP) polymerases, etc. Significant amounts of NHC proteins also function as regulating factors (e.g., DNA-binding proteins, sigma factor, etc.), such as in determining the rate of replication and transcription. It has been known for some time that histones have no specific spacing function in the assembly of nucleosomes in vitro.[63,64] The variable linker DNA (with specific spacing) may be dependent on the availability of NHC proteins, perhaps to limit the sliding of the nucleosome.

Regarding the structural aspect, it has also been found that many NHC proteins are identical in most cell types as revealed by two-dimensional PAGE.[65] Some of these proteins are also abundant in chromatin. These groups of NHC proteins may be involved in a structural aspect rather than a functional role.

However, the two groups of NHC proteins which we have investigated extensively could be involved in both structural and functional aspects, even though their precise role(s) has yet to be delineated. These genes must be transcribed in a concerted fashion in order to maintain the normal physiological state of an organism. NHC proteins may therefore be involved in crucial steps, by affecting conformational changes in chromatin structure, perhaps as well as by modulating the rate of transcription in order to express genes involved in differentiating cell types.

ACKNOWLEDGMENTS

Generous support by grants from the Medical Research Council and the Ontario Heart Foundation, Canada, as well as from the U.S. Public Health Service (HL-18824-01) was greatly appreciated. Dr. M. S. Zhao was a Fellow of the Canadian Heart Foundation and Mr. M. J. Halikowski was supported by an open fellowship from the University of Toronto.

REFERENCES

1. **Defer, N., Ditzis, A., Tichonicky, L., Sabatier, M., Levy, F., and Kruh, J.,** Presence of non-histone proteins in nucleosomes, *Eur. J. Biochem.*, 88, 583, 1978.
2. **Goodwin, G. H., Woodhead, L., and Johns, E. W.,** The presence of high mobility group proteins in isolated nucleosomes, *FEBS Lett.*, 73, 85, 1977.
3. **Liew, C. C. and Chan, P. K.,** Identification of non-histone chromatin proteins in chromatin subunits, *Proc. Natl. Acad. Sci. U.S.A.*, 73, 3458, 1976.
4. **Paul, J. and Malcolm, S.,** A class of chromatin proteins associated with the nonhistone proteins, *Biochemistry*, 15, 3510, 1976.
5. **Chan, P. K. and Liew, C. C.,** Identification of specific phosphoproteins in nuclease-digested chromatin subunits, *Can. J. Biochem.*, 55, 847, 1977.
6. **Chan, P. K. and Liew, C. C.,** Purification of a phosphoprotein from chromatin of rat liver, *Biochem. J.*, 183, 147, 1979a.
7. **Chan, P. K. and Liew, C. C.,** Identification of nonhistone chromatin proteins in chromatin subunits (or mononucleosomes) devoid of histone H_1, *Can. J. Biochem.*, 57, 666, 1979b.
8. **Mardiau, J. K. W., Paton, A. E., Bunick, G. J., and Olins, D. E.,** Nucleosome cores have two specific binding sites for nonhistone chromosomal protein HMG 14 and HMG 17, *Science*, 209, 1534, 1980.
9. **Johns, E. W.,** HMG chromosomal proteins — conformations and interactions, in *HMG Chromosomal Proteins*, Johns, E. W., Ed., Academic Press, New York, 1982, 89.
10. **Goodwin, G. H., Walker, J. M., and Johns, E. W.,** The high mobility group (HMG) nonhistone chromosomal proteins, in *The Cell Nucleus*, Vol. 6 (Part C), Busch, H., Ed., Academic Press, New York, 1978, 181.
11. **Phillips, I. R., Shephard, E. A., Stein, J. L., and Stein, G. S.,** Role of nonhistone chromosomal proteins in selective gene expression, in *Eukaryotic Gene Expression*, Kolodny, G. M., Ed., CRC Press, Boca Raton, Fla., 1980, 129.
12. **Kitzis, A., Tichonicky, L., Defer, N., and Kruh, J.,** Localization of phosphoproteins and of protein kinases in chromatin from hepatoma tissue-cultured cells, *Eur. J. Biochem.*, 111, 237, 1980.
13. **Martinson, H. G., True, R., Burch, J. B. E., and Kunkel, G.,** Semihistone protein A24 replaces H_{2a} as an integral component of the nucleohistone core, *Proc. Natl. Acad. Sci. U.S.A.*, 76, 1030, 1979.
14. **Goldknopf, I. L., Frency, R., Musso, R., and Busch, H.,** Presence of protein A24 in rat liver nucleosomes, *Proc. Natl. Acad. Sci. U.S.A.*, 74, 5492, 1977.
15. **Goldknopf, I. L. and Busch, H.,** Modification of nuclear proteins: the ubiquitin-histone 2A conjugate, in *The Cell Nucleus*, Vol. 6, Busch, H., Ed., Academic Press, New York, 1978, 149.
16. **Mullins, D. W., Jr., Giri, C. P., and Smulson, M.,** Poly (ADP) polymerase: the distribution of a chromosome associated enzyme within the chromatin substructure, *Biochemistry*, 16, 506, 1977.
17. **Yukioka, M., Ikai, Y., Hasuma, T., and Inoue, A.,** Non-preferential localization of poly(ADP-ribose) polymerase activity in transcriptionally active chromatin, *FEBS Lett.*, 86, 85, 1978.
18. **Poirier, G. G., Murcio, G., Jongstro-Bilen, J., Niedergang, C., and Mandel, P.,** Poly(ADP-ribosyl) ation of polynucleosomes cause relaxation of chromatin structure, *Proc. Natl. Acad. Sci. U.S.A.*, 79, 3423, 1982.
19. **Germond, J. E., Ronviere-Yaniv, J., Yaniv, M., and Brutlag, D.,** Nicking-Closing enzyme assembles nucleosome-like structure, *in vitro*, *Proc. Natl. Acad. Sci. U.S.A.*, 76, 3779, 1979.
20. **Laskey, R. A., Honda, B. M., Mills, A. D., and Finch, J. T.,** Nucleosomes are assembled by an acidic protein which binds histones and transfers them to DNA, *Nature (London)*, 275, 416, 1978.

21. **Mills, A. D., Laskey, R. A., Black, P., and DeRabertis, E. M.,** An acidic protein which assembles nucleosomes *in vitro* is the most abundant protein in Xenopus oocyte nuclei, *J. Mol. Biol.*, 139, 561, 1980.

21a. **Baer, B. W. and Rhodes, D.,** Eukaryotic RNA Polymerase II binds to nucleosome core from transcribed genes, *Nature (London)*, 301, 482, 1983.

22. **Zhao, M. S. and Liew, C. C.,** Production of antibody to phosphoprotein associated with nucleosome structure, *Can. J. Biochem.*, 60, 356, 1982.

23. **Thoma, F., Koller, T., and Klug, A.,** Involvement of histone H_1, in the organization of the nucleosome and of the salt-dependent superstructures of chromatin, *J. Cell. Biol.*, 83, 403, 1979.

24. **Finch, J. T., Brown, R. S., Rholdes, D., Richmond, T., Rushton, B., Lutter, L. C., and Klug, A.,** X-ray diffraction study of new crystal form of the nucleosome core showing higher resolution, *J. Mol. Biol.*, 145, 757, 1981.

25. **Bentley, G. A., Finch, J. T., and Lewit-Bentley, A.,** Neutron diffraction studies on crystals of nucleosome cores using contrast variation, *J. Mol. Biol.*, 145, 771, 1981.

26. **Levy, W. B. and Dixon, G. H.,** Partial purification of transcriptionally active nucleosomes from trout testis cells, *Nucleic Acids Res.*, 5, 4155, 1978.

27. **Berkowitz, E. M. and Riggs, E. A.,** Characterization of rat liver oligonucleosomes enriched in transcriptionally active genes: evidence for altered base composition and a shortened nucleosome repeat, *Biochemistry*, 20, 7283, 1981.

28. **Simpson, R. T.,** Structure of the chromotosome, a chromatin particle containing 160 base pairs of DNA and all the histones, *Biochemistry*, 17, 5524, 1978.

29. **Thoma, F. and Koller, T. H.,** Unraveled nucleosomes, nucleosome beads and higher order structures of chromatin: influence of non-histone components and histone H_1, *J. Mol. Biol.*, 149, 709, 1981.

30. **Hamkalo, B. A. and Rattner, J. B.,** Folding up genes and chromosomes, *Q. Rev. Biol.*, 55, 409, 1980.

31. **Zentgraf, H., Muller, U., and Franke, W. W.,** Reversible *in vitro* packaging of nucleosomal filaments into globular supranucleosomal units in chromatin of whole chick erythrocyte nuclei, *Eur. J. Cell. Biol.*, 23, 171, 1980.

32. **Berezney, R.,** Dynamic properties of the nuclear matrix, in *The Cell Nucleus*, Vol. 6, Busch, H., Ed., Academic Press, New York, 1979, 143.

33. **Berezney, R. and Coffey, D. S.,** Nuclear protein matrix: association with newly synthesized DNA, *Science*, 189, 291, 1975.

34. **Chiu, J.-F., Fujitani, H., and Hnilica, L. S.,** Methods for selective extraction of chromosomal proteins, in *Methods in Cell Biology*, Vol. 16, Stein, J., Stein, G. S., and Kleinsmith, L. J., Eds., Academic Press, New York, 1977, 283.

35. **Barrett, T. and Gould, H. J.,** Tissue and species specificity of non-histone chromatin proteins, *Biochim. Biophys. Acta*, 294, 165, 1973.

36. **Catins, J. J., Yeoman, L. C., Mandel, M., and Busch, H.,** Characterization of a DNA binding protein from rat liver chromatin which decreases during growth, *Biochemistry*, 17, 983, 1978.

37. **Spelsberg, T. C.,** The role of nuclear acidic proteins in binding steroid hormones, in *Acidic Proteins of the Nucleus*, Cameron, I. L. and Jeter, J. R., Jr., Eds., Academic Press, New York, 1974, 247.

38. **Thrall, C. L., Webster, R. A., and Spelsberg, T. C.,** Steroid receptor interaction with chromatin, in *The Cell Nucleus*, Busch, H., Ed., Academic Press, New York, 1979.

39. **Spelsberg, T. C., Webster, R., Pikler, G., Thrall, C., and Wells, D.,** Role of nuclear proteins as high affinity sites ("acceptors") for progesterone in the avian oviduct, *J. Steroid Biochem.*, 7, 1091, 1976.

40. **Hardy, K., Chiu, J.-F., Beyer, A. L., and Hnilica, L. S.,** Immunological properties of fractionated avian erythroid nuclei, *J. Biol. Chem.*, 53, 5825, 1978.

41. **Boffa, L. C. and Allfrey, V. G.,** Modifications of nuclear protein structure and function during carcinogenesis, in *The Cell Nucleus*, Vol. 7(Part D), Busch, H., Ed., Academic Press, New York, 1979, 521.

42. **Jackson, D. A., McCready, S. J., and Cook, P. R.,** RNA is synthesized at the nuclear cage, *Nature (London)*, 292, 552, 1981.

43. **Gates, D. M. and Bekhor, I.,** DNA sequence selection by tightly-bound non-histone chromosomal proteins, *Nucleic Acids Res.*, 6, 3411, 1979.

44. **Bekhor, I and Mirell, C. J.,** Simple isolation of DNA hydrophobically complexed with presumed gene regulatory proteins (M_3), *Biochemistry*, 19, 609, 1979.

45. **Norman, G. L. and Bekhor, I.,** Enrichment of selected active human gene sequences in the placental deoxyribonucleic acid fraction associated with tightly-bound nonhistone chromatin proteins, *Biochemistry*, 20, 3568, 1981.

46. **Liew, C.C., Halikowski, M. J., and Zhao, M. S.,** A phosphoprotein is preferentially released by mild micrococcal nuclease digestion of nuclei, *Biochem. J.*, in press.

47. **Leder, R.,** The genetics of antibody diversity, *Sci. Am.*, 246, 5, 1982.

48. **Weinberg, R. A.,** Fewer and fewer oncogenes, *Cell*, 30, 3, 1982.

49. **Pulciani, S., Santos, E., Lauver, A., Long, L., Robbins, K., and Barbacid, M.,** Oncogenes in human tumor cell lines: molecular cloning of a transforming gene from human bladder carcinoma lines, *Proc. Natl. Acad. Sci. U.S.A.*, 246, 2845, 1982.
50. **MacIntyre, I., Hillyard, C. J., Murphy, P. K., Reynolds, J. J., Gaines Das, R. E., and Craig, R. K.,** A second plasma calcium-lowering peptide from human calcitonin precursor, *Nature (London)*, 300, 460, 1982.
51. **Suria, D. and Liew, C.C.,** Characterization of proteins associated with nuclear ribonucleoprotein particles by two-dimensional polyacrylamide gel electrophoresis, *Can. J. Biochem.*, 57, 32, 1979.
52. **Stevenin, J., Gattani, R., Gallinaro-Matringe, H., and Jacob, M.,** Nuclear ribonucleoprotein particles contain specific proteins and unspecific non-histone nuclear proteins, *Eur. J. Biochem.*, 84, 541, 1978.
53. **Karn, J., Vidali, G., Boffa, V. C., and Allfrey, V. G.,** Characterization of the nonhistone nuclear proteins associated with rapidly labelled heterogeneous nuclear RNA, *J. Biol. Chem.*, 252, 7307, 1977.
54. **Beyer, A. L., Christensen, M. E., Walker, B. W., and LeStourgeon, W. M.,** Identification and characterization of the packaging proteins of core HnRNP 40S particles, *Cell*, 11, 127, 1977.
55. **Liew, C. C., Sole, M. J., Silver, M. D., and Wigle, E. D.,** Electrophoretic profiles of nonhistone nuclear proteins of human hearts with muscular subaortic stenosis, *Circ. Res.*, 46, 573, 1980.
56. **Wang, H. Y., Zhao, M. S., and Liew, C. C.,** Immunological studies of proteins which are associated with heterogeneous RNA in rat ventricular nuclei of different ages, Abstract No. 83, 11th Ann. Natl. Meeting Assoc. Aging, New York, 1982.
57. **Liew, C. C. and Sole, M. J.,** Nuclear proteins in the heart of the cardiomyopathic hamster: fractionation of the phenol-soluble nonhistone proteins by two-dimensional polyacrylamide gel electrophoresis, *Circ. Res.*, 42, 628, 1978.
58. **Finch, J. T. and Klug, A.,** Solenoid model for superstructure in chromatin, *Proc. Natl. Acad. Sci. U.S.A.*, 73, 1897, 1976.
59. **Wu, C., Wong, Y. C., and Elgin, S. C. R.,** The chromatin structure of specific genes. II. Disruption of chromatin structure during gene activity, *Cell*, 16, 807, 1979.
60. **Wu, C.,** The 5′ ends of Drosophila heat shock genes in chromatin are hypersensitive to DNase I, *Nature (London)*, 286, 854, 1980.
61. **Wu, C. and Gilbert, W.,** Tissue-specific exposure of chromatin structure at the 5′ terminus of the rat preproinsulin II gene, *Proc. Natl. Acad. Sci. U.S.A.*, 78, 1577, 1981.
62. **Cartwright, I. L., Abmayr, S. M., Fleischman, G., Lowenhaupt, K., Elgin, S. C. R., Keene, M. A., and Howard, G. C.,** Chromatin structure and gene activity: the role of nonhistone chromosomal proteins, in *Crit. Rev. Biochem.*, 13, 1, 1982.
63. **McGhee, J. D. and Felsenfeld, G.,** Nucleosome structure, *Ann. Rev. Biochem.*, 49, 1115, 1980.
64. **Felsenfeld, G.,** Chromatin, *Nature (London)*, 271, 115, 1978.
65. **Liew, C. C.,** Nonhistone chromatin proteins. Chemistry fractionation, and "fingerprints" by two-dimensional polyacrylamide gel electrophoresis, in *Structure and Function of DNA, Chromatin and Chromosomes*, Dion, D. S., Ed., Symposia Specialists, Inc., Miami, 1979, 153.
66. **Teng, C. S., Teng, C. T., and Allfrey, V. G.,** Studies of nuclear acidic proteins: evidence for their phosphorylation, tissue specificity, selective binding to deoxyribonucleic acid and stimulatory effects on transcription, *J. Biol. Chem.*, 246, 3597, 1971.
67. **Reeder, R. H.,** Transcription of chromatin by bacterial RNA polymerase, *J. Mol. Biol.*, 80, 229, 1973.

Chapter 3

NONHISTONE CHROMOSOMAL PROTEIN FRACTIONS OF REGULATORY CHARACTER SYNTHESIZED IN SPLEEN CELLS DURING THE IMMUNE RESPONSE

A. Horst and E. M. Rakowicz-Szulczyńska

TABLE OF CONTENTS

I. INTRODUCTION: THE MODEL

The regulation of genes and differentiation programs remains one of the central unresolved problems in modern biology. Sequencing and cloning of genes changed views not only on the structure and function of genes, but also on gene regulation. Previous views on gene regulation were principally based on the action of nuclear proteins, i.e., histones and non-histones. Nonhistone chromatin (NHC) proteins were generally considered the main gene regulators. Because there is a great number of eukaryotic genes (some hundreds of thousands), it has been suggested that there must be a similar great number of particular NHC proteins as specific gene regulators. According to modern views, specific chromatin structure, recognized by the regulatory proteins (principally enzymes) in interaction with specific gene inducers (hormones and antigens) may play the decisive role in gene regulation. With this model of gene regulation there is no need for an enormous number of regulatory proteins, because the same enzymes in interaction with specific gene inducers may activate particular genes for various functions.

Many models have been established to study gene regulation. One of the most attractive models is based on the activation of genes by steroid hormones. It is well proven that the early steps depend on hormones binding with specific cytoplasmic receptors. Structural changes of the hormone-receptor complex are necessary for entering into the nucleus, where a series of not entirely known events occur, causing an increased rate of transcription of specific genes with consequent increase of the level of specific mRNAs, coding for the synthesis of specific proteins.[1]

Much more complicated events must occur during the differentiation process in developing organisms leading to heterogeneous cell populations. The biochemical events differ during cell differentiation both in substances produced and in timing. Therefore when studying gene regulations in this respect it is important to work (if possible) with homogeneous and synchronized cell populations.

Changes in gene structure are important not only for gene induction, but also for its repression. Modifications of DNA, especially methylation of cytosine residues, seem to play an important role in repression of genes. According to modern views, methylation of cytosine residues occurs nonrandomly, which supports the idea of a regulatory character to this process. It has been proven that methylation of cytosine residues in prokaryotes inhibits synthesis of enzymes probably as a result of gene repression. The regulatory character of cytosine methylation in eukaryotes refers to the symmetrical methylation of both DNA chains, its clonal inheritability and tissue specificity, and methylation of other cytosine residues in differentiated cells.[2]

As a model for studying the NHC proteins, during cell differentiation we have chosen the induction of the immune response by injection of various antigens to animals. The advantage of this model is its physiological character in which distinct gene regulation is proceeding step by step until efficient genes for the immunological reactions are expressed. The first steps toward differentiation of these cells are based on the rearrangement of embryonic DNA with the consequence that V and C genes being separated by an unknown stretch of the embryonic DNA are moved closer together in the genome.[2-9] Differentiation consists of the recombination of genes coding for the variable part of the immunoglobulin light and heavy chains, securing the variability of the synthesized immunoglobulins.[10,11] These steps of the differentiation of immunocompetent cells are realized probably without participation of antigens. The newly rearranged V-genes seem to be responsible for the synthesis of receptors specific for the particular antigens.

The next steps in differentiation of immunocompetent cells are realized under the action of antigens able to interact with the products of V-genes of the future antibody molecule. As the consequence of the action of antigens, proliferation of antigen-reactive cells occurs with V-genes specific for the given antigen. During the immune reaction further differentiation is taking place; i.e., selection of immunoglobulin classes in which the V-part of the immunoglobulin molecule of both light and heavy chains remains, and the C-parts of the heavy chains change.[6] This selection is realized through excision of consequent genes coding for particular C-parts of the heavy chains, organized in clusters. It is suggested that the excision of particular genes with consequent joining of the remaining ones may be executed by specific recombinases.[7,8] The general outlines of the differentiation process in immunocompetent cells is shown in Figure 1.

The main disadvantage of our model is the complexity of the immune reaction, where a great number of cells participate in this reaction, e.g., B-cells, antigen-producing macrophages, and different subsets of T-cells. It can be anticipated that the majority of reactive proliferative cells are the B-cells. Some of the cells participating in our model may be active in a restricted number like the helper T-cells, the suppressor T-cells, and to a lesser amount, the macrophages. The main proliferating population of the induced cells during the immune reaction are the B-cells. The single B-cells with adequate V-parts of the synthesized immunoglobulins form clones synthesizing great amounts of antibodies. These cells, in addition to other cells of the immune reaction cause the enlargment of lymphatic organs, principally the spleen which can double in size. In consequence we presume that the bulk of active cells analyzed by us during the phase of antibody synthesis are clones of B-cells induced by various antigens to proliferate and synthesize immunoglobulins. However, to define exactly the cells synthesizing the NHC protein fractions described by us during the immune reaction is actually not possible.

II. MATERIALS AND METHODS

The experiments were performed with mice of CFW/bog strain and rats of the Wistar strain.

Mice were immunized i.p. with different antigens: 0.3 mℓ of 10% suspension of sheep red blood cells, 10 mg of aggregated human gamma globulin, 8 to 200 μg of pure human IgG in complete Freund's adjuvant (1:1), or 20 μg of lipopolisaccharides from *Escherichia coli* per mouse.

Rats were immunized with 3.5 × 10^7 of mouse spleen cells of CH3 strain, 0.5 to 1.0 mg of pure IgG in complete Freund's adjuvant (1:1) or 200 μg of lipopolisaccharides from *E. coli* per rat.

On the particular days of the immune reaction spleens were removed. Spleen cells were obtained by dissecting the tissues with scissors and pressing them through fine nylon mesh sieves. The spleen cells (3 to 7 × 10^7/mℓ) were incubated 12 hr in Eagle's medium supplemented with 20% of fetal calf serum, L-glutamine 30 μg/mℓ, kanamycin 60 μg/mℓ and one of the labeled amino acids (2 μCi/mℓ): [^{3}H]-tryptophan (spec. act. 8.8 Ci/mmol); [^{3}H]-methionine (spec. act. 12 Ci/mmol) and [^{3}H]-leucine (spec. act. 50 Ci/mmol).

After the incubation time chromatin was isolated as described previously.[12] Chromatin pellet obtained by centrifugation through 1.7 *M* sucrose, 10 m*M* Tris-HCl (pH 7.9) was dissociated in 5 *M* urea-2 *M* NaCl-0.1 M Tris-HCl (pH 7.9)-12 m*M* mercaptoethanol-1 m*M* phenylmethanosulfonylfluoride (PMSF). To each chromatin preparation sodium dodecyl

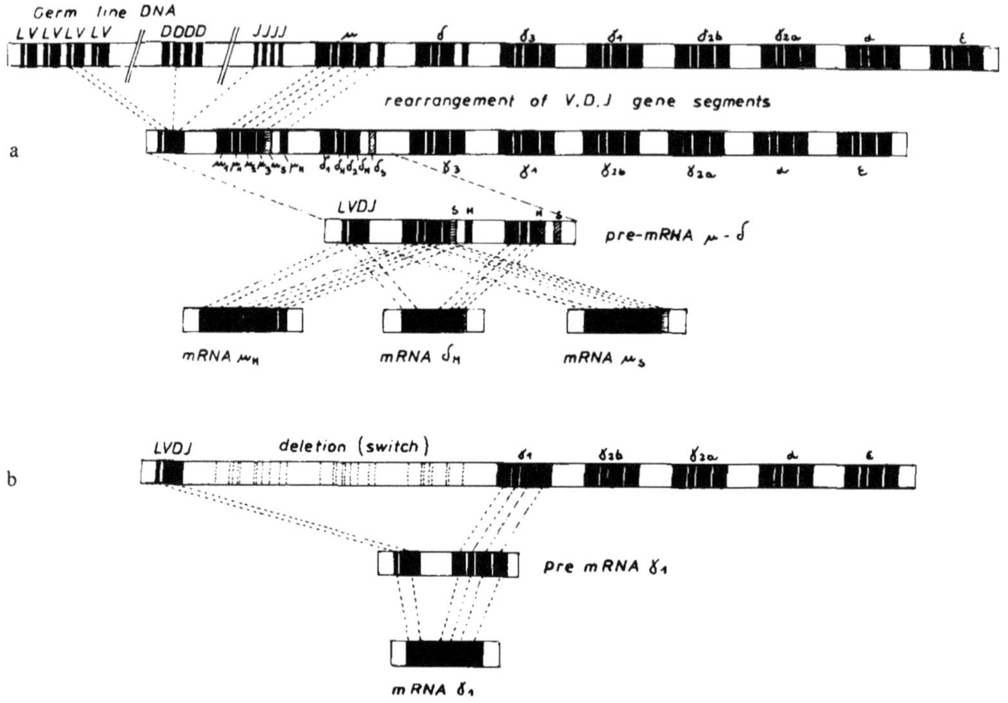

FIGURE 1. Model for gene rearrangements in generation of genes coding various heavy chain immunoglobulin molecules. (a) Germ line DNA (first line) carries particular clusters of V-genes (V, D, J) and C-genes of immunoglobulin heavy chain, interrupted by shorter or longer DNA sequences (L = leader sequences). Somatic recombination of the germ line DNA brings together the particular segments (V, D, J) coding the variable part of the heavy chain (second line). After rearrangement transcription yields pre-mRNA (third line). Processing of mRNA (splicing by excision of intervening sequences and several modifications) yields mature mRNAs: $mRNA_{\mu M}$ (membrane-bound IgM), $mRNA_{\delta M}$ (membrane-bound IgD), $mRNA_{\mu s}$ (secretory IgM) (fourth line). This step of differentiation of immuncompetent cells is done without stimulation by antigens. The synthesized molecules are probably the cell-receptors for circulating antigens. (b) The second step occurs after stimulation of B lymphocytes by antigens. Lymphocytes B, stimulated by thymus-dependent antigens synthesize in the first phase of the immune reaction IgM molecules by translation of $mRNA_{\mu s}$. In further phases of the immune reaction there occur sequential synthesis of different classes of immunoglobulins (IgG: four subclasses, IgA, and IgE). The sequential synthesis of immunoglobulin classes occurs by switch of the synthesis from one class of immunoglobulin to another until the most adequate class is synthesized. The switch of immunoglobulin classes is done by deletion of preceding genes (DNA sequences). Our experiments were executed at the events of immuoglobulin synthesis presented in (b).

sulfate (SDS) was added to 1% and the mixture dialyzed overnight at room temperature against 0.1 *M* phosphate buffer pH 7.2, containing 0.1% SDS.

Each chromatin preparation was characterized by monitoring absorbance at 260, 280, and 320 nm. The ratio of protein to DNA for each preparation was calculated from the ratio of A_{260} to A_{280} and from chemical analyses. The ratio of A_{260} to A_{280} differed at the particular days of the immune reaction. The ratio of histone to DNA is constant (about 1.0); therefore changes of the protein-to-DNA ratio after immunization must be ascribed to the changes of NHC protein contents. Protein was measured by the method of Lowry et al.[13] and DNA according to Burton.[14]

NHC proteins were analyzed by electrophoresis of chromatin preparation in 7.5% polyacrylamide gel with 0.1% SDS, according to the method of Bhorjee and Pederson[15] with the modification consisting in elimination of urea, as described previously.[12] Slices of gels (1.8 mm) were solubilized in H_2O_2 and radioactivity was measured in triton/toluene (1:1) scintillation fluid.

III. RESULTS

A. NHC Protein Fractions in Nonimmunized Mice and Rats

The A_{260} to A_{280} ratio of chromatin preparation was about 1.5 and the protein to DNA ratio 1.19 in mice and 1.58 in rats (Tables 1 and 2). After labeling with [^{3}H]-tryptophan, [^{3}H]-leucine, or [^{3}H]-methionine, chromatin proteins were separated by electrophoresis. The profile of nonhistone proteins from nonimmunized mice, labeled with [^{3}H]-tryptophan is shown in Figure 2a with the characteristic fraction A (molecular weight about 160,000). After incubation with [^{3}H]-methionine the highest incorporation occurred into fractions E_2 and E_3 (molecular weights about 35,000 and 20,000, respectively). After incubation with [^{3}H]-leucine no peaks were observed.

The electrophoretic profiles of nonhistone proteins from nonimmunized rats, labeled with [^{3}H]-tryptophan or [^{3}H]-methionine, did not show characteristic higher peaks; with [^{3}H]-leucine two peaks could be distinguished: fraction B (molecular weight about 120,000) and fraction E_1 (molecular weight about 40,000). (See Figure 2b.)

B. NHC Protein Fractions in Immunized Mice and Rats

1. Immunization with Cellular Antigens

a. Immunization of Mice with Sheep Red Blood Cells (Chromatin Proteins Labeled with [^{3}H]-Tryptophan)

The ratio of protein to DNA increased from 1.19 in control, nonimmunized mice, to 1.75 on the fourth day of immunization and later (in 7 weeks) decreased to the normal values (Table 1). Fraction A, synthesized in nonimmunized spleen cells, was not found 1 day after immunization of mice.

A newly synthesized fraction F (molecular weight 12,000) was found 2 days after immunization (Figure 3a). In addition to fraction F, another new fraction H (molecular weight 3000) was synthesized (Figure 3b) 4 days after immunization. Both fractions F and H were still present 8 days after immunization but were depleted at 14 days after immunization, at which time fraction A reappeared (Figure 3c).

b. Immunization of Rats with Mouse Spleen Cells (Chromatin Proteins Labeled with [^{3}H]-Tryptophan, [^{3}H]-Methionine, or [^{3}H]-Leucine)

The protein-to-DNA ratio increased during the first 4 days from 1.58 in control, nonimmunized rats to 2.38 in immunized, and later decreased slowly (during 7 weeks) to the control values (Table 2).

All of the three amino acids were incorporated in small amounts into all fractions (without distinct peaks) 1 day after immunization. Two days after immunization (Figure 3d) [^{3}H]-tryptophan was incorporated in greater amounts into fraction F (molecular weight 12,000) [^{3}H]-leucine into fraction E_2 (molecular weight 35,000). Radioactivity of [^{3}H]-tryptophan in fraction F increased 3 days after immunization at which time the new fraction H (molecular weight 3000) appeared. Fractions F and H were highly labeled with [^{3}H]-tryptophan 4 days after immunization (Figure 3e). Incorporation of [^{3}H]-methionine into these fractions was also observed (Figure 3e). Fraction F was depleted 5 days after immunization and radioactivity of fraction H decreased slowly (data not shown). Fraction A (molecular weight 160,000) appeared 14 days after immunization (Figure 3f), although in control nonimmunized cells this fraction was absent.

2. Immunization with Noncellular Antigens

a. Immunization of Mice and Rats with IgG or Aggregated Human Gamma Globulin

Mice and rats were immunized with high doses of aggregated human gamma globulin or with low doses of pure IgG in complete Freund's adjuvant. All the profiles obtained with

Table 1
TIME COURSE OF PROTEIN TO DNA RATIO IN CHROMATIN OF MOUSE SPLEEN CELLS AFTER IMMUNIZATION WITH SHEEP RED BLOOD CELLS, AGGREGATED HUMAN GAMMA GLOBULIN, IgG, OR LIPOPOLYSACCHARIDES FROM *E. COLI*[a]

Ratio	Days after immunization									
	0	1	2	3.5 (3)	4	8	14	21	28	49
Sheep red blood cells										
A_{260}/A_{280}	1.50 ± 0.005	1.46 ± 0.01	1.44 ± 0.02	1.40 ± 0.02	1.38 ± 0.02	1.44 ± 0.02	1.46 ± 0.02	1.49 ± 0.01	1.50 ± 0.005	1.50 ± 0.005
A_{320}/A_{260}	0.08 ± 0.005	0.09 ± 0.005	0.09 ± 0.005	0.08 ± 0.005	0.09 ± 0.005	0.09 ± 0.005	0.08 ± 0.005	0.09 ± 0.005	0.08 ± 0.01	0.08 ± 0.005
Protein/DNA	1.19 ± 0.01	1.32 ± 0.02	1.42 ± 0.03	1.58 ± 0.03	1.75 ± 0.05	1.42 ± 0.04	1.33 ± 0.04	1.25 ± 0.02	1.20 ± 0.01	1.19 ± 0.01
Aggregated human gamma globulin										
A_{260}/A_{280}	1.50 ± 0.005	1.46 ± 0.01	1.45 ± 0.02	1.35 ± 0.02	1.38 ± 0.02	1.44 ± 0.02	1.46 ± 0.02	1.49 ± 0.01	1.50 ± 0.005	1.50 ± 0.005
A_{320}/A_{260}	0.08 ± 0.005	0.08 ± 0.005	0.08 ± 0.005	0.08 ± 0.005	0.08 ± 0.005	0.07 ± 0.01	0.09 ± 0.005	0.09 ± 0.005	0.08 ± 0.005	0.07 ± 0.005
Protein/DNA	1.19 ± 0.01	1.32 ± 0.02	1.40 ± 0.03	1.78 ± 0.05	1.75 ± 0.04	1.42 ± 0.03	1.34 ± 0.02	1.25 ± 0.02	1.19 ± 0.01	1.19 ± 0.01
IgG										
A_{260}/A_{280}	1.50 ± 0.005	1.46 ± 0.02	1.45 ± 0.02	1.36 ± 0.02	1.35 ± 0.02	1.45 ± 0.02	1.46 ± 0.01	1.46 ± 0.01	1.46 ± 0.01	1.48 ± 0.01
A_{320}/A_{260}	0.08 ± 0.005	0.08 ± 0.005	0.08 ± 0.005	0.09 ± 0.015	0.08 ± 0.005	0.09 ± 0.01	0.07 ± 0.005	0.08 ± 0.01	0.08 ± 0.005	0.07 ± 0.01
Protein/DNA	1.19 ± 0.01	1.35 ± 0.03	1.38 ± 0.03	1.55 ± 0.03	1.80 ± 0.04	1.38 ± 0.02	1.36 ± 0.02	1.36 ± 0.02	1.34 ± 0.02	1.28 ± 0.02
Lipopolysaccharides from *E. coli*										
						(5)	(8)	(14)	(21)	(49)
A_{260}/A_{280}	1.50 ± 0.005	1.46 ± 0.01	1.43 ± 0.01	1.44 ± 0.02	1.46 ± 0.02	1.40 ± 0.02	1.49 ± 0.01	1.50 ± 0.005	1.50 ± 0.005	1.50 ± 0.005
A_{320}/A_{260}	0.08 ± 0.005	0.08 ± 0.01	0.08 ± 0.005	0.08 ± 0.05	0.07 ± 0.01	0.07 ± 0.01	0.08 ± 0.005	0.08 ± 0.01	0.07 ± 0.01	0.08 ± 0.005
Protein/DNA	1.19 ± 0.01	1.33 ± 0.02	1.40 ± 0.02	1.38 ± 0.02	1.34 ± 0.02	1.50 ± 0.02	1.22 ± 0.01	1.20 ± 0.01	1.19 ± 0.01	1.19 ± 0.01

[a] The ratio of protein to DNA was calculated from protein and DNA chemical analyses. Values present the mean ± SE from three experiments.

Table 2

[ME COURSE OF THE RATIO OF PROTEIN TO DNA IN CHROMATIN OF RAT SPLEEN CELLS AFTER IMMUNIZATION WITH MOUSE SPLEEN CELLS AND LIPOPOLYSACCHARIDES FROM *E. COLI*[a]

	Days after immunization									
tio	0	1	2	3	4	8	14	21	28	49
80	1.51 ± 0.01	1.49 ± 0.01	1.48 ± 0.01	1.42 ± 0.02	1.36 ± 0.02	1.39 ± 0.02	1.43 ± 0.02	1.47 ± 0.01	1.50 ± 0.01	1.51 ± 0.01
60	0.07 ± 0.01	0.08 ± 0.005	0.08 ± 0.005	0.08 ± 0.01	0.08 ± 0.005	0.07 ± 0.01	0.07 ± 0.01	0.08 ± 0.005	0.09 ± 0.005	0.07 ± 0.01
DNA	1.58 ± 0.01	1.68 ± 0.02	1.75 ± 0.03	1.98 ± 0.03	2.38 ± 0.05	2.10 ± 0.03	1.94 ± 0.02	1.70 ± 0.02	1.58 ± 0.03	1.58 ± 0.02
Lipopolysaccharides from *E. coli*										
						(5)	(8)	(14)	(21)	(49)
80	1.51 ± 0.0	1.49 ± 0.01	1.48 ± 0.01	1.42 ± 0.02	1.42 ± 0.02	1.40 ± 0.02	1.43 ± 0.02	1.45 ± 0.02	1.51 ± 0.01	1.51 ± 0.01
50	0.07 ± 0.0	0.08 ± 0.005	0.07 ± 0.005	0.08 ± 0.005	0.07 ± 0.005	0.08 ± 0.005	0.08 ± 0.01	0.07 ± 0.01	0.07 ± 0.01	0.08 ± 0.005
DNA	1.58 ± 0.0	1.64 ± 0.01	1.74 ± 0.02	1.98 ± 0.02	1.98 ± 0.03	2.20 ± 0.04	2.00 ± 0.03	1.80 ± 0.03	1.58 ± 0.02	1.58 ± 0.01

: ratio of protein to DNA was calculated from protein and DNA chemical analyses. Values presen: the mean ±SE from three experiments.

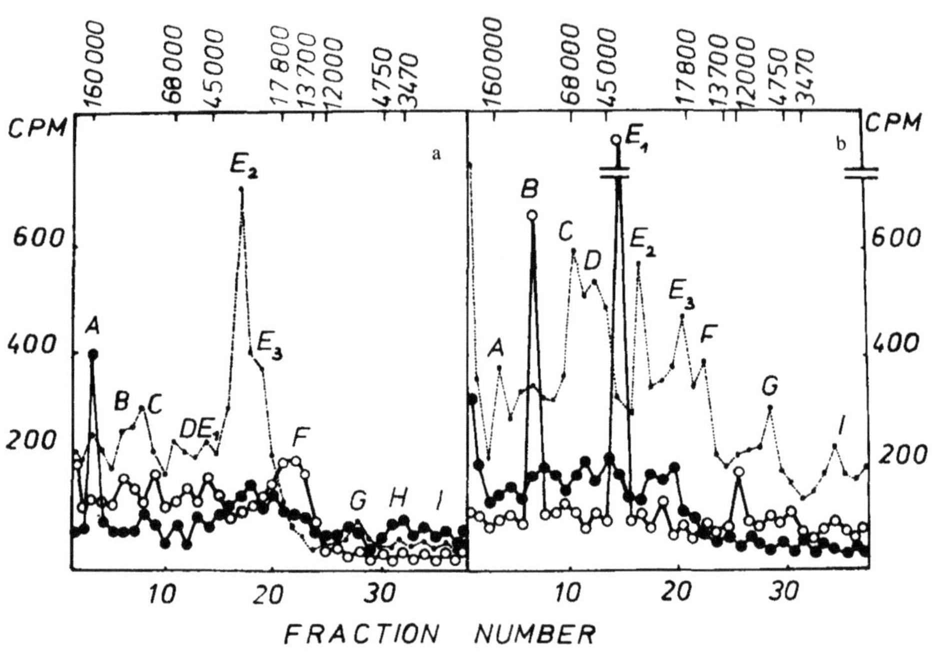

FIGURE 2. Electrophoretic profiles of chromatin proteins from spleen cells of nonimmunized (a) mice and (b) rats labeled with [^{3}H]-tryptophan (●——●), [^{3}H]-methionine (........), and [^{3}H]-leucine (○——○). The molecular weights of the protein fractions were estimated using the following marker proteins: human IgG (160,000), bovine serum albumin (68,000), ovalbumin (45,000), myoglobulin (17,800), ribonuclease (14,700), cytochrome c (12,000), insulin (4750), and glucagon (3470).

the high doses of aggregated human gamma globulin were similar to that obtained with pure IgG in complete Freund's adjuvant.[12,16] Therefore we show only profiles obtained in mice immunized with IgG in complete Freund's adjuvant. For testing the dose effect of antigen, we injected mice i.p. with 8, 20, 100, and 200 μg of pure IgG in complete Freund's adjuvant (Figure 4). On the second day, the highest incorporation of [^{3}H]-tryptophan in fraction I occurred at a dose of 200 μg per mouse. Fraction I is characteristically synthesized at the second day of immunization of mice with this antigen. In follow-up experiments mice were therefore immunized with pure human IgG (200 μg per mouse in complete Freund's adjuvant).

The protein-to-DNA ratio of chromatin preparation increased from 1.19 in control, non-immunized mice, to 1.8 during the first 4 days and later decreased (over 7 weeks) to 1.28.

At the first day after immunization the electrophoretic profile of NHC proteins of spleen cells was similar, as it was in the control mice with the exception that fraction A was absent. Fraction I (molecular weight below 3000) appeared 2 days after the immunization, highly labeled with [^{3}H]-tryptophan and low labeled with [^{3}H]-methionine, but not labeled with [^{3}H]-leucine (Figure 5a). An increase of fraction C was also observed. Fraction I was still present 3 days after the immunization and fraction B (molecular weight 120,000), labeled with [^{3}H]-tryptophan and [^{3}H]-methionine, appeared (Figure 5b). Four days after immunization radioactivity of fraction I decreased and that of fraction B increased; 8 days after immunization fraction I and B were absent, and 14 days after immunization fraction A (molecular weight 160,000) was present as observed in nonimmunized control mice (electrophoretic profile similar to the profile in Figure 2a). The same fractions (I and B), labeled with [^{3}H]-tryptophan, were also observed in mice immunized with aggregated human gamma globulin and in rats immunized with IgG (Figure 5c and 5d).

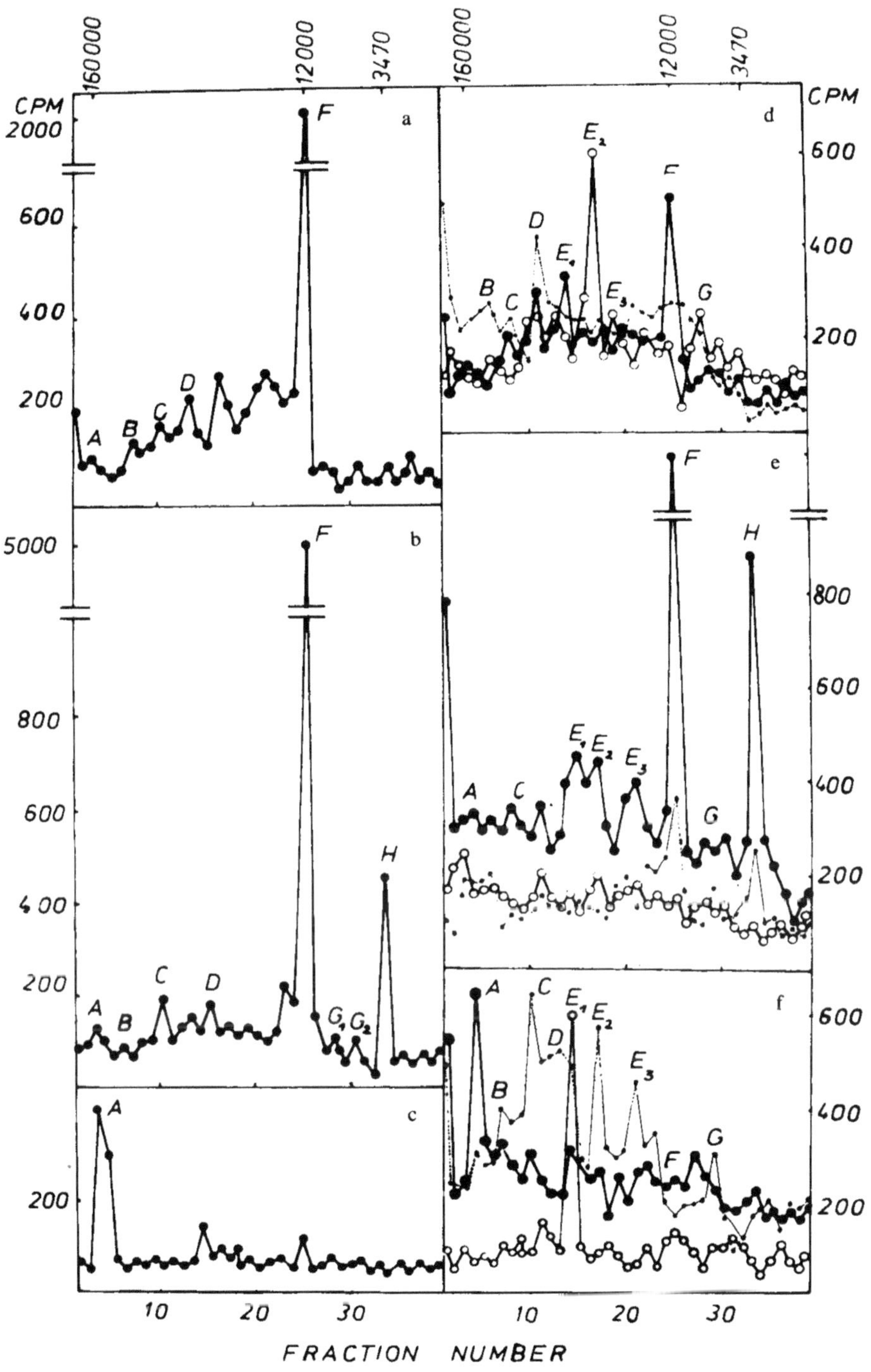

FIGURE 3. Electrophoretic profiles of chromatin proteins from spleen cells of mice immunized with sheep red blood cells (a) 2 days after immunization, (b) 4 days after immunization, and (c) 14 days after immunization; and rats immunized with mouse spleen cells (d) 2 days after immunization, (e) 4 days after immunization, and (f) 14 days after immunization. Proteins were labeled with [^{3}H]-tryptophan (●——●), [^{3}H]-methionine (.......) and [^{3}H]-leucine (○——○).

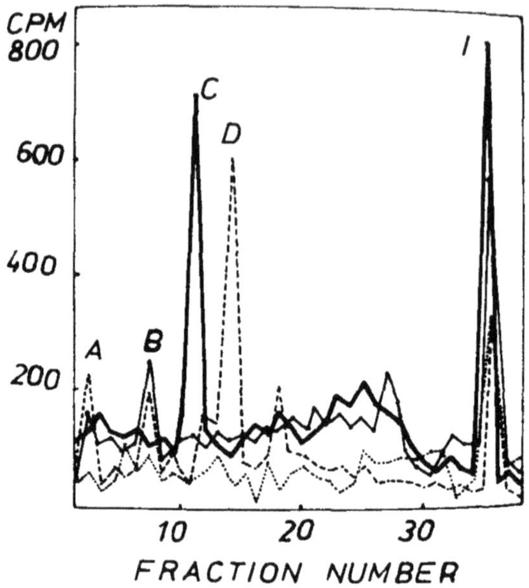

FIGURE 4. Electrophoretic profiles of NHC proteins labeled with [^{3}H]-tryptophan synthesized in spleen cells of mice 2 days after immunization with 8 μg (.......), 20 μg (------), 100 μg (————), or 200 μg (▬▬▬) of IgG.

b. Immunization of Mice and Rats with Lipopolisaccharides from E. coli as the Thymus-Independent Antigen

From the first to the fourth day after immunization only small quantitative changes of the particular fractions of NHC proteins occurred, being in general, similar in both species. In the chromatin of both mice and rats, a new fraction I (molecular weight below 3000) appeared 5 days after immunization (Figure 6). This was highly labeled in mice and less so in rats, but in both species it was clearly distinguished from other fractions. After the eighth day in mice and after the ninth day in rats fraction I was depleted.

C. Evidence for the Regulatory Character of the Described NHC Protein Fractions

The presence of the described fractions at specific days following the immune reaction, its antigen, and species nonspecificity suggests a regulatory character for these fractions. For ascertainment of the regulatory character of these fractions in further experiments we analyzed the:

1. Synthesis of the described NHC protein fractions in active spleen cells.
2. Correlation of the time of the synthesis of the described NHC protein fractions with the time of RNA synthesis.
3. Presence of the described NHC protein fractions in transcriptionally active chromatin fractions.

1. Presence of NHC Protein Fractions in Particular Subpopulations of Spleen Cells

Spleen cells were separated on the basis of two principles: centrifugation on 1.095 ficoll:sodium metrizoate density gradient and adhesiveness to plastic (Figure 7).[12,19] The experiments were performed on mice immunized with aggregated human gamma globulin. The characteristic fractions of NHC proteins of spleen cells immunized with aggregated human gamma globulin were synthesized only in the fraction 2b of fractionated spleen cells,

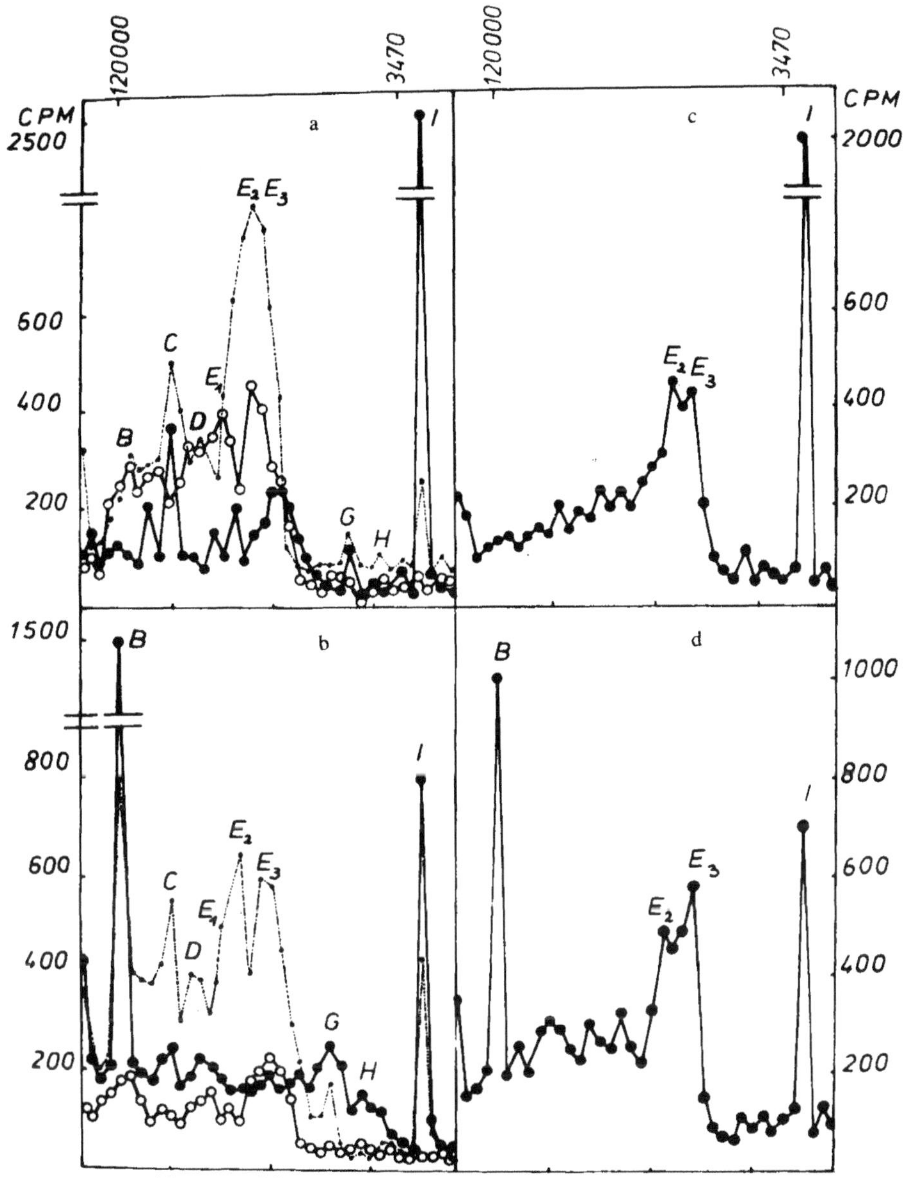

FIGURE 5. Electrophoretic profiles of NHC proteins from spleen cells of mice and rats immunized with IgG (a) 2 days after immunization of mice, (b) 4 days after immunization of mice, (c) 2 days after immunization of rats, and (d) 4 days after immunization of rats.

i.e., adherent macrophage-like cells plus activated lymphocytes (blast cells) (Figures 8a and 8b). Electrophoretic profiles of the other fractions of fractionated spleen cells (Figures 8c, 8d, and 8e) were similar to those in the control (Figure 2a) with only one difference, namely, the absence of fraction A. The experiments proved that the new fractions of NHC proteins were synthesized only by activated cells.

2. Synthesis of "Rapidly Labeled" RNA

"Rapidly labeled" RNA was determined according to the method of Albrecht and Hemminki.[20] Spleen cells isolated at specific days following the immune reaction were incubated

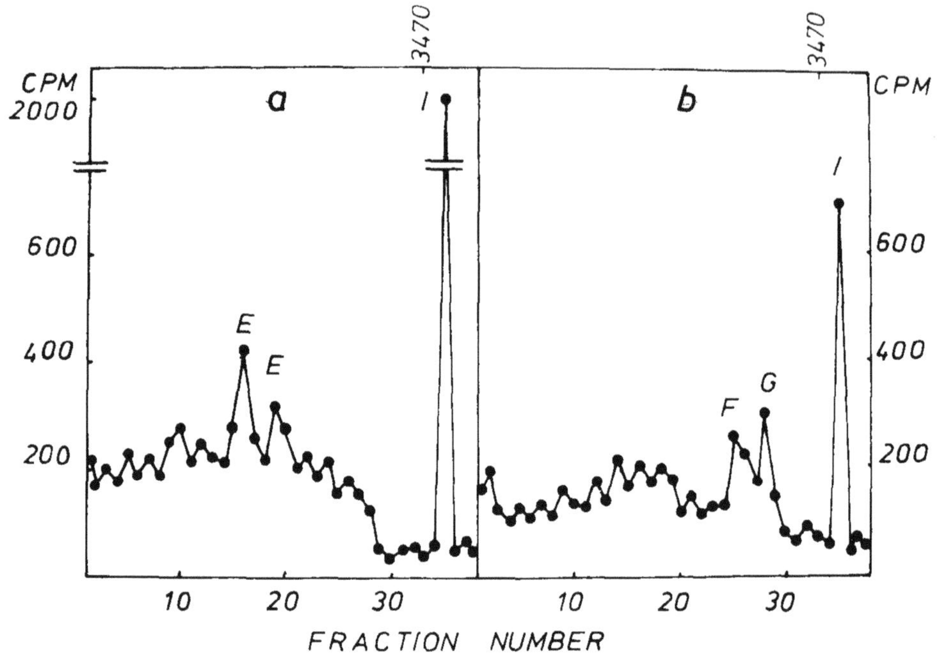

FIGURE 6. Electrophoretic profiles of NHC proteins synthesized in spleen cells of (a) mice and (b) rats 5 days after immunization with lipopolysaccharides from *E. coli*.

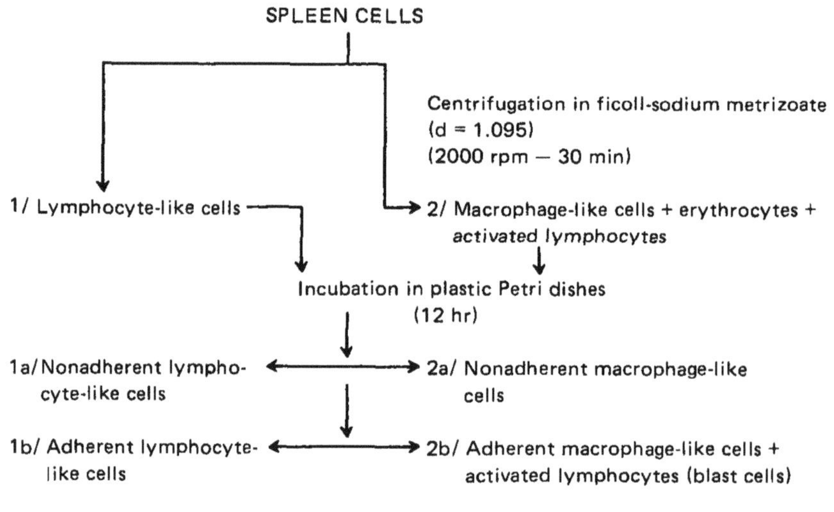

FIGURE 7. Scheme of spleen cells fractionation.

for 0.5 hr in the medium used for NHC proteins labeling, containing 5.6 μCi [^{3}H]-uridine (spec. act. 49 Ci/mmol) instead of the labeled amino acids. The curves of [^{3}H]-uridine incorporation into rapidly labeled RNA associated with chromatin are shown in Figure 9. In the case of immunization with thymus-dependent antigens, when two new NHC protein fractions were synthesized, two peaks of [^{3}H]-uridine incorporation were observed. The first one was present on the second day after immunization, when fraction I in the case of immunization with IgG (Figures 5a and 5c) or fraction F in the case of immunization with sheep red blood cells (Figure 3a) and mouse spleen cells (Figure 3d) were synthesized. The

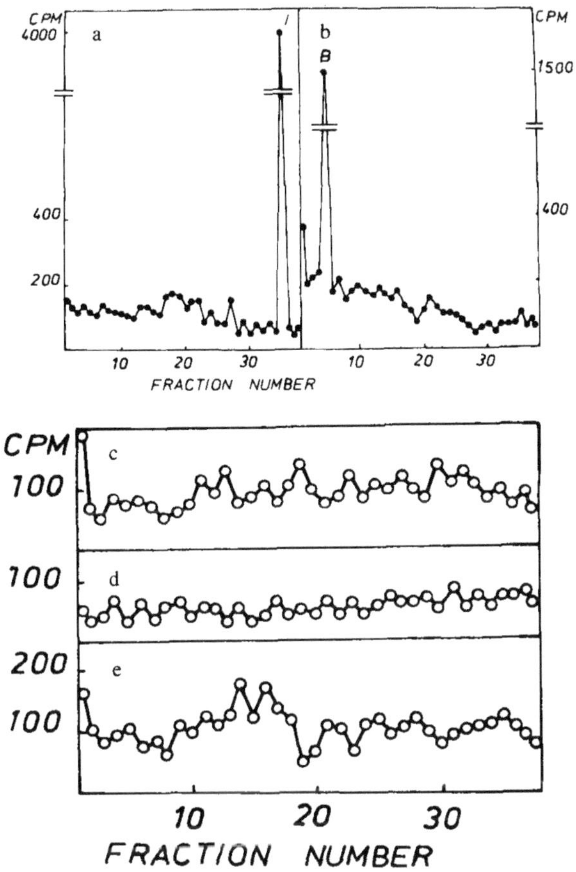

FIGURE 8. Electrophoretic profiles of NHC proteins labeled with [^{3}H]-tryptophan, isolated from four subpopulations of spleen cells of mice immunized with IgG (a) adherent macrophage-like cells and blast forms 2 days after immunization with IgG, (b) adherent macrophage-like cells and blast forms 5 days after immunization, (c) nonadherent lymphocyte-like cells 2 or 5 days after immunization, (d) adherent lymphocyte-like cells 2 or 5 days after immunization, and (e) nonadherent macrophage-like cells 2 or 5 days after immunization.

second peak was present on the fourth day after immunization, when fraction B in the case of immunization with IgG (Figures 5b and 5d) or fraction H in the case of immunization with sheep red blood cells (Figure 3b) or mouse spleen cells (Figure 3e) was synthesized. In the case of immunization with lipopolysaccharides from *E. coli*, only one peak of incorporation of [^{3}H]-uridine on the fifth day after immunization was observed; also, only one NHC protein fraction at the same day (Figure 6) after immunization with lipopolysaccharides was synthesized.

The synthesis of fractions F, H, I, and B at the time of maximal RNA synthesis favors the regulatory character of these fractions, although it cannot be excluded that besides the synthesis of mRNAs coding for the synthesis of particular antibodies, other kinds of RNAs could also be synthesized.

3. Presence of the Described Fractions of NHC Proteins in Transcriptionally Active Chromatin Fractions

Two kinds of procedures were used for evaluation of transcriptionally active chromatin: fractionation at increasing salt concentrations and limited digestion with DNAse II.

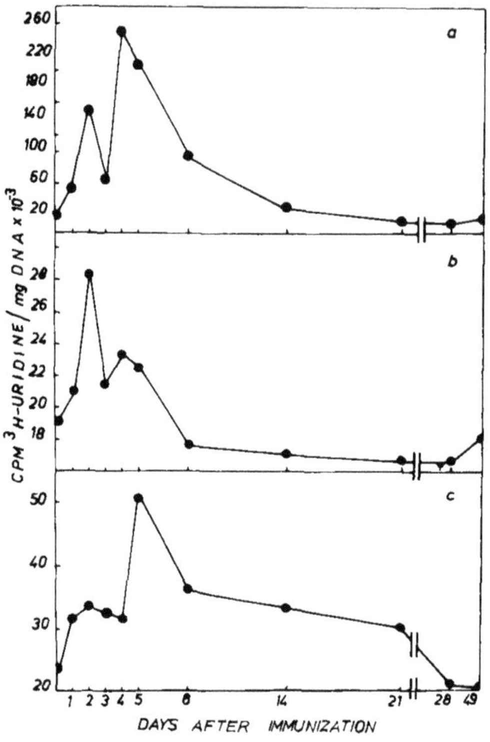

FIGURE 9. Incorporation of [^{3}H]-uridine into "rapidly labeled" chromatin RNA at various days after immunization; (a) mice with IgG, (b) rats with mouse spleen cells, (c) mice with lipopolisaccharides from *E. coli*.

a. Fractionation of Chromatin at Increasing Salt Concentrations

Isolated chromatin was fractionated by the previously described method[21] as shown in Figure 10. Chromatin was fractionated into three fractions: 0.35 *M* NaCl-soluble, 2 *M* NaCl-soluble and -insoluble, and a residual fraction. All extractions were performed in 0.1 *M* Tris-HCl (pH 7.9), 0.1% 2-mercaptoethanol, and 1 m*M* PMSF at 0°C. In some experiments histones have been removed from chromatin by extraction with 0.15 *M* H_2SO_4 (twice) and 0.4 *M* H_2SO_4 (twice) and the remaining chromatin was fractionated into 0.35 *M* NaCl-soluble, 2 *M* NaCl-soluble, and residual fractions.

All of the chromatin fractions isolated from the spleen cells of immunized mice have been characterized by ultracentrifugation at 4°C in their own solvent. The homogenous residual fraction showed a sedimentation coefficient of S_{apt} = 1.98 S. The 2 *M* NaCl-soluble fraction was less homogenous and dominated by molecules with a sedimentation coefficient of S_{apt} = 4.97 S. The 0.35 *M* NaCl-soluble fraction was highly heterogeneous, and was dominated by high molecular weight molecules (over 50,000,000).

All of the three fractions of chromatin contained DNA-protein complexes. The amounts of DNA and proteins in these fractions were similar in chromatin isolated from both immunized and nonimmunized mice and rats (Tables 3 and 4). The bulk of DNA and proteins was isolated in the 0.35 *M* NaCl-soluble fraction (approximately 50% of the total DNA and proteins). The 2 *M* NaCl-soluble fraction contained less than 30% of DNA and proteins. The residual fraction contained only up to 25% of DNA and approximately 20% of the proteins. Proteins of the residual fraction were labeled with the highest amounts of [^{3}H]-tryptophan, [^{3}H]-methionine, and [^{3}H]-leucine. H_2SO_4 extraction showed the presence of

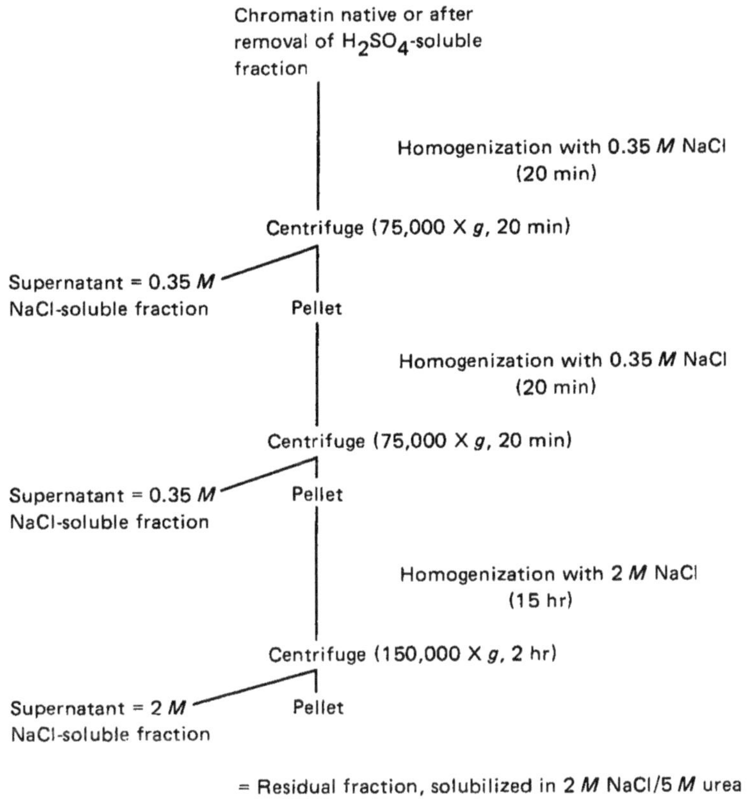

FIGURE 10. Scheme of chromatin fractionation at increasing salt concentrations.

histones in the 0.35 *M* NaCl-soluble and 2 *M* NaCl-soluble chromatin fractions. The residual fraction was practically histone free (it contained only 1 to 3% of the total amount of histones).

b. Template Activity of the 0.35 or 2.0 M NaCl-Extracted Chromatin Fractions from Immunized and Nonimmunized Mice Spleen Cells

The activity was determined by the method of Marushige and Bonner.[22] The 0.75-mℓ incubation medium for RNA synthesis contained 30 μmol Tris-HCl (pH 7.5), 3 μmol $MgCl_2$, 0.75 μmol $MnCl_2$, 9 μmol 2-mercaptoethanol, 0.3 μmol each of ATP, GTP, CTP, and UTP, 1 μCi [^{3}H]-UTP (16.6 Ci/mmol) and 3 to 4 units of RNA-polymerase from *E. coli* per 10 μg chromatin. RNA synthesis was carried out at 37°C for 2 to 30 min. After incubation, the reaction mixture was chilled on ice, 100 μg bovine serum albumin (BSA) was added, and the reaction was stopped by adding an equal volume of 10% trichloracetic acid. After 30 min, the precipitates were collected on Whatman® GF/C filters. The paper disks were subsequently treated with 10% cold trichloracetic acid, 5% trichloracetic acid (three times), ethanol/diethyl ether (1:1) and diethyl ether. The dry paper disks were placed in vials containing 10 mℓ toluene scintillation fluid. Radioactivity was measured in a scintillation counter.

The template activity of native chromatin and of chromatin extracted either with 0.35 or 2.0 *M* NaCl of nonimmunized and immunized mice (3 days after immunization with IgG) is shown in Figures 11a and 11b, respectively. The endogenous template activity of the 0.35 *M* NaCl-extracted chromatin, isolated from nonimmunized mice, increased 1.4 times and with RNA-polymerase from *E. coli*, 1.5 times. The template activity of the 2 *M* NaCl-extracted chromatin (residual fraction), assayed with endogenous RNA-polymerase was three

Table 3
THE CONTENT OF PROTEIN, DNA, AND RADIOACTIVITIES OF [^{3}H]-TRYPTOPHAN, [^{3}H]-METHIONINE, OR [^{3}H]-LEUCINE IN PARTICULAR CHROMATIN FRACTIONS ISOLATED FROM SPLEEN CELLS OF NONIMMUNIZED AND IMMUNIZED MICE (3 DAYS AFTER IMMUNIZATION WITH IgG)

Tissue	Chromatin fraction	% DNA	% Protein	[^{3}H]-Tryptophan %	[^{3}H]-Tryptophan cpm (mg protein × 10^{-3})	[^{3}H]-Methionine %	[^{3}H]-Methionine cpm (mg protein × 10^{-3})	[^{3}H]-Leucine %	[^{3}H]-Leucine cpm (mg protein × 10^{-3})
Spleen cells	0.35 *M* NaCl	47.4 ± 5.5	50.5 ± 2.7	33.7 ± 4.2	13.3	46.7 ± 3.8	34.7	53.2 ± 3.9	8.0
from nonim-	2 *M* NaCl	26.1 ± 2.6	28.9 ± 3.1	10.9 ± 0.5	7.6	27.7 ± 2.1	36.1	13.7 ± 1.2	3.4
munized mice	Residual	26.5 ± 4.2	20.6 ± 0.8	55.4 ± 3.8	54.0	25.6 ± 3.1	46.8	33.1 ± 3.2	12.0
	H_2SO_4	32.6 ± 1.2	81.7 ± 0.8						
	0.35 *M* NaCl	28.3 ± 1.4	0.6 ± 0.2						
	2 *M* NaCl	1.4 ± 0.4	0.7 ± 0.3						
	Residual	37.7 ± 1.3	17.0 ± 0.6						
Spleen cells	0.35 *M* NaCl	48.1 ± 3.1	53.8 ± 3.1	27.7 ± 5.1	10.3	45.0 ± 2.5	31.4	32.8 ± 0.4	4.6
from immu-	2 *M* NaCl	27.8 ± 1.9	28.9 ± 1.5	5.8 ± 0.6	4.0	4.0 ± 0.4	5.2	7.4 ± 2.2	2.0
nized mice	Residual	24.1 ± 2.3	17.3 ± 2.3	66.5 ± 4.4	78.2	51.0 ± 2.7	115.5	59.8 ± 3.0	26.3
	H_2SO_4	31.0 ± 0.9	66.5 ± 1.5						
	0.35 *M* NaCl	0.9 ± 0.2	5.2 ± 0.9						
	2 *M* NaCl	1.2 ± 0.2	4.3 ± 1.0						
	Residual	66.9 ± 1.3	24.0 ± 0.5						

Table 4
THE CONTENT OF PROTEIN, DNA, AND RADIOACTIVITIES OF [^{3}H]-TRYPTOPHAN, [^{3}H]-METHIONINE, AND [^{3}H]-LEUCINE IN PARTICULAR CHROMATIN FRACTIONS ISOLATED FROM SPLEEN CELLS OF NONIMMUNIZED AND IMMUNIZED RATS (3 DAYS AFTER IMMUNIZATION WITH MOUSE SPLEEN CELLS)

				[^{3}H]-Tryptophan	
Tissue	**Chromatin fraction**	**% DNA**	**% Protein**	**%**	**cpm (mg protein × 10^{-3})**
Spleen cells from nonimmunized rats	0.35 *M* NaCl	53.6 ± 3.3	56.8 ± 2.3	35.8 ± 3.6	5.0
	2 *M* NaCl	24.4 ± 1.8	27.3 ± 1.5	14.2 ± 0.7	4.2
	Residual	22.0 ± 2.0	21.0 ± 3.3	50.0 ± 2.9	19.0
	H_2SO_4	16.7 ± 1.1	72.2 + 2.0		
	0.35 *M* NaCl	35.8 ± 2.7	4.8 ± 0.9		
	2 *M* NaCl	4.5 ± 2.0	3.0 ± 0.6		
	Residual	43.0 ± 3.1	20.1 ± 1.1		
Spleen cells from immunized rats	0.35 *M* NaCl	52.2 ± 0.5	44.7 ± 2.1	28.1 ± 3.7	6.2
	2 *M* NaCl	26.0 ± 3.7	33.8 ± 1.3	11.7 ± 1.1	3.3
	Residual	21.9 ± 4.0	21.5 ± 3.4	60.3 ± 2.5	27.4
	H_2SO_4	21.8 ± 2.3	51.0 ± 1.0		
	0.35 *M* NaCl	28.9 ± 3.1	12.0 ± 2.1		
	2 *M* NaCl	8.0 ± 2.0	8.5 ± 1.9		
	Residual	49.0 ± 2.8	28.5 ± 2.3		

[a] Values present the mean ±SE from four experiments.

times higher, and with RNA-polymerase from *E. coli* six times higher than that of unfractionated chromatin (Figure 11a).

In immunized mice, the template activity of the various chromatin fractions differed only quantitatively from that in control mice. After removal of the 0.35 *M* NaCl-soluble chromatin fraction, the template activity of the remaining chromatin assayed with endogenous RNA-polymerase increased 1.4 times and with RNA-polymerase from *E. coli* about 1.8 times. After removal of the 2 *M* NaCl-soluble chromatin fraction the template activity of residual chromatin with endogenous polymerase was four times higher and with RNA-polymerase from *E. coli*, approximately seven times higher than that of native chromatin (Figure 11b).

c. Chromatin Fractionation by DNase II

Fractionation of chromatin was performed according to the method of Gottesfeld et al.[23] based on preferential sensitivity of active transcribed chromatin to DNase II. Chromatin was digested for various times at 37°C with DNase II (350 to 400 μg chromatin per milliliter in 25 m*M* sodium acetate pH 6.6). The reaction was stopped by cooling in ice and raising the pH to 7.5 with 0.1 *M* Tris-HCl (pH 11.0). Undigested chromatin was removed by centrifugation yielding a pellet, P_1. The supernatant S_1 was further fractionated by the addition of $MgCl_2$, up to 2 m*M*. After stirring at 0°C for 30 min, the insoluble material was removed by centrifugation yielding a pellet P_2, and supernatant S_2. The same procedure was used for digestion of the residual fraction of chromatin.

Chromatin isolated from spleen cells of nonimmunized and immunized mice was highly sensitive to DNAase II. Rapid digestion of the chromatin was caused by 100 units of DNAase II per 10 A_{260} units of DNA; therefore for limited chromatin digestion we used 0.1 unit per 10 A_{260} unit of DNA (Figure 12). Chromatin from immunized mice was more sensitive to

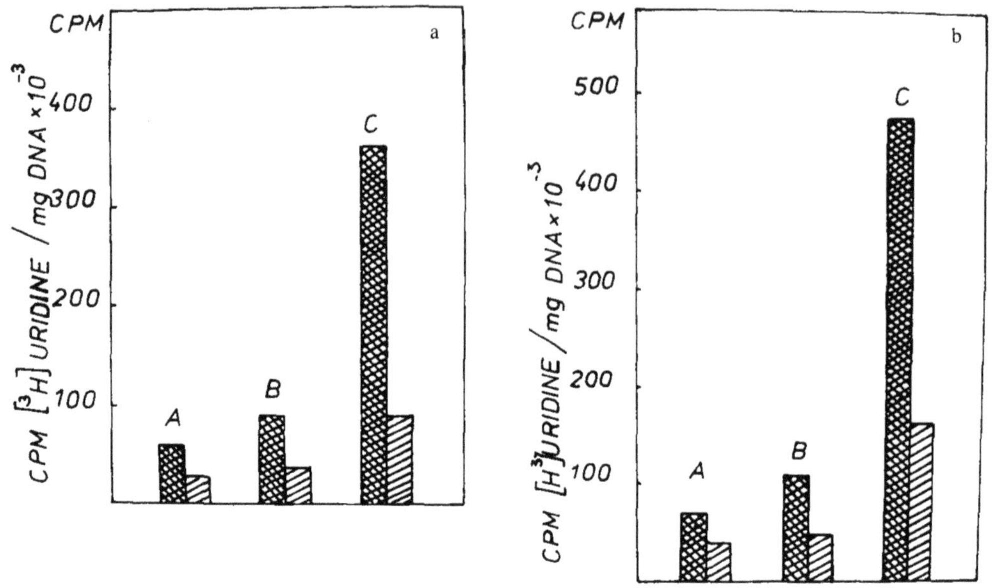

FIGURE 11. Template activity of native chromatin (a), 0.35 *M* NaCl extracted chromatin (b), and residual fraction (C) from (a) nonimmunized and (b) immunized mice (3 days after immunization with IgG) with endogeneous RNA-polymerase (▨) and *E. coli* RNA-polymerase (▩).

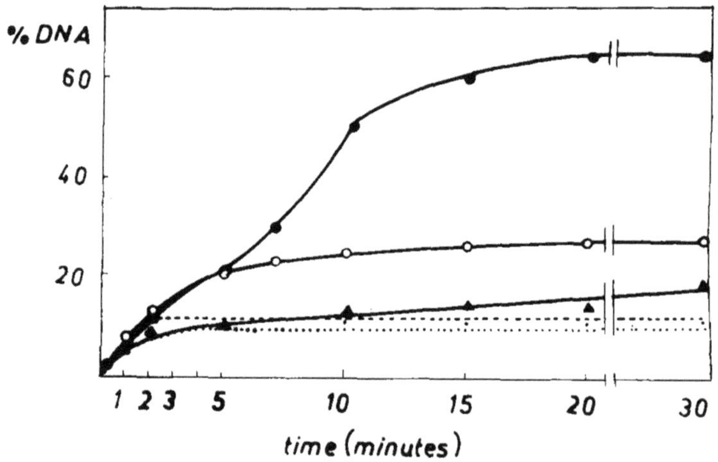

FIGURE 12. Time course of mouse chromatin digestion with DNAase II (0.1 units per 10 A_{260} units of chromatin DNA). Nonimmunized mice: S_1 (▲——▲), S_2 (.....); 3 days after immunization with IgG:S_1 (○——○), S_2 (------); 4 days after immunization: S_1 (●——●), S_2 (------).

DNAse II digestion than that from nonimmunized mice. Chromatin isolated from immunized mice at the fourth day after immunization was more sensitive to DNAase II digestion than that isolated at the third day after immunization. Chromatin isolated from spleen cells of nonimmunized mice contained approximately 10% Mg^{2+}-soluble sequences, and chromatin from immunized mice approximately 12% Mg^{2+}-soluble sequences (Figure 12).

The residual chromatin fraction from spleen cells of nonimmunized mice was less sensitive to DNAse II digestion than that of immunized mice (Figure 13). The residual fraction from nonimmunized mice contained only 30% active transcribed sequences and that from immunized mice, 40% active transcribed sequences (Mg^{2+}-soluble).

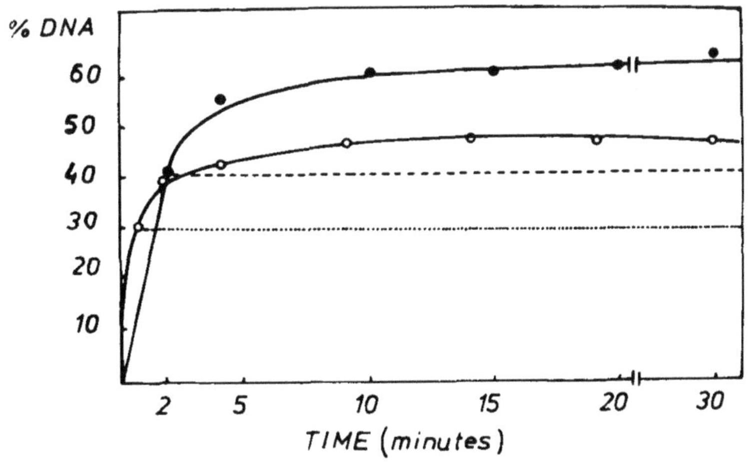

FIGURE 13. Time course of the residual chromatin fractions digestion with DNAase II (0.1 units per 10 A_{260} units of chromatin DNA. Nonimmunized mice: S_1 (○——○), S_2 (.....); 3 days after immunization with IgG: S_1 (●——●), S_2 (——).

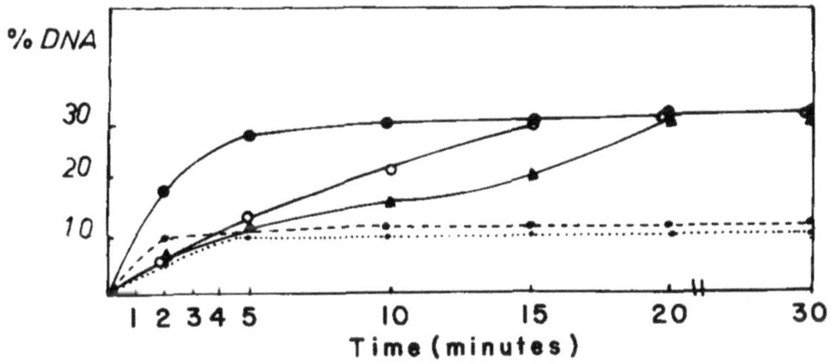

FIGURE 14. Time course of chromatin digestion with DNAase II. Nonimmunized rats: S_1 (▲——▲), S_2 (.....) (0.5 unit per 10 A_{260} units of chromatin DNA); 3 days after immunization: S_1 (●——●), S_2 (------) (0.5 unit DNAase II per 10 A_{260} units of chromatin DNA), S_1 (○——○) (0.1 unit DNAase II per 10 A_{260} units of chromatin DNA).

Chromatin from immunized rats was in general less sensitive to DNAase II digestion than that from immunized mice spleen cells, but the amount of Mg^{2+}-soluble sequences was the same as in mice, i.e., 10% Mg^{2+}-soluble sequences in nonimmunized rats and 12% in immunized rats (Figure 14).

d. Presence of Active Transcribed Sequences in Chromatin Fractions: 0.35 M NaCl-Soluble, 2 M NaCl-Soluble, and Residual Chromatin

For evaluation of which of the three isolated chromatin fractions contained the active transcribed sequences, chromatin from immunized mice and rats was digested to 11% of sequences in mice and 16% of sequences in rats and then fractionated at increasing salt concentrations.

In the case of mice, chromatin digested to 11% of sequences showed decreasing amounts of DNA in the residual fraction, with equal amounts of DNA in the 0.35 *M* NaCl-soluble and 2 *M* NaCl-soluble fractions (Table 5).

In the case of rats, digestion of chromatin to approximately 16% of sequences revealed only 10% of sequences being Mg^{2+}-soluble (Table 6). Thus, besides the diminished amounts

Table 5
DNA AND PROTEIN CONTENT IN CHROMATIN FRACTIONS OF MOUSE SPLEEN CELLS BEFORE AND AFTER DNAase II TREATMENT (3 DAYS AFTER IMMUNIZATION WITH IgG)[a]

	% DNA		% Protein	
Chromatin fraction	Before DNAase II treatment	After DNAase II treatment	Before DNAase II treatment	After DNAase II treatment
$S_1 = S_2$	—	11.0 ± 1.0	—	22.5 ± 1.5
0.35 *M* NaCl	48.1 ± 3.1	48.7 ± 4.0	53.8 ± 3.1	42.2 ± 3.5
2 *M* NaCl	27.8 ± 1.9	29.0 ± 2.5	28.9 ± 1.5	26.2 ± 1.8
Residual	24.1 ± 2.3	11.3 ± 2.6	17.3 ± 2.3	9.1 ± 1.4

[a] Values present the mean ±SE from three experiments.

Table 6
DNA AND PROTEIN CONTENT IN CHROMATIN FRACTIONS OF RAT SPLEEN CELLS BEFORE AND AFTER DNAase II TREATMENT (3 DAYS AFTER IMMUNIZATION WITH MOUSE SPLEEN CELLS)[a]

	% DNA		% Protein	
Chromatin fraction	Before DNAase II treatment	After DNAase II treatment	Before DNAase II treatment	After DNAase II treatment
S_1	—	16.5 ± 0.4	—	26.1 ± 0.3
S_2	—	10.0 ± 0.2	—	22.0 ± 0.2
0.35 *M* NaCl	51.8 ± 0.1	42.1 ± 0.8	45.1 ± 1.0	35.5 ± 0.9
2 *M* NaCl	23.2 ± 1.0	26.3 ± 1.7	34.1 ± 0.7	28.1 ± 0.5
Residual	25.0 ± 1.1	15.1 ± 0.3	20.8 ± 0.4	10.3 ± 0.4

[a] Values present the mean ±SE from three experiments.

of the DNA in the residual fraction from 25 to 15% of sequences, a decrease of the DNA in the 0.35 *M* NaCl-soluble fraction was also observed.

On the basis of data obtained by limited DNA digestion with DNAase II a consequent fractionation of the remaining chromatin at increasing salt concentrations, we may conclude that the residual chromatin fraction contains the active transcribed DNA sequences.

The solubility of the histone-free residual fraction in magnesium buffer, approximately 10% of the total DNA (Figure 13), may suggest that it had been done by the higher DNAase II sensitivity of the histone-free DNA. However, this possibility may be excluded on the basis of the observation of Gottesfeld et al.[23] who, using purified DNA as a substrate, demonstrated that at pH 6.6 (pH of the reaction) the DNAase II exhibits only 6 to 8% of its maximal activity.

For determining whether fraction A, synthesized in nonimmunized spleen cells of mice, and fractions F, H, I, and B, synthesized in spleen cells of mice immunized with IgG or rats immunized with mouse spleen cells, are present in active or inactive chromatin fractions, we fractionated chromatin before and after DNAase II treatment into three fractions: 0.35 *M* NaCl-soluble, 2 *M* NaCl-soluble, and residual (Figures 15 to 17).

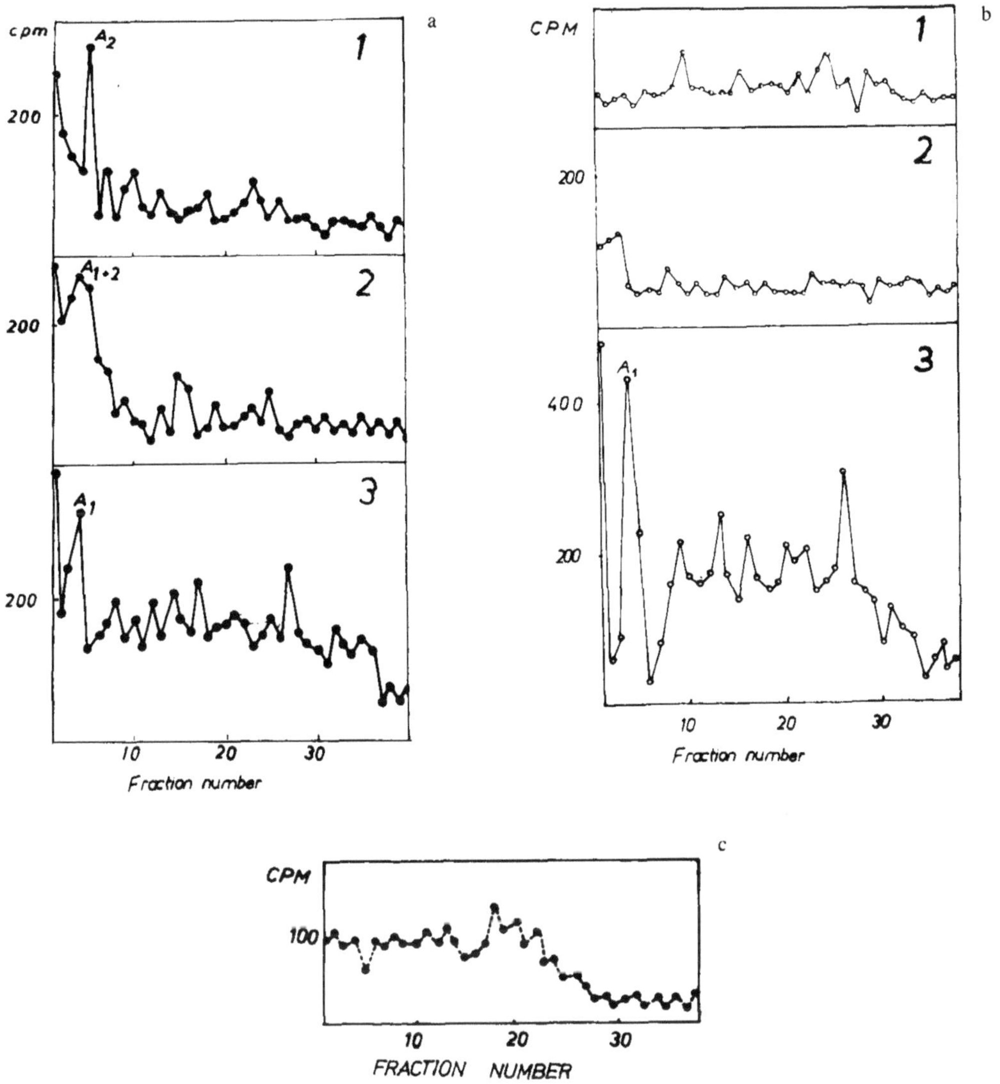

FIGURE 15. Electrophoretic profiles of [^{3}H]-tryptophan labeled NHC proteins from spleen cells of nonimmunized mice: (1) 0.35 *M* NaCl-soluble fraction, (2) 2 *M* NaCl-soluble fraction, (3) residual fraction; (a) before DNAase II treatment, (b) after DNAase II treatment, (c) Mg^{2+}-soluble fraction (S_2).

In the case of spleen cells from nonimmunized mice, fraction A (molecular weight 160,000), found to be homogeneous as shown by the electrophoretic profile of nonhistone proteins (Figure 2a) from unfractionated chromatin,[21] was found to be composed of two subfractions (Figure 15a); A_1 (molecular weight 160,000) and A_2 (molecular weight 150,000).

In chromatin fractionated after digestion with DNAase II (Figure 15b), subfraction A_2 was not found in the 0.35 *M* NaCl-soluble and 2 M NaCl-soluble chromatin fractions while subfraction A_1 decreased in the 2 *M* NaCl-soluble chromatin fraction. In the residual chromatin fraction subfraction A_1 was still present. In the Mg^{2+}-soluble chromatin fraction the NHC protein fraction A was absent (Figure 15c).

After digestion of chromatin with DNAase II the NHC protein subfraction A_2 was not found in the 0.35 *M* NaCl-soluble and 2 *M* NaCl-soluble chromatin fractions. Absence of this NHC subfraction in the residual and Mg^{2+}-soluble chromatin fractions suggests that the

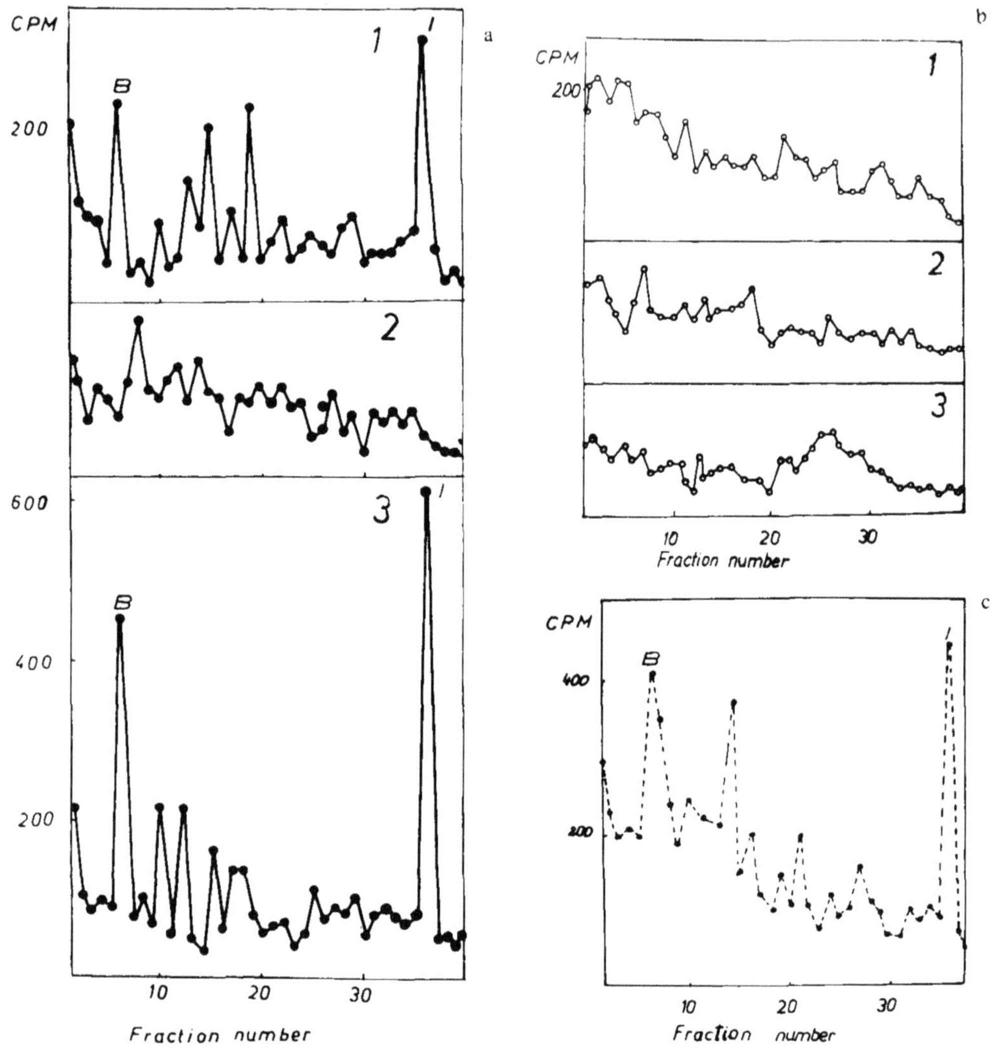

FIGURE 16. Electrophoretic profiles of [^{3}H]-tryptophan labeled NHC proteins from mice immunized with IgG (3 days after immunization): (1) 0.35 *M* NaCl-soluble fraction, (2) 2 *M* NaCl-soluble fraction, (3) residual fraction; (a) before DNAase II treatment, (b) after DNAase II treatment, (c) Mg^{2+}-soluble fraction (S_2).

subfraction A_2 is a loosely bound protein, and is extracted together with the bulk of loosely bound NHC proteins in the 0.35 M NaCl-soluble fraction of chromatin.

After fractionation of chromatin isolated from spleen cells of mice immunized with IgG, fractions I and B, present at the third day after immunization, were present mainly in the residual chromatin fraction and in smaller amounts in the 0.35 M NaCl-soluble chromatin fraction (Figure 16a). The NHC protein fractions I and B were absent in the 2.0 *M* NaCl-soluble chromatin fraction. After DNAase II digestion (approximately 11% DNA) the nonhistone chromatin protein fractions I and B were not found in the 0.35 *M* NaCl-soluble and residual fractions of chromatin (Figure 16b), but simultaneously these fractions were present in the Mg^{2+}-soluble fraction of chromatin (Figure 16c).

Similar results were obtained in the case of chromatin from rat spleen cells immunized with mouse spleen cells. Fractions F (molecular weight 12,000) and H (molecular weight

3000), specific for cellular antigens (Figure 3e) were present in the residual fraction and in smaller amounts in the 0.35 *M* NaCl-soluble fraction (Figure 17a). In the 2 *M* NaCl-soluble fraction these NHC protein fractions were absent. After DNAase II digestion (approximately 16% digested DNA and 12% Mg^{2+}-soluble), NHC protein fractions F and I were not found in the 0.35 *M* NaCl-soluble and residual chromatin fractions (Figure 17b), but were present in Mg^{2+}-soluble fraction (together with fractions E_1 and E_3 (Figure 17c).

e. NHC Proteins in Myeloma Cells Synthesizing Immunoglobulins

Of special interest seemed to be whether myelomas (neoplastic cells synthesizing immunoglobulins), exhibit synthesis of the same or similar NHC protein fractions as found in the immunocompetent cells during the immune reaction. In these experiments we used RPC5 myeloma cells synthesizing γ_2A immunoglobulins and myeloma cells ABPC 22 synthesizing IgM immunoglobulins. Balb/c mice were injected i.p. with 3×10^6 of RPC5 or ABPC 22 myeloma cells, and isolated from the peritoneal exudate on the 8th day after myeloma cell injection.[24,25]

Tumor cells and spleen cells from myeloma-bearing mice, isolated at the eighth day after tumor cell injection, were incubated for 12 hr in culture medium supplemented with labeled amino acids [^{3}H]-tryptophan, [^{3}H]-methionine, or [^{3}H]-leucine. After the incubation, chromatin was isolated and prepared for electrophoresis as described for chromatin of normal spleen cells.

Differences in the protein-to-DNA ratio of chromatin preparations are shown in Table 7. The protein-to-DNA ratio of myeloma cells exceeded 3.0, while in chromatin from normal spleen cells from Balb/c mice the same ratio was 1.80, and in normal spleen cells from mice of the CFW/bog strain only 1.18. The protein-to-DNA ratio in chromatin from spleen cells of tumor-bearing mice of Balb/c strain was also elevated (Table 7). The high protein-to-DNA ratio in neoplastic cells may be the consequence of increased metabolism of proliferating neoplastic cells.

The electrophoretic profiles of NHC proteins of myeloma cells did not exhibit any of the described NHC protein fractions[25] (fractions F, H, I, and B described in antibody-producing cells).[12,16-19] The profiles of myeloma cells resemble to some degree the profiles of NHC proteins in normal, nonimmunized spleen cells and in stimulated proliferating cells.[27]

In RPC 5 and ABPC 22, the 0.35 *M* NaCl-soluble chromatin fraction contained approximately 30% DNA and the residual fraction 58 to 65% DNA (Table 8), while in antibody-synthesizing spleen cells of mice and rats approximately 50% DNA was found in the 0.35 *M* NaCl-soluble fraction and only 20 to 25% DNA in the residual fraction (Tables 3 and 4). Thus, in myeloma cells there was a dramatic increase in DNA in the residual fraction, shown by our data as the fraction containing actively transcribed DNA sequences (genes) during the immune reaction.[21]

As shown in Figure 18, the dramatic increase in DNA in the residual fraction of chromatin from myeloma cells was not in accordance with sensitivity to DNAase II digestion. Chromatin from myeloma cells RPC5 and ABPC 22 when digested with 0.1 unit DNAse II per 10 A_{260} of chromatin DNA, exhibited low sensitivity to digestion (10% sequences). As shown in Figure 12 and 14, chromatin from mice spleen cells (4 days after immunization with IgG) digested with 0.1 unit DNAase II per 10 A_{260} of chromatin DNA, exhibited digestion to 60% during 30 min (30% in the case of rat spleen cells). The same dramatic difference has been shown by fractionation with Mg^{2+}. The Mg^{2+}-soluble sequences, in the case of myeloma chromatin, were approximately 3% (Figure 18) and in the case of chromatin from immunized mice and rats, the amount of Mg^{2+}-soluble sequences was approximately 12% (Figures 12 and 14). Similarly, in the residual chromatin fraction from myeloma cells, approximately 8% DNA was digested, and only 4% of the sequences were Mg^{2+}-soluble (Figure 19). In comparison, the residual fraction from nonimmunized mice spleen cells

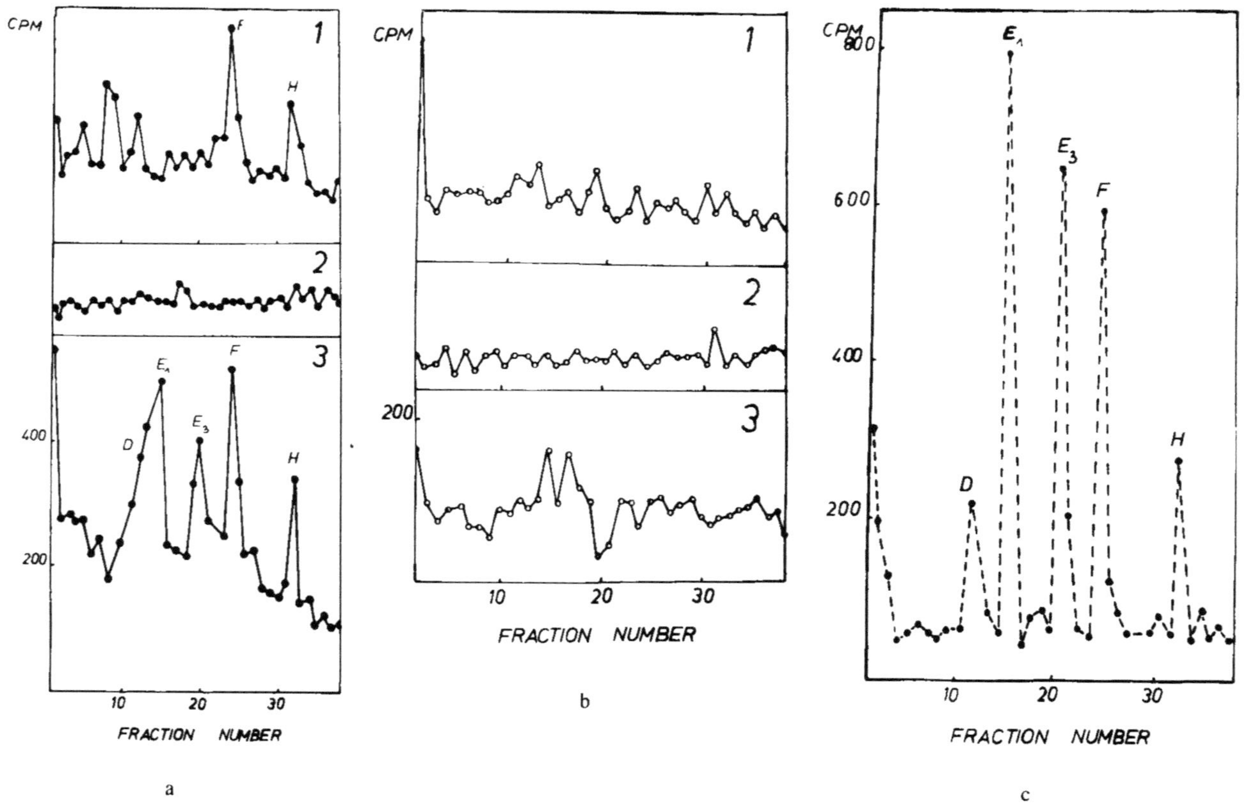

FIGURE 17. Electrophoretic profiles of [^{3}H]-tryptophan labeled NHC proteins of rats immunized with mouse spleen cells (3 days after immunization): (1) 0.35 *M* NaCl-soluble fraction, (2) 2 *M* NaCl-soluble fraction, (3) residual fraction; (a) before DNAase II treatment, (b) after DNAase II treatment, (c) Mg^{2+}-soluble fraction (S_2).

Table 7
A_{260} TO A_{280} RATIO AND THE PROTEIN TO DNA RATIO OF CHROMATIN FROM MYELOMA CELLS AND SPLEEN CELLS OF CONTROL MICE AND MICE BEARING TUMORS[a]

Tissue	A_{260}/A_{280}	Protein/DNA
RPC5	1.42 ± 0.04	3.50 ± 0.20
ABPC 22	1.45 ± 0.04	3.30 ± 0.20
Spleen cells from Balb/c mice bearing myeloma RPC5	1.49 ± 0.02	2.64 ± 0.05
Spleen cells from Balb/c mice bearing myeloma ABPC 22	1.50 ± 0.02	2.35 ± 0.04
Spleen cells from Balb/c mice	1.68 ± 0.01	1.28 ± 0.02

[a] Values present the mean ±SE from three experiments.

contained 30% of Mg^{2+}-soluble sequences and from immunized mice, 40% Mg^{2+}-soluble sequences.

These results suggest that in spite of the increase in the amount of the residual chromatin fraction of myeloma cells, the amount of actively transcribed sequences was diminished. After removal of the 0.35 *M* NaCl-soluble fraction and the 2 *M* NaCl-soluble fraction, the template activity of the residual chromatin fraction in RPC5 myeloma decreased and in ABPC 22 myeloma was the same as that of native chromatin (Figures 20a and 20b).

IV. DISCUSSION

A. Newly Synthesized Nonhistones

Experiments[12,16,17] suggest that immunization with thymus-dependent antigens causes the synthesis of two fractions of nonhistones: fraction F (molecular weight 12,000) in mice immunized with sheep red blood cells, and fraction H (molecular weight 3000) in rats immunized with mouse spleen cells. In the case of immunization with noncellular antigens (aggregated human gamma globulin, pure human IgG) of mice and rats, two fractions of NHC proteins were also synthesized, i.e., fraction I (molecular weight below 3000) and fraction B (molecular weight 120,000).

In the case of immunization of mice and rats with thymus-independent antigen (lipopolysaccharides from *E. coli*), only one fraction I (molecular weight below 3000) was synthesized.[18]

All the newly synthesized NHC protein fractions appeared only during the immune reaction but were not found after termination of the reaction. Therefore, we assume that their synthesis must be in some way related to the immune reaction, principally with the synthesis of antibodies. The last hypothesis is strongly supported by the results shown in this communication. In the case of thymus-dependent antigens, when two new classes of antibodies were synthesized (antibodies of the IgM class and antibodies of the IgG class), two fractions of NHC proteins were synthesized. In the case of thymus-independent antigen, when only one class of IgG antibodies was synthesized, only one fraction of NHC proteins was synthesized, namely the fraction I appearing at the later phase of the immune reaction, where a switch in the synthesis of antibodies from the IgM class to the IgG class occurs.

Table 8

THE CONTENT OF PROTEIN, DNA, AND RADIOACTIVITIES OF [^{3}H]-TRYPTOPHAN AND [^{3}H]-LEUCINE IN PARTICULAR CHROMATIN FRACTIONS ISOLATED FROM MYELOMA CELLS RPC5 AND ABPC 22[a]

				[^{3}H]-Tryptophan		[^{3}H]-Leucine	
Tissue	Chromatin fraction	DNA	Protein	%	cpm (mg protein × 10^{-3})	%	cpm (mg protein × 10^{-3})
RPC 5	0.35 *M* NaCl	30.1 ± 4.1	26.2 ± 0.1	31.0 ± 0.8	4.2	32.5 ± 1.1	3.3
	2 *M* NaCl	5.2 ± 2.7	37.7 ± 0.7	34.3 ± 1.0	3.6	33.1 ± 1.0	2.6
	Residual	64.8 ± 6.9	36.1 ± 0.6	34.7 ± 1.8	3.8	34.4 ± 2.1	2.7
	H_2SO_4	—	50.4 ± 3.9				
	0.35 *M* NaCl	—	11.3 ± 1.5				
	2 *M* NaCl	2.5 ± 0.5	1.0 ± 0.5				
	Residual	97.5 ± 0.5	37.3 ± 2.1				
ABPC 5	0.35 *M* NaCl	27.4 ± 4.2	33.3 ± 1.0	44.3 ± 2.2	6.1	45.6 ± 2.9	6.9
	2 *M* NaCl	14.7 ± 3.1	28.1 ± 2.1	15.7 ± 1.3	3.7	23.0 ± 1.6	4.1
	Residual	57.9 ± 1.3	38.6 ± 3.1	40.0 ± 3.2	6.0	31.4 ± 1.9	4.0

[a] Values present the mean ±SE from three experiments.

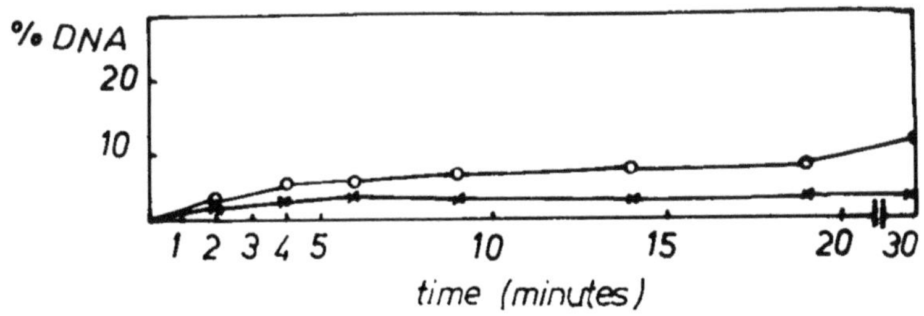

FIGURE 18. Time course of myeloma chromatin digestion with DNAase II 0.1 to 1 unit per 10 A_{260} units of chromatin DNA: S_1 (○——○), S_2 (x————x).

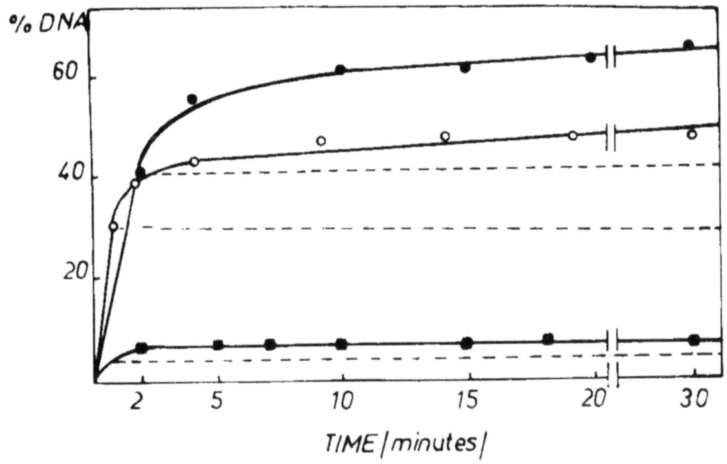

FIGURE 19. Time course of digestion of the residual chromatin fraction with DNAase II (0.1 unit per 10 A_{260} units of chromatin DNA); residual fraction from myeloma cells: S_1 (■——■), S_2 (------), residual fraction from nonimmunized mouse spleen cells: S_1 (○——○), S_2 (------), and residual fraction from mouse spleen cells immunized with IgG: S_1 (●——●), S_2 (-------).

The synthesis of the same fractions after immunization with various antigens in mice and rats (e.g., fraction I after immunization with aggregated human gamma globulin, IgG, and lipopolysaccharides from *E. coli*) suggests that these nonhistones were not specifically synthesized in response to a specific antigen nor appear to be specific relative to the animal species used for raising these antibodies.

B. Regulatory Role for the Newly Synthesized Nonhistones

It is important to determine which cells participating in the immune reaction synthesize new NHC proteins: those cells actively engaged in the immune reaction or those cells accompanying this reaction. The performed experiments confirmed the synthesis of these fractions in only one kind of spleen cells, namely macrophage-like and activated lymphocytes (blast cells), which suggest that the newly synthesized NHC protein fractions may be functionally related to the regulation of the immune reaction.[12,19]

One of the arguments favoring the regulatory function of the described NHC protein fractions may be the correlation between their synthesis with *de novo* synthesis of RNA. This was strongly supported by the simultaneous appearance of peaks of rapidly labeled RNAs together with the newly synthesized nonhistones.

We observed two peaks in the case of synthesis of two fractions of NHC proteins (in mice or rats immunized with thymus-dependent antigens) and only one peak in the case of

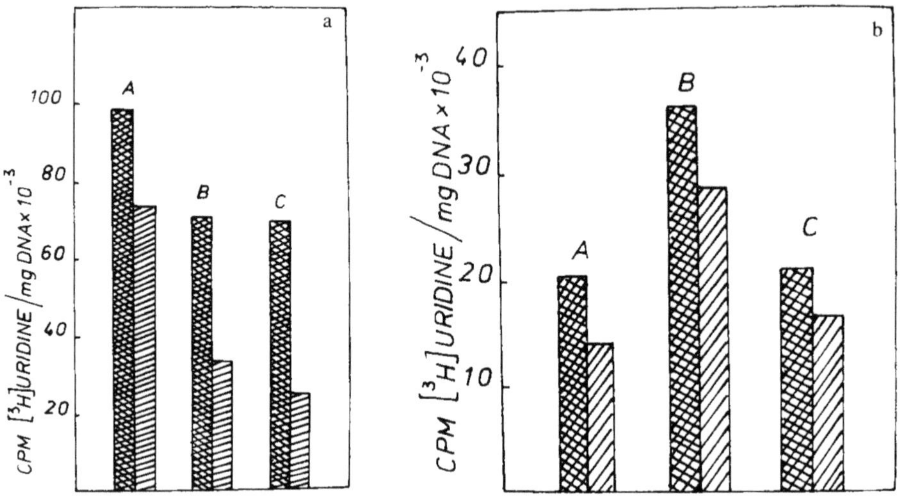

FIGURE 20. Template activity of native chromatin (A) after removal of 2 *M* NaCl-soluble fraction (B) and residual (C) from myeloma cells RPC 5 (a) and ABPC 22 (b) with endogeneous RNA-polymerase (▨) and *E. coli* (▩) RNA-polymerase.

synthesis of only one fraction of nonhistone chromatin (in mice or rats immunized with thymus-independent antigens). See Figures 3, 5, and 9.

The most important argument supporting the regulatory character of the described NHC fractions may be their presence in actively transcribed chromatin sequences during the immune reaction.[17,21] For this purpose the following experiments were performed: fractionation of chromatin at increasing salt concentrations, measurement of the template activity in particular chromatin fractions from spleen cells of immunized and nonimmunized mice, chromatin fractionation by DNAase II, and evidence for the presence of active transcribed sequences in particular chromatin fractions. Our experiments on the residual chromatin fraction containing tightly bound nonhistone chromatin proteins support the proposed regulatory role for these proteins. In this respect, fractions I and B of NHC proteins, synthesized by immunized mouse spleen cells, were present mainly in the residual chromatin fraction of the highest DNAase II sensitivity and highest template activity. The presence of these fractions in the 0.35 *M* NaCl-soluble chromatin fraction differed quantitatively, which may be caused by unspecific loose binding of these proteins to this fraction of chromatin. It may be that there is also a state of equilibrium between high-affinity binding of these fractions to the residual chromatin fraction and low-affinity binding to the 0.35 *M* NaCl-soluble chromatin fraction. The loss of these NHC protein fractions after DNAase II treatment suggests that these fractions interact with actively transcribed DNA sequences (genes) during the immune reaction.

Special attention should be paid as well to the NHC protein fraction A, depleted during the immune reaction (Figures 3 and 5). On the basis of our experiments, fraction A of NHC proteins may suppress gene expression. This hypothesis is supported by the presence of fraction A in normal, nonimmunized spleen cells, its loss at the start of the immune reaction (no matter what kind of antigen was used), and its reappearance after the termination of the immune reaction. This hypothesis is also supported by the absence of fraction A in actively transcribed DNA sequences (this fraction is not solubilized from residual chromatin after DNase II digestion).

Further evidence supporting the repressive character of fraction A results from the experiments on both immunized and nonimmunized rats, where this fraction was absent. After termination of the immune reaction this fraction appeared, which may be interpreted that in normal, nonimmunized rats its level is so low that it was undetectable, and during the termination of the immune reaction the synthesis of this fraction increased so that it could be detected.

C. Nonhistone Proteins in Myeloma Cells

Changes in nuclear proteins of neoplasms have been described by many authors. Murphy et al.[28] showed differences between nonhistone proteins of active chromatin and inactive heterochromatin from mouse myelomas. Ruoslahti et al.,[29] using immunological methods, found an absence of one specific NHC protein fraction in hepatomas, which was present in normal liver cells.

In our experiments the main difference in chromatin of myeloma cells as compared to normal lymphoid spleen cells was an increase in the total amount of NHC proteins, without the presence of characteristic NHC protein fractions found in chromatin of spleen cells from immunized mice and rats.[24-26]

Salt solubility, DNAase II sensitivity, and template activity showed significant differences between chromatin from tumor-myeloma cells synthesizing spontaneously monoclonal immunoglobulins and chromatin from normal spleen cells, and also spleen cells synthesizing immunoglobulins after immunization with various antigens. In myeloma cells approximately 60% of the DNA was tightly bound with NHC proteins (isolated in the residual fraction of chromatin).

DNAase II sensitivity of myeloma-cell chromatin was very low as compared to the immunized spleen cells synthesizing immunoglobulins. In normal, nonimmunized spleen cells, after the removal of the 0.35 *M* NaCl-soluble and 2 *M* NaCl-soluble fractions, an increase of the template activity was observed. Contrary to this, the template activity of the residual fraction of chromatin from myeloma cells was low, which suggests that the low amount of active chromatin (DNAase sensitive) and the low template activity of the residual DNA-nonhistone protein complex may be characteristic for neoplastic (myeloma) cells.[26]

D. Summary and Conclusions

The experiments on chromatin of spleen cells of immunized mice and rats revealed the synthesis of some new nonhistone protein fractions absent in normal, nonimmunized chromatin of spleen cells and depleted after termination of the immune reaction. Simultaneously, a characteristic high molecular weight fraction, present in chromatin of spleen cells of nonimmunized mice, was always depleted after immunization, and resynthesized at the termination of the immune reaction.

A problem arises about the role of these immunization-induced NHC protein fractions. NHC proteins have been considered as the main gene regulators since the era of molecular biology. Recent discoveries in molecular biology, contrary to the preceding views, revealed that DNA is not a stable and unalterable structure, but a structure undergoing various changes like rearrangements of genes (e.g., genes for the synthesis of immunoglobulins[3-5]) and switching of genes (e.g., genes for the synthesis of globin from embryonic to adult hemoglobin,[30] and genes for the synthesis of immunoglobulins from IgM to IgG[6-8]). In addition, moveable and transposable elements have been discovered, at first in prokaryotes and viruses, then also in eukaryotes, including man.[31] Finally, nucleotide signals for initiation and termination of transcription have been detected, including signals for mRNA processing (splicing) prior to its maturation for programming protein biosynthesis.

All these activities require the presence in the genetic code of signals that must be recognized for the correct function of the genetic code, and sites for the specific binding of regulatory proteins leading to the temporal transcription of specific genes. Therefore, a great number of nuclear proteins, especially NHC proteins, which comprise a number of enzymes necessary for DNA[31] and RNA processing, include gene activators and repressors. Our suggestion is that the described NHC protein fractions may act as gene regulators.

The regulatory character of the described NHC protein fractions as gene activators is indicated by their (1) appearance during the active phase of the immune reaction, especially at the time of immunoglobulin switch from IgM to IgG class of antibodies (Figures 3 and 5); (2) synthesis by a specific active fraction of spleen cells, i.e., adherent to plastic macrophage-like cells and blast forms (Figures 8a and 8b); (3) interaction with the transcriptionally active sequences of the residual chromatin fraction (Figures 16 and 17); (4) decrease or loss at various times of the immune reaction — the occurrence of specific fractions of nonhistones during the time of a switch in the synthesis from one class of immunoglobulins to another (Figures 3 and 5), and their total depletion after the reaction was completed; (5) synthesis of two fractions after immunization with thymus-dependent antigens, producing the two classes of antibody IgM and IgG (Figures 3 and 5), and synthesis of only one fraction after immunization with thymus-independent antigen (lipopolysaccharides from *E. coli* as shown in Figure 6), producing only one class of antibody; i.e., IgM; and (6) nonspecies specificity, where the same fractions are synthesized after immunization of both mice and rats with cellular antigens — sheep red blood cells in mice (Figures 3a to 3c) or spleen cells of mice in rats (Figures 3d to 3f).

The repressive character of fraction A (molecular weight 160,000), synthesized in normal, nonimmunized mice, is suggested by its (1) depletion on the second day after immunization, irrespective of the type of antigen used for immunization;[12,16,18] (2) synthesis at termination of the immune reaction (Figure 3b); (3) presence in the residual fraction of chromatin, containing proteins tightly bound with DNA (Figure 15); and (4) binding with transcriptionally inactive DNA sequences (in nonimmunized mice as shown in Figure 15). This fraction was not detected in normal, nonimmunized rats, but was found in chromatin of immunized rats after termination of the immune reaction (Figure 6). We suspect that the amount synthesized in normal, nonimmunized rats may be so minute that it is not detectable, but at the phase of termination of the immune reaction its synthesis must be greater, and therefore detectable.

In chromatin from myeloma cells spontaneously producing immunoglobulins, any characteristic fraction of NHC proteins, described previously in immunoglobulin-producing spleen cells,[12,16,19] could not be detected.[24,25] There were also distinct changes in chromatin from myeloma cells, namely low DNAase II sensitivity (Figures 18 and 19) with low template activity (Figure 10), as compared with normal spleen cells. On the basis of our results we suppose that in neoplastic cells, there is a decrease in active transcribed genes with the exception of gene or genes coding for synthesis of neoplastic proteins. In this respect, myeloma cell genes coding for synthesis of monoclonal immunoglobulin are of the highest activity.

Proteins of fractions F, H, I, and B were highly labeled with [^{3}H]-tryptophan and lower with [^{3}H]-methionine, but only very low (or not at all) with [^{3}H]-leucine (Figure 3 and 5), [^{3}H]-valine, [^{14}C]-isoleucine,[31] [^{3}H]-serine, and [^{3}H]-proline (not published). High labeling with tryptophan and methionine and low labeling with the other amino acids suggests a specific role of amino-acid residues in the regulatory processes of gene expression. The dominant presence of tryptophan suggests a specific role of residues of this amino acid in binding of NHC proteins with DNA. Histones, in which tryptophan residues are absent, can bind DNA only very loosely. Contrary to this, the NHC proteins bind DNA very tightly, especially the transcriptionally active fraction, i.e., the residual fraction of chromatin.[21] The strong interaction of NHC proteins with DNA is probably done by stacking of tryptophan with purine and pyrimidine bases of DNA.[33] Therefore we suggest that the specific absence of tryptophan in histones may be not only the result of the formation of histones in evolution at the time when tryptophan was absent, but conserved because of the specific role of histones in the nucleosomal structure of chromatin.

Finally, we would like to make some suggestions about the functional character of the described NHC fractions. As seen in Figure 1, our experiments with immunized animals happen at the time of immunoglobulin heavy chain-gene rearrangements to form new classes of immunoglobulins; e.g., switch from IgM, synthesized at the beginning of the immune reaction, to the synthesis of IgG and later other classes of immunoglobulins. For the process of switching the synthesis from IgM to IgG, enzymes would be needed to first restrict the DNA and later join the remaining ends. Enzymes of this competence have been described in prokaryotes, e.g., rec A protein in *E. coli* which has the ability to catalyze recombination of the DNA and regulate its own synthesis.[34] Recently Sakano et al.[8], Early et al.,[7] and others have postulated the presence of specific recombinases for the immunoglobulin heavy chain genes during the course of the immune reaction. The described NHC protein fractions appear to be synthesized during the period when IgM synthesis is switched to IgG as a result of specific DNA recombination events. Therefore, we suggest as a hypothesis that the described nonhistone fractions may be the postulated recombinases. This hypothesis is strongly supported by the absence of the described NHC protein fractions in chromatin of myeloma cells, where monoclonal immunoglobulins are synthesized without switching from one class of antibodies to another.

Another hypothesis may be established for fraction A (molecular weight 160,000), found in chromatin of normal, nonimmunized spleen cells, where it was depleted at the beginning of the immune reaction and again synthesized at the termination of the reaction. For this reason we suggested it to be a repressor.[12,21] Enzymes causing repression of genes are well known, and the greatest attention has been paid to methylases.[2] Data on methylases appears to be somewhat controversial. There are investigators who showed that the pattern of methylation of endogenous virus sequences is specifically altered during tumorigenesis.[35,36] Demethylation caused activation of the oncogenic viruses and tumorigenesis, and methylation had an opposite effect, i.e., inactivation of the virus and lack of tumorigenesis. On the other hand, the opposite effect of methylation on gene activity was found to occur in X chromosomes isolated from normal human cells.[37] A molecular mechanism for inactivating X chromosomes was suggested.[37] Nevertheless, we suggest that the fraction A described by us may be a specific methylase or an enzyme possessing functions capable of modifying DNA structure in a specific manner that prevents gene activation.

E. Prospects for Future Investigations

In order to elucidate the structure and function of the described NHC protein fractions, we must be aware that regulatory proteins are present in cell nuclei only in very small amounts (hundreds to thousands of molecules per cell). Such small amounts are insufficient either for biochemical or functional studies. For these reasons we have initiated studies on immunization of mice with chromatin isolated from nonimmunized and immunized mice as described in this report. After immunization, sera from immunized mice have been tested by complement fixation. In low dilution of antiserum, antibodies from "nonimmunized" and "immunized" mice caused about 80% complement fixation with the antigen used for immunization. A complement-fixation test of antibodies from mice immunized with "immunized chromatin", tested against nonimmunized chromatin, revealed only about 45% fixation. The same result has been found in the case of serum from mice immunized with nonimmunized chromatin tested against immunized chromatin. These results suggest differences in the antigenic property of nonimmunized and immunized chromatin, caused by specific proteins present in each one.

Next, serum from mice immunized with immunized chromatin was absorbed with chromatin from nonimmunized mice for the purpose of eliminating antibodies against constituents of chromatin from normal, nonimmunized mice. In this way the remaining serum should be enriched in antibodies against new protein fractions present in immunized chromatin and absent in nonimmunized ones. The results so far look promising, however much laboratory work is needed to obtain certain data, as we are also attempting to look for and isolate specific antibodies to precipitate polysomes for additional work on mRNA.

ACKNOWLEDGMENTS

The authors wish to thank Mrs. Elzbieta Wojtasiewicz for the preparation of the manuscript, and Mrs. Małgorzata Kapcińska and Mr. Waldemar Kotowski for excellent technical assistance and drafting of figures. This work was supported by the Polish Academy of Sciences within the project 09.7.4.1.6. and 09.7.3.2.1.

REFERENCES

1. **Coffino, P.,** Hormonal regulation of cloned genes, *Nature (London)*, 292, 492, 1981.
2. **Razin, A. and Riggs, A. D.,** DNA methylation and gene function, *Science*, 210, 604, 1980.
3. **Hozumi, N. and Tonegawa, S.,** Evidence for somatic rearrangements of immunoglobulin genes coding for variable and constant regions, *Proc. Natl. Acad. Sci. U.S.A.*, 73, 3628, 1976.
4. **Seidman, J. G., Max, E. L., and Leder, P.,** A k-immunoglobulin gene is formed by site specific recombination without further somatic mutation, *Nature (London)*, 280, 370, 1979.
5. **Rabbits, T. H.,** Evidence for splicing of interrupted immunoglobulin variable and constant region sequences in nuclear RNA, *Nature (London)*, 275, 291, 1978.
6. **Davis, M. M., Kim, S. K., and Hood, L.,** Immunoglobulin class switching: developmentally regulated DNA rearrangements during differentiation, *Cell*, 22, 1, 1980.
7. **Early, P., Huang, H., Davis, M., Calame, K., and Hood, L.,** An immunoglobulin heavy chain variable region is generated from three segments of DNA: V_{HJ}, D and J_H, *Cell*, 19, 891, 1980.
8. **Sakano, H., Maki, R., Kurosawa, Y., Roeder, W., and Tonegawa, S.,** Two types of somatic recombination are necessary for the generation of complete immunoglobulin heavy-chain genes, *Nature (London)*, 286, 676, 1980.
9. **Siebenlist, U., Ravetch, J. V., Korsmeyer, S., Waldmann, T., and Leder, P.,** Human immunoglobulin D segments encoded in tandem multigenic families, *Nature (London)*, 294, 631, 1981.
10. **Robertson, M. and Hobart, M.,** Antibodies, introns and biosynthetic versatility, *Nature (London)*, 290, 543, 1981.
11. **Robertson, M.,** Genes of lymphocytes. I. Diverse means to antibody diversity, *Nature (London)*, 290, 625, 1981.
12. **Horst, A., Rakowicz-Szulczyńska, E. M., and Wiland, E.,** Non-histone chromatin proteins during various stages of activity of immunocompetent cells, *Mol. Cell. Biochem.*, 38, 3, 1981.
13. **Lowry, O. H., Rosenbrough, N. J., Farr, A. L., and Randal, R. J.,** Protein measurement with the Folin phenol reagent, *J. Biol. Chem.*, 193, 26, 1951.
14. **Burton, K.,** A study of the conditions and mechanism of the diphenylamine reaction for the colorimetric estimation of deoxyribonucleic acid, *Biochem. J.*, 62, 315, 1956.
15. **Bhorjee, J. S. and Pederson, T.,** Chromatin: its isolation from cultured mammalian cells with particular reference to contamination by nuclear ribonucleoprotein particles, *Biochemistry*, 12, 2766, 1973.
16. **Rakowicz-Szulczyńska, E. M. and Horst, A.,** Incorporation of various amino acids into non-histone chromatin protein fractions of spleen cells of mice immunized with IgG, *Mol. Cell. Biochem.*, 37, 13, 1981.
17. **Rakowicz-Szulczyńska, E. M. and Horst, A.,** Specific non-histone protein fractions associated with "active chromatin" in immunized rats, *Mol. Cell. Biochem.*, in press.
18. **Horst, A. and Rakowicz-Szulczyńska, E. M.,** Species nonspecific fraction of non-histone chromatin proteins in mice and rats immunized with bacterial lipopolisaccharides, *Mol. Cell. Biochem.*, 37, 21, 1981.
19. **Rakowicz-Szulczyńska, E. M. and Horst, A.,** The occurrence of new non-histone chromatin protein (NHP) fractions induced by immunization with aggregated human gamma globulin only in one subpopulation of spleen cells, *Bull. Acad. Pol. Sci., Ser. Sci. Biol.*, 27, 823, 1979.
20. **Albrecht, J. and Hemminki, K.,** Protein composition of chromatin subfraction prepared from chick embryos, *Biochem. Biophys. Acta*, 475, 152, 1977.

21. **Rakowicz-Szulczyńska, E. M. and Horst, A.,** Non-histone proteins in fractionated chromatin before and after DNAse II treatment, *Biochim. Biophys. Acta,* 653, 69, 1981.
22. **Marushige, K. and Bonner, J.,** Fractionation of liver chromatin, *Proc. Natl. Acad. Sci. U.S.A.,* 68, 2941, 1971.
23. **Gottesfeld, J. M., Bagi, G., Berg, B., and Bonner, J.,** Sequence of the template-active fraction of rat liver chromatin, *Biochemistry,* 15, 2472, 1976.
24. **Rakowicz-Szulczyńska, E. M.,** Analysis of NHP fractions of myeloma cells RPC 5 synthesizing gamma2A, and ABPC 22 synthesizing IgM immunoglobulins, *Acta Anthropogen.,* 3, 45, 1979.
25. **Rakowicz-Szulczyńska, E. M. and Horst, A.,** Synthesis of nonhistone chromatin proteins of mice spleen cells and myeloma cells RPC 5 and ABPC 22, *Mol. Cell Biochem.,* 33, 115, 1980.
26. **Rakowicz-Szulczyńska, E. M. and Horst, A.,** Differences in salt solubility, DNAase II sensitivity and template activity of chromatin from antibodies synthesizing spleen cells and myeloma cells RPC 5 and ABPC 22, *Mol. Cell. Biochem.,* in press.
27. **Wiland, E. and Horst, A.,** Non-histone chromatin proteins (NHP) of mice thymocytes after in vivo immunization and during in vitro proliferation by PHA, *Bull. Acad. Pol. Sci. Ser. Sci. Biol.,* 28, 3, 1980.
28. **Murphy, E. C., Jr., Hall, H., Shepherd, J. H., and Weiser, R. S.,** Fractionation of mouse myeloma chromatin, *Biochemistry,* 12, 3843, 1973.
29. **Ruoslahti, E., Engvall, E., Jalanko, H., and Comings, D. E.,** Antigenic differences in nuclear proteins of normal liver and hepatoma, *J. Exp. Med.,* 146, 1054, 1977.
30. **Stalder, J., Groudine, M., Dodgson, J. B., Engel, J. D., and Weintraub, H.,** Hb switching in chickens, *Cell,* 19, 973, 1980.
31. **Newmark, P.,** Cancer genes — processed genes — jumping genes, *Nature (London),* 296, 393, 1982.
32. **Horst, A. and Rakowicz-Szulczyńska, E. M.,** New fractions of nonhistone proteins in spleen cells of mice immunized with various antigens, *Bull. Acad. Pol. Sci. Ser. Sci. Biol.,* 25, 211, 1977.
33. **Lancelot, G. and Hélène, C.,** Model studies of interaction between nucleic acids and proteins: hydrogen binding of amides with nucleic acid bases, *Nucleic Acids Res.,* 6, 1063, 1979.
34. **Sedgwick, S.,** Roles of Rec. A revealed, *Nature (London),* 287, 676, 1980.
35. **Groudine, M., Eisenman, R., and Weintraub, H.,** Chromatin structure of endogenous retroviral genes and activation by an inhibitor of DNA methylation, *Nature (London),* 292, 311, 1981.
36. **Breznik, T. and Cohen, J. C.,** Altered methylation of endogenous viral promoter sequences during mammary carcinogenesis, *Nature (London),* 295, 255, 1982.
37. **Wolf, S. F. and Migeon, B. R.,** Studies of X chromosome methylation in normal human cells, *Nature (London),* 295, 667, 1982.

Chapter 4

CHARACTERIZATION OF THE 2 *M* NaCl-RESISTANT CHROMATIN FRACTION FROM CHICKEN ERYTHROID CELLS

Peter C. Hentzen and Isaac Bekhor

TABLE OF CONTENTS

I. INTRODUCTION

The histone and nonhistone chromosomal proteins play important roles in determining the structure and function of the eukaryotic genome. The relatively small basic proteins called histones are the major structural proteins in chromatin, localized in the nucleosomes. In addition to their role in chromatin structure, the histones also repress the transcriptional activity of chromatin.[48]

The nonhistone chromosomal proteins are acidic nuclear proteins, indicating an excess of the acidic amino acids, glutamate and aspartate, over the basic amino acids, arginine, lysine, and histidine.[19] The nonhistone chromosomal proteins consist of structural elements, many different enzymatic activities, and other proteins which serve to modulate the expression of the eukaryotic genome.[58]

As the histone proteins lack the ability to specifically regulate gene expression in chromatin, attention has turned to the nonhistone proteins in the search for specific gene regulatory proteins. The nonhistone proteins represent an extremely diverse family of proteins as shown by two-dimensional polyacrylamide gel electrophoresis (PAGE). This technique has resolved from 69 and 84 proteins in normal rat liver and Novikoff hepatoma cells, respectively, to 470 protein components in HeLa cells.[57,91] Many nuclear enzymes are included in the nonhistone protein fraction, thereby adding to the heterogeneity of these proteins.[58] The nonhistones also contain proteins serving a structural function, while others serve to regulate gene activity.[58]

In sharp contrast to the histones, the nonhistone proteins display marked tissue- and species-specific differences when analyzed by one- or two-dimensional gel electrophoresis. Considerable variability was observed between nonhistones extracted from rat liver, kidney, spleen, and brain.[77] A similar study of nonhistones from rat liver, spleen, kidney, lung, and brain found that each tissue had a recognizable pattern.[90] In addition, the brain nonhistone proteins included many new bands and consisted of many high molecular weight proteins when compared to other tissues. A comparison of Friend and HeLa cell nonhistone proteins found approximately 50% of the proteins to be cell-specific with a resolution of 200 proteins.[57] During differentiation of Friend cells the nonhistone protein content decreased by 50%.[49] Nonhistone protein fractions have been found to contain proteins which form tissue-specific immunochemical complexes with homologous DNA from both Novikoff hepatoma cells and chicken erythroid cells.[31,82] Duck erythroid cells from different stages of development reveal a general decrease in kinds and amounts of nonhistone proteins during differentiation into red blood cells.[79]

When nonhistone proteins are fractionated by elution of chromatin with increasing salt concentrations, the electrophoretic analysis is greatly facilitated. Thus, when rabbit brain, heart, liver, and submandibular salivary gland nonhistone proteins are studied by this method, a specific set of nonhistones is associated with each tissue.[6,7] From the above data, it is evident that the nonhistone proteins are very heterogeneous and some are tissue-specific.

The great diversity of nonhistones and the extent to which their structure and function appears to change during gene activation strongly suggests that they play an extremely important role in determining the transcriptional state of cells.[58] Implicit in this role is the assumption that this control is mediated by nonhistone proteins interacting with and binding to the DNA, resulting in altered transcriptional activity. Many DNA-binding nonhistone proteins have been characterized in the hope of finding a specific nonhistone which regulates gene expression. To date, however, no evidence exists which demonstrates that a specific nonhistone protein can regulate the expression of a defined gene sequence in chromatin.[42]

The binding of nonhistone proteins from rat liver to DNA has been studied using the membrane filter technique.[71] This experiment showed that maximum binding occurred at 0.05 *M* salt, and was pH and Mg^{2+} dependent. Using the techniques of tandem DNA columns and filter binding to assay the specificity of rat DNA-binding nonhistone proteins, it has been shown that these proteins bind preferentially to rat DNA and in particular to middle-repetitive sequences of the genome.[69] A nonhistone fraction with a high affinity for A-T rich DNA has been characterized.[78] Nonhistone proteins with high affinity to single-copy and repetitive sequences have also been isolated.[33]

A great deal of work using DNase I digestion of chromatin (or nuclei) has yielded evidence supporting the idea that active chromatin exists in a more open or accessible conformation.[21,84] The genes coding for adult globin chains were preferentially cleaved when nuclei from 18-day chick embryo red blood cells (containing adult globin chains) were digested 15 to 20% by DNase I.[84] DNase I digestion of 5-day embryo red blood cells (containing embryonic globin chains) was specific only for embryonic sequences. When brain or fibroblast cells were digested 15 to 20% with DNase I, neither the embryonic nor the adult globin coding sequences were sensitive to cleavage. The two red blood cell types and the fibroblast cells failed to be preferentially digested when ovalbumin cDNA was used as a probe. Thus, these experiments have shown that the globin genes are preferentially sensitive to DNase I digestion in nuclei from cells which are actively transcribing these genes.

Other active genes have also been discovered to be preferentially sensitive to DNase I digestion: several endogenous retroviral genes,[29,84] the ovalbumin gene in the hen oviduct,[10,12,22,23] the albumin gene in Xenopus hepatocytes,[20] and the protamine gene of trout testes nuclei.[41] Furthermore, DNase I sensitivity is evident in chromatin sequences coding for total nuclear RNA[84] and a scarce, complex class of polyadenylated mRNA.[23]

The ability of DNase I to preferentially digest active gene sequences as well as structural genes which are potentially transcribable has been utilized extensively to investigate hemoglobin switching during erythropoiesis in chick embryo nuclei.[30,73] These experiments are performed by digesting the nuclei from embryonic and adult red blood cells with DNase I followed by specific cleavage using restriction enzymes. In primitive chick erythroblasts, both the embryonic (expressed) and the adult (nonexpressed) β-globin genes are in the DNase I sensitive, open chromatin conformation. Stalder et al.[73] suggest that the adult β-globin gene exists in a preactivated chromatin state, awaiting the actual transcriptional event. As erythropoiesis proceeds, the adult β-globin gene remains equally sensitive to DNase I while the embryonic β-globin becomes increasingly resistant to digestion, suggesting that the embryonic β-globin gene structure changes during development.

Several factors appear to affect the degree to which chromatin is sensitive to DNase I digestion. It appears that chromatin containing hyperacetylated histones is digested much more rapidly by DNase I.[47,68] Furthermore, active sequences which are preferentially cleaved by DNase I are found to be undermethylated.[37,51] Thus, it has been shown by many experiments that active genes exist in an altered, open chromatin conformation.

The high mobility group (HMG) proteins are extracted from chromatin or nuclei by 0.35 *M* NaCl (for review see Reference 27). These nonhistone proteins have molecular weights below 30,000 daltons and contain approximately 25% basic and 30% acidic amino acids. HMG-1 and -2 are soluble in 2% trichloroacetic acid (TCA) while HMG 14 and -17 are soluble in 10% TCA.[61] These proteins are present in isolated mononucleosomes, and two binding sites for HMG-14 and -17 have been identified on core nucleosomes.[46,67] The HMG proteins are released from chromatin when preferentially digested with DNase I,[41] DNase II,[40] and micrococcal nuclease.[17]

Chromatin extracted with 0.35 *M* NaCl loses its preferential DNase I sensitivity.[26,86] If either the 0.35 *M* NaCl-extracted fraction or HMG-14 and -17 from this fraction is reconstituted back to the depleted chromatin, preferential DNase I sensitivity is restored. The eluted HMG proteins do not appear to exhibit tissue specificity.[85] These experiments suggest that HMG-14 and -17 are involved in maintenance of the active chromatin conformation. HMG-14 and -17 bind specifically to active nucleosomes and prefer single-stranded to double-stranded DNA.[85,88] The factor which confers the HMG-14 and -17 binding specificity to only active nucleosomes has not yet been determined, but may be a tightly bound nonhistone protein.

Additional evidence in support of the importance of the nonhistone proteins has come from studies of chromatin domains. These domains consist of 35- to 85-kb loops of chromatin which are anchored to a supporting structure of the nucleus consisting of nonhistone proteins.[2,50,70] The domains have been termed nuclear matrix,[11] the nuclear scaffold of metaphase chromosomes,[1] among others (see Reference 2). The structures are usually isolated by treating nuclei with Triton®X-100, then 2 *M* salt, followed by digestion with DNase I and RNase A.[70] This domain structure consists of 98.2% protein (very tightly bound), 0.1% DNA, 1.2% RNA, and 0.5% phospholipid.[11] The nuclear protein matrix consists primarily of three polypeptides with molecular weights in the range 60,000 to 70,000 daltons.[11,63] Electron microscopy of metaphase protein scaffold shows a fibrous structure resembling whole paired chromatids (see Reference 1 and also other articles in this volume).

The nuclear protein matrix has been shown to contain high concentrations of hnRNA;[44] nuclear globin RNA;[65] polyoma T-antigen;[13] and sex steroid hormone binding sites.[4] DNA replication sites have also been localized in the nuclear protein matrix.[55,80]

Studies investigating the nature of DNA sequences tightly bound to the nuclear protein matrix are of particular interest. It has been reported that the nuclear protein matrix is enriched in satellite and middle repetitive DNA sequences[63] and rapidly reassociating sequences.[34] However, some results indicate that neither repetitive nor unique sequences were enriched in the nuclear protein matrix,[5,54] while ribosomal DNA genes were significantly enriched. Other workers have found the nuclear matrix of SV40-infected 3T3 cells to be enriched three- to sevenfold in SV40 sequences,[52] while the ovalbumin gene has been preferentially associated with the nuclear protein matrix isolated from chicken oviduct cells.[64]

Another group of nonhistone proteins of particular interest are the proteins which remain tightly bound to DNA after chromatin is treated with either 2 *M* NaCl or 2.5 *M* NaCl-5 *M* urea. Bekhor and associates[7,9,39,66] have devised a method of fractionating nonhistone proteins using 5 *M* urea and stepwise increments of NaCl concentration (0, 1, and 3 *M* NaCl) to extract chromatin. These experiments showed that when histones are present in DNA, all of the nonhistone classes except for fraction M3 (3 *M* NaCl) bind DNA to an extent one magnitude greater than in the absence of the histones. Bekhor's group found that fraction M3 bound most tightly to DNA whether or not the histones were present. Fraction M0 (0 *M* NaCl) consisted of protein bound hydrophobically to DNA. The M1 proteins (1 *M* NaCl) include the histones as well as the majority of the nonhistone proteins. Fraction M3 associated very strongly with DNA ($K_{ass} = 10^{10}/M$) and bound to 25 to 50 base pairs (bp) of DNA. Bekhor and Samal[9] observed that fractions M0, M1, and M3 stimulated transcription from DNA histone complexes with nonhistone fraction M1 being the least active. Other workers have used the 5 *M* urea-2.5 *M* NaCl residual DNA-nonhistone protein complex to demonstrate that the complex is immunologically tissue- and species-specific[15,59] and the proteins contain organ-specific androgen acceptor proteins.[35]

The search for a group of nonhistone proteins which might be associated with active gene sequences resulted in Bekhor's group using 2 *M* NaCl to extract all the histones and approximately 95% of the nonhistones from chromatin. This results in a tightly associated DNA-nonhistone protein (DNA-P) complex resistant to 2 *M* NaCl extraction and represents 1 to 5% of the total DNA, depending on the tissue.[8] This residual DNA-P obtained from chicken reticulocytes was found to be enriched 20-fold in globin gene sequences when compared to total DNA. Liver DNA-P showed little hybridization to the globin gene probe as would be expected for a nonerythroid tissue.

Further work has shown that chicken reticulocyte DNA-P is depleted in repetitive sequences and consists primarily of unique DNA sequences.[24] Gates and Bekhor[25] reported that chicken liver DNA-P hybridized to 25% of a cDNA probe complementary to total chicken liver polyadenylated mRNA. These results suggest that the DNA-P proteins are distributed nonrandomly, and in part are tightly associated with active gene sequences. Results supporting this claim found DNA-P from human placenta to be enriched 22-fold in active human placental lactogen gene sequences.[53] In addition, Norman and Bekhor[53] observed that DNA-P from human placenta hybridized to 57% of the total polyadenylated mRNA. The authors concluded that these tight bound proteins may play an active role in controlling gene expression.

In this communication we present a study of the DNA sequences and proteins which constitute the tightly associated complex referred to as DNA-P as isolated from chickens. The results show that DNA-P nonhistone proteins bind tightly to DNA in a nonrandom manner, they selectively enrich for globin gene sequences in chicken erythrocyte DNA-P, and they confer added specificity to DNase I digestion of DNA-P sequences.

II. EXPERIMENTAL PROCEDURES

A. Preparation of Erythroid Nuclei

Reticulocytes were induced in adult chickens by injection of phenylhydrazine (1 mg/kg) for a period of 7 days. After this time, the chickens were decapitated and bled. All subsequent procedures were carried out at 5°C unless stated otherwise. Cells were collected by centrifugation at 3000 rpm for 10 min in a Sorvall GSA rotor. The cellular pellet was washed with deionized water to lyse contaminating red blood cells, and centrifuged at 3000 rpm for 10 min. The cellular pellet was further washed with NET (0.1 *M* NaCl, 1m*M* EDTA, 10 m*M* Tris-HCl, pH 8.0) and repelleted. Erythrocytes were collected from untreated decapitated chickens into NET containing 0.1% heparin and pelleted as above. The NET wash of the pellet was repeated twice. The cells were lysed in RSB (0.1 *M* NaCl, 0.5 m*M* $MgCl_2$, 0.1 m*M* PMSF, 0.2 m*M* DTT, 10 m*M* Tris-HCl, pH 8.0) containing 0.2% Triton®X-100, and nuclei were obtained by centrifugation at 10,000 rpm for 10 min. This step was repeated until no red pigment remained in the pelleted nuclei. The nuclei were washed three times with RSB without Triton®X-100, and centrifuged at 10,000 rpm for 10 min. The final nuclear pellet was lysed as described below to yield chromatin.

B. Preparation of Nuclei from Chicken Liver

Frozen chicken livers were placed into grinding buffer (0.25 *M* sucrose, 1.5 m*M* $MgCl_2$, 10 m*M* Tris-HCl, pH 8.0) and mixed in a blender at low speed. Cells were pelleted at 5000 rpm for 10 min in the GSA rotor. The cellular pellets were resuspended in grinding buffer and mixed at high speed to rupture the cells. Crude nuclei were pelleted by centrifugation. The crude nuclear pellets were homogenized into 5 volumes of 2.2 *M*-sucrose in TPD (10 m*M* Tris-HCl, pH 8.0, 0.1 m*M* PMSF, 0.2 m*M* DTT), layered over 5-mℓ cushions of 2.2 *M*-sucrose in TPD, and centrifuged at 26,000 rpm for 90 min in a Beckman SW27 rotor to pellet purified nuclei.

C. Isolation of Chromatin

The final nuclear pellets were homogenized into TPD to lyse the nuclei. Nuclear lysis was confirmed by phase contrast microscopy. Crude chromatin was layered on 25-mℓ cushions of 1.1-*M* sucrose in TPD, and nucleoli were pelleted by centrifugation at 2500 rpm for 15 min in a SW27 rotor. Bulk chromatin was collected from the interphase, layered on 20-mℓ cushions of 1.7-*M* sucrose in TPD and centrifuged at 26,000 rpm for 90 min to pellet purified chromatin. The final chromatin pellet was used either for extraction of proteins or for the separation of DNA tightly bound to nonhistones (DNA-P) and DNA free of protein (DNA-S) as described below.

D. Separation of DNA-P and DNA-S from Chromatin

Purified chromatin from above was homogenized into 2 *M* NaCl/TPD (at 3 to 5 A_{260}), allowed to extract for 30 min with periodic homogenization, and layered on 10-mℓ cushions of 1.0 *M* sucrose-2 *M* NaCl-TPD.[8] Centrifugation was performed at 60,000 rpm for 24 hr in the Beckman type 60Ti rotor. The extraction and centrifugation was performed twice to ensure histone removal. All of the histones and about 95% of the nonhistones were solubilized from chromatin-DNA into the 2 *M* NaCl-TPD solution. Centrifugation yielded a clear DNA pellet (DNA-S) underlaid by an opaque DNA-protein complex (DNA-P). The combined pellet was homogenized into TPD, and was allowed to swell overnight by dialysis against 4.0 ℓ of TPD. After dialysis, a clear fraction and a diffuse DNA-protein precipitate was visually evident. This dialyzed mixture was collected and centrifuged at 10,000 rpm for 15 min in the Sorvall SS-34 rotor. The opaque DNA-P was in the pellet, and DNA-S in the supernatant. The DNA-P pellet was washed several times with TPD until the supernatant A_{260} approached zero.

E. Extraction of Chromatin Proteins

Proteins from chromatin and DNA-P were extracted by phenol-Tris-HCl-EDTA buffer (pH 8.4).[72,75,77] Both basic and acidic chromatin proteins were extractable in the phenol buffer phase, whereas nucleic acids remained in the interphase and the aqueous phase. The chromatin proteins extracted in 2 *M* NaCl-TPD were first dialyzed against TPD to remove the salt, and later lyophilyzed for solubilization into the phenol-Tris-HCl-EDTA buffer.

F. Isolation of Core Histones

Nuclei were suspended into TPD by homogenization. Histone H1 was removed by blending the nuclear suspension with an equal volume of 1.2 *M* NaCl-TPD to attain a final salt concentration of 0.6 *M* NaCl. This suspension was dialyzed overnight against 0.6 *M* NaCl-TPD at 5°C. The dialyzed material was spun down at 10,000 rpm for 20 min in a GSA rotor. The pellet was resuspended in 0.25 *N* HCl-TPD and blended at high speed.[74] The acid-extracted material was spun down at 10,000 rpm for 20 min in a GSA rotor. The supernatant containing the core histones was dialyzed against 0.01 *N* HCl, 0.2 m*M* DTT, 0.1 m*M* PMSF. The core histones were kept at −70°C.

G. Isolation of 0.35 *M* NaCl Proteins

Nuclei were homogenized into RSB and the suspension was made 0.35 *M* NaCl by the addition of an equal volume of 0.7 *M* NaCl-RSB. This suspension was blended until smooth and centrifuged at 10,000 rpm for 15 min in a GSA rotor. The supernatant was kept while the pellet was resuspended in RSB, blended with an equal volume of 0.7 *M* NaCl-RSB, and spun down at 10,000 rpm for 15 min. The two supernatants were pooled and contain the HMG proteins.

H. HMG Purification by TCA Precipitation

Nuclei were suspended in 5 m*M* Na phosphate buffer, pH 7.0, 0.2 m*M* PMSF and blended until a smooth suspension was obtained.[87] An equal volume of 0.7 *M* NaCl, 5 m*M* Na phosphate pH 7.0, 0.2 m*M* PMSF was added slowly to the suspension while blending to extract proteins. The extracted nuclei were dialyzed against 0.35 *M* NaCl, 5 m*M* Na phosphate pH 7.0, 0.2 m*M* PMSF overnight at 5°C. The dialyzed material was centrifuged in a GSA rotor at 10,000 rpm for 15 min. The supernatant was made 2% TCA and the precipitate was collected by centrifugation at 10,000 rpm for 15 min. The supernatant was then made 10% TCA and the precipitate collected at 10,000 rpm for 15 min. The resultant supernatant was neutralized with NaOH, dialyzed against 0.01 *N* HCl, 0.2 m*M* PMSF, and stored at −70°C.

I. Purification of DNA

All DNA samples were made 1% in sodium dodecyl sulfate (SDS) to which 50 μg/mℓ proteinase K was added, and incubated for 2 hr at 37°C. The DNA was extracted twice with an equal volume of chloroform to isopentyl alcohol (24:1) for 20 min and the phases resolved by centrifugation. The aqueous phase was made 0.3 *M* in NaCl, and 2.5 volumes of cold ethanol was added to precipitate the DNA at −20°C. The precipitated DNA was dissolved in distilled water, 20 μg/mℓ RNase A was added, and it was incubated for 2 hr at 37°C. Proteinase K (50 μg/mℓ) and SDS (1%) were later added and incubation continued for 2 hr at 37°C. The DNA solution was extracted twice with an equal volume of chloroform to isopentyl alcohol as above, and once with an equal volume of phenol to chloroform to isopentyl alcohol (25:24:1) and resolved by centrifugation. The DNA was precipitated from the final aqueous phase as above. The precipitated DNA was dissolved in distilled water and dialyzed against 4 ℓ of 0.3 *M* NaCl, 10 m*M* Tris-HCl, pH 8.0. The purified DNA(A_{260}/A_{230} = 2.4, A_{260}/A_{280} = 1.8 to 1.9) was stored at −70°C.

J. Reconstitution

DNA samples (DNA tightly bound to nonhistones [DNA-P], purified unfractionated DNA [DNA-T], and DNA purified from DNA-P) were solubilized in 5 *M* urea, 2 *M* NaCl, TPD, and reconstituted with 1.0 mg core histones per milligram of DNA (resulting in optimum reconstitution as shown by varying the amount of histones per milligram of DNA) by step dialysis.[74] The final DNA concentration prior to dialysis was 0.4 to 0.5 mg/mℓ. The dialysis steps were as follows: 5 *M* urea, 2 *M* NaCl, TPD; 5 *M* urea, 1 *M* NaCl, TPD; 5 *M* urea, 0.8 *M* NaCl, TPD; 5 *M* urea, 0.6 *M* NaCl, TPD; and 5 *M* urea, 0.4 *M* NaCl, TPD. This was followed by dialysis against 0.4 *M* NaCl, TPD to remove the urea. Urea is required during reconstitution to solubilize the DNA samples containing the tightly bound nonhistone proteins. At this point, either purified HMGs or 0.35 *M* NaCl-extracted proteins were added to the appropriate DNA samples at a concentration of 0.2 mg/mg DNA. Dialysis continued against 0.4 *M* NaCl, TPD, and finally against TPD.

K. DNase I Digestion

DNA samples including nuclei, Hl-depleted nuclei, 0.35 *M* NaCl-extracted nuclei, DNA reconstituted with core histones and the 0.35 *M* NaCl-soluble chromosomal proteins, and the minor DNA associated with tightly bound nonhistones (DNA-P) were suspended in 10 m*M* Tris-HCl, pH 7.4, 5 m*M* $MgCl_2$ to which 10 μg of DNase I was added per milligram of DNA.[84] Digestion was carried out at either room temperature or 37°C until 15% (or to the extent stated) of the DNA was rendered acid soluble (as determined by a kinetic curve). The reaction was terminated by the addition of SDS and EDTA to 1% and 10 m*M*, respectively. The DNA was purified for hybridization as described above.

L. Micrococcal Nuclease Digestion

Reconstituted DNA samples were suspended in 10 m*M* Tris-HCl, pH7.4, 10 m*M* NaCl, 3 m*M* $MgCl_2$, 0.2 m*M* $CaCl_2$ and 300 units per milligram of DNA micrococcal nuclease added.[53] Digestion occurred at 37°C and aliquots were taken at various time points. The reaction was stopped by the addition of SDS and EDTA to 1% and 5 m*M*, respectively. DNA was purified as described above.

M. Amplification of pHB1001 and pOV230 (Isolation of Plasmid DNA)

The plasmids pHB1001 (cDNA to chicken β-globin mRNA obtained from W. Salser, UCLA) and pOV230 (cDNA to chicken ovalbumin mRNA, obtained from B. W. O'Malley, Baylor College of Medicine) were amplified and purified by standard procedures.

N. Nick Translation of DNA

Nick translation of purified DNA (DNA-P and DNA-T) was done according to Norman and Bekhor.[53] The specific activity of nick-translated DNA using (^{3}H)-dCTP (^{24}Ci/mmol, New England Nuclear) ranged from 1 to 2 × 10^6 cpm/μg DNA.

Plasmid pHB1001 and pOV230 were nick translated using the New England Nuclear nick translation system and (α-^{32}P)dCTP (800 Ci/mmol). Specific activity was 1 to 2 × 10^7 cpm/μg DNA.

O. Shearing of DNA

The procedure of Davis and Philips[16] was used to shear the DNA for the reassociation experiments to a length of 500 bp. DNA was dissolved at a concentration of 0.25 mg/mℓ in 10 m*M* Tris-HCl, pH 7.4, 1.0 *M* NaCl and cooled in an ice-water bath. The DNA was sheared by sonicating for a total of 3 min in 5-sec bursts at 350 W with a Braunsonic 1510 sonicator using the intermediate probe. To precipitate the sheared DNA at −20°C, 2.5 volumes cold ethanol was added. The precipitated, sheared DNA was collected by centrifugation at 10,000 rpm for 15 min in an SS-34 rotor, redissolved in 10 m*M* Tris-HCl, pH 7.4, and passed through a 5-mℓ Chelex®-100 (Bio-Rad®) column in order to remove heavy metals.

P. Reassociation

The reassociation of the probes with the various driver DNA samples was performed at 68°C in 0.6 *M* NaCl, 50 m*M* Tris-HCl, pH 7.9, 0.1% SDS as routinely done in our laboratories.[53]

Q. Dot Hybridization

The DNA concentration of all DNA samples was determined by the diphenylamine assay.[14] All samples were diluted to 0.24 mg DNA per milliliter, after which they were denatured and any residual RNA hydrolyzed by the addition of 1/10 volume 1.0 *N* NaOH at 95°C for 10 min, cooled in an ice-water bath and neutralized by the addition of 1/10 volume 1.8 *M* Tris-HCl, 0.2 *M* Tris-base. The final DNA concentration of all samples was 0.2 mg/mℓ.

Nitrocellulose BA85 filters (10 × 15 cm) were washed in 20xSSC (1xSSC = 0.15 *M* NaCl, 0.015 *M* Na citrate) at room temperature for 30 min prior to use, blotted between Whatman® no.1 filters, and spotted with DNA as below. Each DNA spot was applied 1 μg at a time and respotted until each spot contained 4 μg of DNA. Filters were prehybridized in 50% formamide, 0.1 *M* Hepes®, pH 7.0, 0.8 *M* NaCl, 3x Denhardt's solution (1x = 0.2% polyvinylpyrrolidine, 0.02% ficoll, 0.02% bovine serum albumin), 0.1% SDS, and 500 μg/mℓ denatured, sheared salmon sperm DNA for 4 hr at 42°C with constant shaking.[81] Hybridization was done in prehybridization solution containing 10% dextran sulfate and 1 to 2 × 10^7 cpm of denatured ^{32}P-dCTP labeled plasmid pHB1001 (β-globin) or pOV230 (ovalbumin) DNA for 24 hr at 42°C with constant shaking. Filters were washed two times for 5 min at room temperature with 2xSSC, 0.1% SDS, 0.1% Na pyrophosphate with constant agitation. Filters were then washed four times with 0.2xSSC, 0.1% SDS, 0.1% Na pyrophosphate over 2 hr at 50°C in a shaking water bath. The nitrocellulose filters were blotted between Whatman® no.1 filters, covered with Saran® Wrap and exposed to Kodak® XAR-5 film with a Du Pont lightning-plus intensifying screen at −70°C. The film was developed, printed, and the intensity of each dot on the film was determined by scanning in a Quick-Scan/Flur-Vis instrument (Helena Laboratories).

R. Agarose Gel Electrophoresis

DNA samples were electrophoresed on 1 or 3% agarose gels using 50 m*M* Tris-HCl, pH 8.3, 20 m*M* Na acetate, 2 m*M* EDTA as the running buffer.[32] Samples contained 1% SDS and 1 μℓ of 1% bromophenol blue as the tracking dye. Gels were run for 6.5 to 8 hr at 50 V, stained with 0.5 μg/mℓ ethidium bromide and photographed with ultraviolet illumination. Size standards consisted of HaeIII digested ϕx174 and EcoRI digested λ DNA.

S. Polyacrylamide Gel Electrophoresis

One-dimensional PAGE was carried out on 10 or 18% gels as described by Laemmli.[38] The protein samples (75 μg) which were solubilized in urea buffer[72] were mixed with an equal volume of SDS-buffer containing 2.3% SDS, 10% glycerol, 5% 2-mercaptoethanol, 62.5 m*M* Tris-HCl, pH 6.8, and incubated at 37°C for 30 min prior to electrophoresis.

In addition, isoelectrofocusing gel electrophoresis was done as previously described[75] following dialysis of nonhistone proteins in urea buffers containing 0.01 *M* glycine to prevent possible carbamylation during dialysis.

Two-dimensional PAGE was also carried out as previously described.[75] Molecular weights of nonhistone proteins were estimated with protein markers: chymotrypsinogen (25,000), ovalbumin (45,000), bovine serum albumin (67,000), and phosphorylase A (91,000). For amino acid analysis of the electrophoresed proteins, the spots were extracted from the gels in urea-SDS buffer, dialyzed against distilled water, hydrolyzed in 1 *N* HCl, and the amino acid composition analyzed on an amino acid analyzer programmed to yield the data in absolute values as to content of each amino acid under the peak.

T. Other Methods

Electron microscope visualization was done according to Griffith and Christiansen.[28] The preparation of the samples for tungsten shadowing without the use of cytochrome *c* or cross-linking with glutaraldehyde and the electron microscopic examination were carried out entirely by Dr. Jack Griffith of the University of North Carolina on samples obtained from us. Concentration of DNA was determined by the method of Burton,[14] RNA by the method of Dische,[18] and protein by the method of Lowry et al.[45]

III. RESULTS

A. Properties of Chromatin, DNA-S, and DNA-P

Chromatin, DNA-S, and DNA-P were isolated from various chicken cell types as described in Section II. The results shown in Table 1 represent the spectrophotometric ratios A_{260}/A_{230} and the yields of chromatin, DNA-S, and DNA-P from the three chicken cell types used in this work. Chromatin A_{260}/A_{230} ratios varied from 0.95 for liver to 1.4 for erythrocytes. The yield of DNA-S varied from approximately 93% for reticulocytes and liver to almost 99% for erythrocytes. By spectrophotometric criteria, the DNA-S fractions were virtually protein-free as shown by the high A_{260}/A_{230} ratios of 2.08-2.18.

The DNA-P yield for the two transcriptionally active cell types, liver and reticulocytes, was approximately 6.7%, while the inactive erythrocyte DNA-P yield was only 1.2%. The DNA-P yield for each cell type was remarkably constant from preparation to preparation. The DNA to protein ratios for the three types of DNA-P reflect the high amount of protein found insoluble in 2 *M* NaCl. The DNA-P nonhistone proteins can be dissociated in high salt and urea, suggesting that hydrophobic interactions rather than ionic forces are responsible for the tenacious binding of these nonhistone proteins to DNA and to each other.

One-dimensional SDS-gel electrophoretic analysis of proteins isolated from chicken reticulocytes is shown in Figure 1. It is evident that 2 *M* NaCl extraction of chromatin removed all of the histones and the majority of the nonhistone proteins. The nonhistone proteins isolated from the DNA-P fraction revealed three major proteins: G1 (molecular weight 80,000), G2 (molecular weight 50,000), and G3 (molecular weight 58,000). These three major nonhistone proteins were also present in reticulocyte chromatin, as seen in Figure 1, lane RC.

Table 1
SPECTROPHOTOMETRIC RATIOS AND YIELDS OF VARIOUS CHICKEN DNA FRACTIONS

Cell Type	Component	A_{260}/A_{230}	DNA yield (%)
Reticulocyte	Chromatin	1.06	100
	DNA-S	2.08	98.15
	DNA-P	0.67	6.85
Erythrocyte	Chromatin	1.40	100
	DNA-S	2.13	98.78
	DNA-P	0.72	1.22
Liver	Chromatin	0.95	100
	DNA-S	2.18	93.4
	DNA-P	0.50	6.6

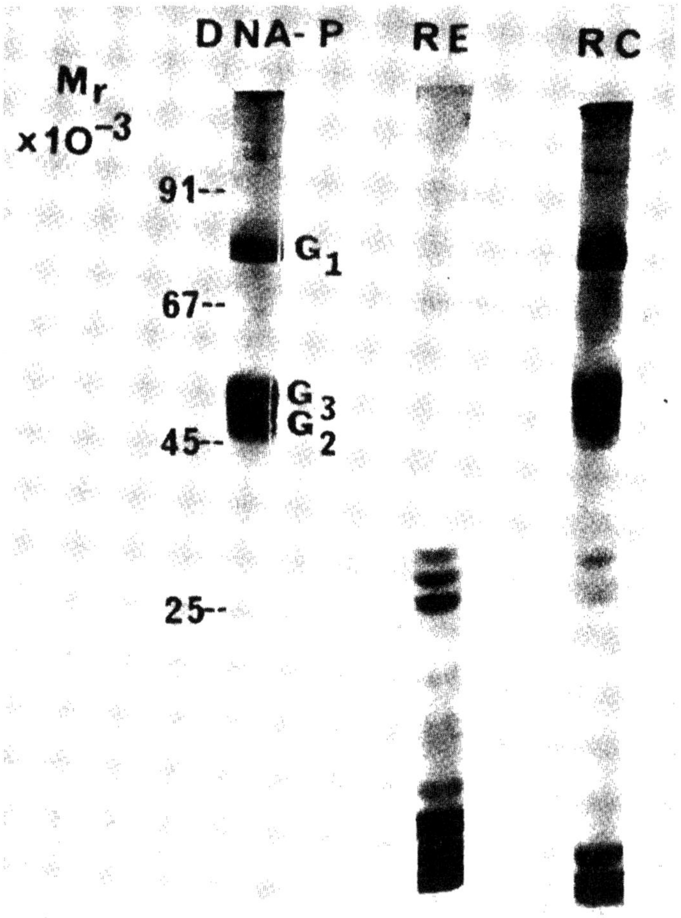

FIGURE 1. Proteins (75 μg) of reticulocyte DNA-P, 2 *M* NaCl chromatin extract (RE), and unfractionated chromatin (RC) were dissolved in SDS buffer and separated by 10% SDS-PAGE as described in Section II. Proteins G1, G2, and G3 are found in reticulocyte and DNA-P, but not in the 2 *M* NaCl-soluble fraction.

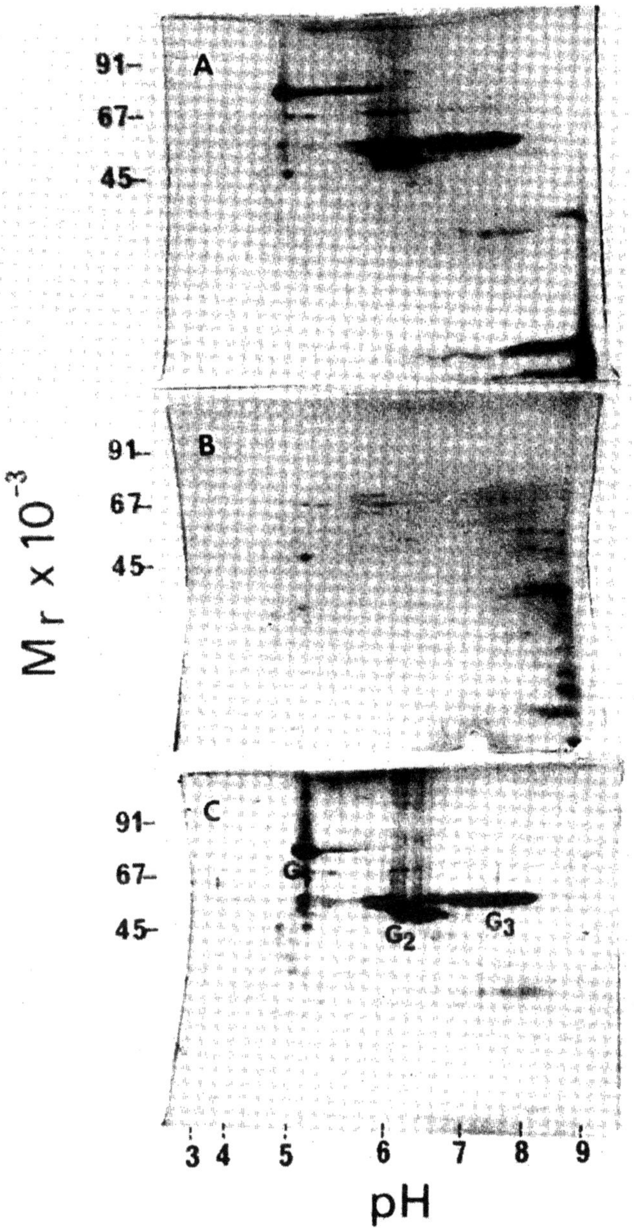

FIGURE 2. Proteins (200 μg) of reticulocyte chromatin (A), 2 *M* NaCl chromatin extract (B), and DNA-P (C) were analyzed by isoelectrofocusing in the first dimension followed by SDS-polyacrylamide slab gel electrophoresis in the second dimension as described in Section II. Proteins with pI > 9 were focused off the gels.

Two-dimensional gel electrophoresis of the reticulocyte proteins gave the results shown in Figure 2. The three major proteins (G1, G2, and G3) are clearly visualized in both chromatin and DNA-P. Other proteins were seen on overloaded gels, but obviously represent a very minor component of the DNA-P nonhistone protein fraction. Figure 2B shows the 2 *M* NaCl-extracted proteins and the complete absence of any of the tightly bound nonhistone proteins G1, G2, and G3.

Table 2
RELATIVE AMINO ACID COMPOSITION OF G1, G2, AND G3 PROTEINS

	G1	G2	G3
Asp	7.7	9.5[a]	9.1[a]
Thr	5.3	5.0	4.5
Ser	9.5	12.3	10.8
Glu	10.8	14.0	13.5
Gly	17.6	17.6	17.6
Ala	4.2	10.0	12.3
Val	1.5	0.5	1.1
Met	9.7	0.2	2.3
Ile	3.0	2.2	2.6
Leu	5.3	5.5	5.4
Tyr	7.2	3.8	3.8
Phe	3.8	3.4	2.8
Hist	1.1	1.4	1.0
Lys	10.4	11.1	10.0
Arg	2.8	4.1	3.4
$\frac{\text{Asp + Glu}}{\text{Hist + Lys + Arg}}$	1.29	1.42	1.57
pI	5.0	6.2	7.5

[a] For comparisons, % composition of amino acids in G2 and G3 was calculated relative to the glycine content found in G1, the pI of which was 5.0. At higher pIs the glycine content increases due to contamination. This amino acid is used during dialysis to prevent carbamylation of proteins in the course of isolation of chromatin proteins.

The reticulocyte DNA-P nonhistone proteins G1 (pI = 5.0), G2 (pI = 6.2), and G3 (pI = 7.5) were extracted from two-dimensional gels as described in Section II. An amino acid analysis of the 1 *N* HCl hydrolysate of these proteins appears in Table 2, as done by Dr. C. C. Liew, University of Toronto. The results show that the acidic to basic amino acid residue ratio for the three nonhistone proteins increases in the order G1 < G2 < G3, while the pI value for each protein increases in the same order (Figure 2). This information suggests that the nonpeptidyl acidic groups of glutamic and aspartic acid may have been modified. Perhaps these groups were methylated, thus conferring upon these nonhistone proteins increased basic and hydrophobic properties.[60a] The amino acid tryptophan is not contained in any of the nonhistone proteins G1, G2, and G3. The three major nonhistone proteins reveal minor variances in the composition of most of the amino acids. The nonhistone proteins G1, G2, and G3 do exhibit large differences in the content of the amino acids methionine, tyrosine, and alanine (Table 2).

The erythrocyte proteins were analyzed on one-dimensional gels as shown in Figure 3. The erythrocyte DNA-P proteins represent a much more heterogeneous population of nonhistones than those associated with reticulocyte DNA-P. As seen in Figure 3, none of the erythrocyte DNA-P proteins appear to predominate. This is in sharp contrast to the nonhistones associated with reticulocyte DNA-P. Many differences are evident upon comparison of the chromatin-associated proteins from erythrocytes and reticulocytes. Similarly, many differences are noted between the 2 *M* NaCl-extractable proteins from reticulocyte and erythrocyte chromatin.

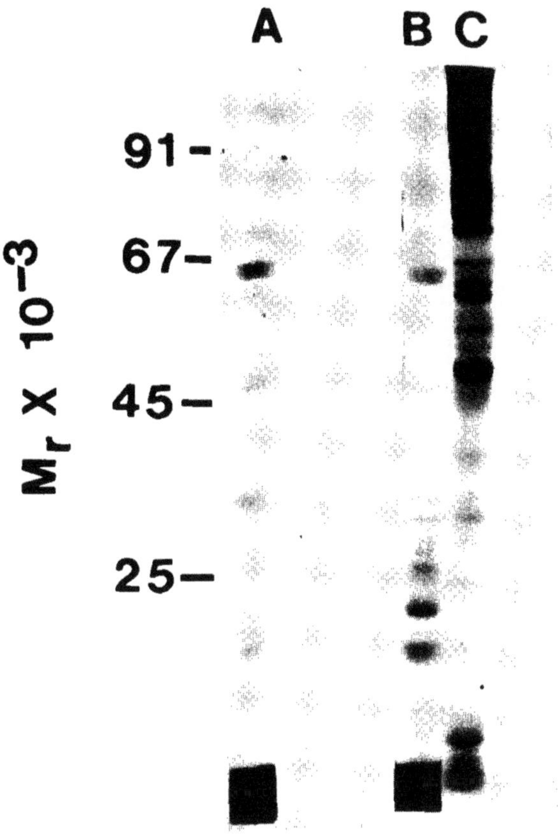

FIGURE 3. Proteins (100 μg) of erythrocyte chromatin (A), 2 *M* NaCl chromatin extract (B), and DNA-P (C). The presence of trace amounts of histones in DNA-P is due to insufficient extraction of DNA-P with 2 *M* NaCl. The latter salt concentration has been shown to extract all of the histones from chromatin.[53] One-dimensional electrophoresis was done as described in Section II.

Two-dimensional gel electrophoresis of the erythrocyte proteins revealed marked differences between reticulocyte and erythrocyte DNA-P nonhistone proteins (results not shown). The erythrocyte DNA-P gel (results not shown) indicates that one protein focused at pI 5 with a molecular weight of 80,000, which is identical to the position of the reticulocyte DNA-P protein G1. The remainder of the erythrocyte DNA-P proteins appear to have pIs around 5, and molecular weights greater than 45,000.

It was of interest to obtain clear evidence that the DNA-P nonhistone proteins contain proteins bound to DNA. Samples of the various DNA subfractions were prepared and sent to Dr. Griffith (UNC) for electron microscopic visualization. The samples were tungsten shadowed without the use of cytochrome *c* or glutaraldehyde as in Section II. DNA-P from chicken reticulocytes appears in Figure 4. Three different domains of protein-DNA binding are evident. The regions designated by the letter "A" show periodic binding of protein to the DNA every 50 to 100 bp. The "B" regions reveal proteins binding in a clustered fashion, with greater frequency per unit length of DNA. The regions represented by the letter "C" are relatively free of bound protein over long stretches of DNA. Some free protein can be seen in the background of the micrograph, suggesting that not all of the proteins isolated with DNA-P are necessarily bound to DNA. Erythrocyte nucleosomes were compared to

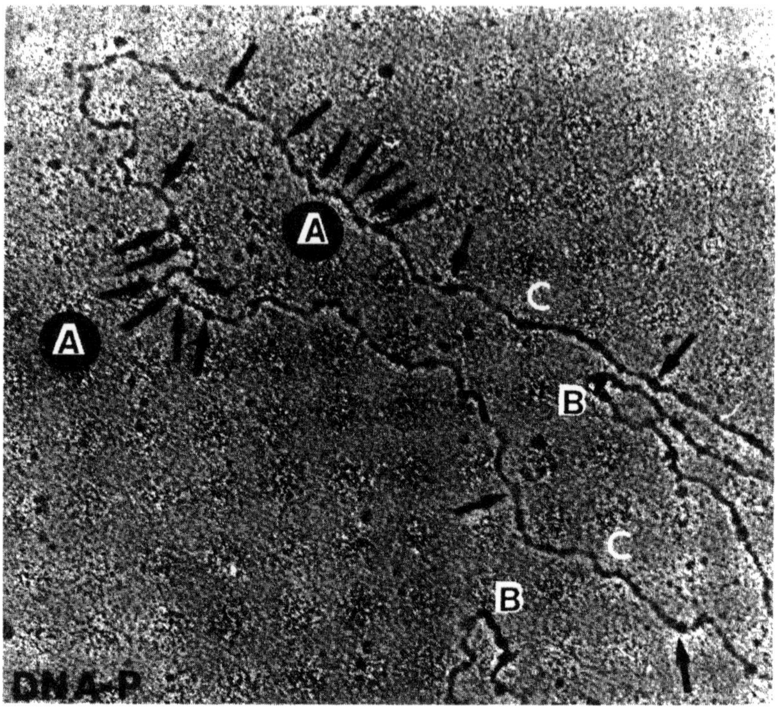

FIGURE 4. Electron micrograph of DNA-P isolated from chicken reticulocyte chromatin as described in Section II. Arrows designate protein beads occurring at various intervals in regions (A) where the protein beads occur at regular intervals of 50 to 100 bp. (B) where the proteins are clustered, and (C) where the protein beads are found at sites separated by long segments of DNA free of protein. (Magnification × 200,000.)

erythrocyte DNA-P (data not shown). The erythrocyte DNA-P once again revealed the different protein-DNA binding domains. Comparison of the nucleosomes and the protein-DNA complexes clearly demonstrated the difference in the size of each complex. Thus it is very unlikely, particularly in view of the fact that DNA-P was twice extracted with 2 *M* NaCl, that the DNA-P protein complexes represent a nucleosomal artifact or residue. This is evident from a comparison of the size and periodicity of each protein-DNA complex. The results from these analyses can be summarized as follows:

1. The tightly bound nonhistone proteins are associated with a small subfraction (1 to 7%) of the total chicken genome and are distributed in a nonrandom manner. These proteins may play a role in the organization of DNA in chromatin.
2. The protein content of DNA-P is relatively high.
3. The nucleosomes present on chromatin are 2 to 3 times larger than the protein-DNA complexes in DNA-P.
4. Most of these proteins have molecular weights greater than 45,000 daltons.

B. DNA-P Renaturation Kinetics

Previous work in this laboratory has shown that DNA-P isolated from chicken reticulocytes,[8] chicken liver,[25] and human placenta[53] is markedly enriched in active gene sequences. It was of interest to compare the sequence homology of the DNA-P isolated from chicken livers, reticulocytes, and erythrocytes. Liquid hybridization was performed as in Section II.

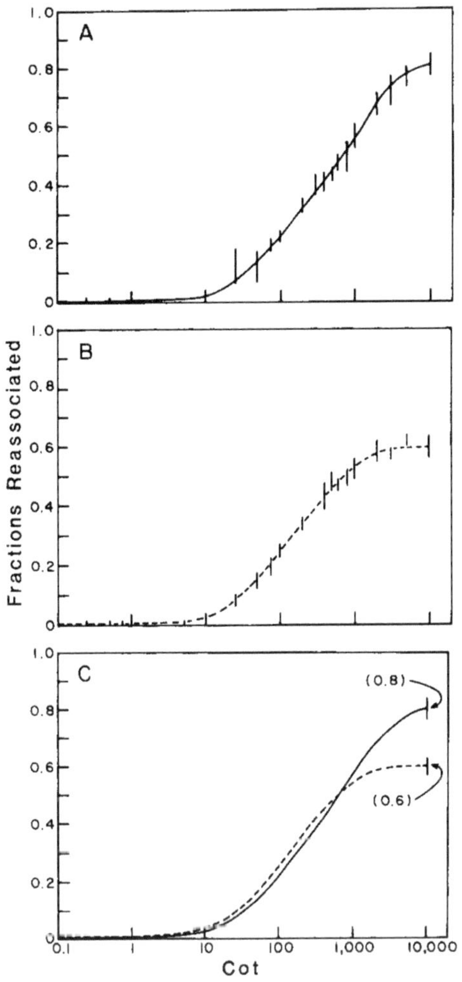

FIGURE 5. Reassociation kinetics of nick translated DNA isolated from chicken liver DNA-P reassociated with either total unfractionated chicken DNA (A), or driver DNA isolated from reticulocyte DNA-P (B). Panel C shows the composite data. Bars represent deviation of measurements between triplicate samples.

The extent of homology of the DNA-Ps from chicken liver, reticulocytes, and erythrocytes was examined. Figure 5 shows the reassociation kinetics of the nick translated liver DNA-P renaturing with total unfractionated DNA (DNA-T) (panel A) and reticulocyte DNA-P (panel B). The more rapidly reassociating sequences (Cot < 30 *M*·s) seen in the self-hybridization of total DNA (data not shown) appear to be completely absent from liver and reticulocyte DNA-P. The composite results shown in panel C reveal the slopes of reassociation to be quite similar. However, the liver DNA-P sequences have a greater extent of homology with total DNA (80%) than with the reticulocyte DNA-P sequences (60%). The $Cot_{1/2}$ for the liver DNA-P/total DNA reaction was approximately 400 M·s, while that for the liver DNA-P/reticulocyte DNA-P approximated 130 M·s. The 20% sequence heterogeneity represented approximately 1.3% of the total chicken genome taking into consideration the liver DNA-P yield of 6.6%.

The reassociation curves of nick translated erythrocyte DNA-P with total unfractionated DNA (panel A) and reticulocyte DNA-P (panel B) are shown in Figure 6. Once again, the

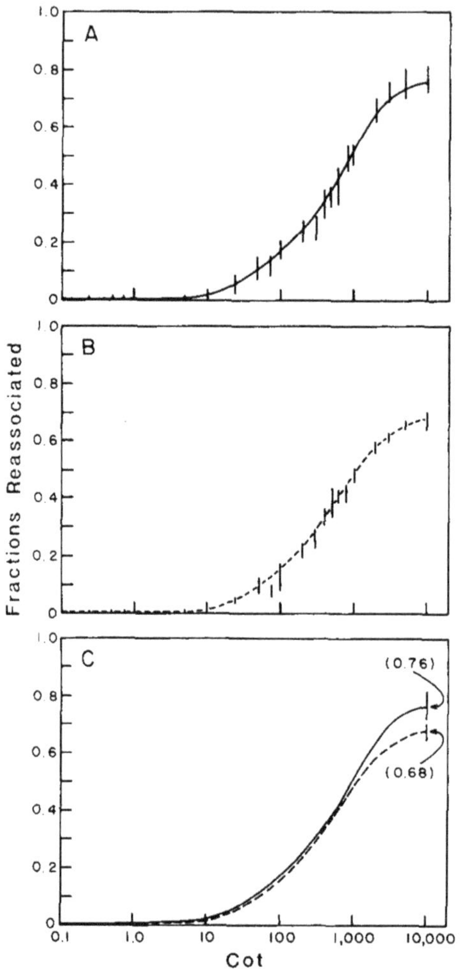

FIGURE 6. Extent of homology between nick translated DNA isolated from erythrocyte DNA-P and driver DNAs from unfractionated erythrocyte chromatin (A), and reticulocyte DNA-P (B). The composite data is shown in (C). Bars represent deviation of measurements between triplicate samples.

complete absence of repetitive sequences is apparent. The total DNA sequences saturated to erythrocyte DNA-P sequences when 76% of the DNA-P had reassociated to total DNA with a $Cot_{1/2}$ of approximately 500 M·s. Reticulocyte DNA-P sequences hybridized to only 68% of the erythrocyte DNA-P sequences with a $Cot_{1/2}$ approximating 400 M·s. The composite curve in panel C reveals a 2 to 8% sequence heterogeneity which represents between 0.13 and 0.5% of the total chicken genome.

C. Reconstitution Experiments

Recent work from this laboratory has shown that the minor chromatin subfraction called DNA-P is significantly enriched in tissue-specific gene sequences.[8,25,53] It was of interest to confirm these results using the technique of dot hybridization to assay for gene enrichment. Furthermore, this technique could be used to ascertain to what extent, of any, the 2 *M* NaCl-resistant proteins in DNA-P help to confer specificity to DNase I digestion in the presence of HMG-14 and -17. In order to determine the contribution of each component to DNase I sensitivity, it was necessary to perform reconstitution experiments using the various purified components of the system.

Core histones were obtained from nuclei, which had been extracted with 0.6 *M* NaCl to remove histone H1, by eluting with 0.25 *N* HCl as described in Section II. Chicken erythrocytes were used in this work.

Two methods were used to provide a source of HMG-14 and -17 as described in Section II. The first procedure was simply extracting nuclei with 0.35 *M* NaCl to elute these proteins off the chromatin. This preparation was quite crude, and included many loosely bound nonhistone proteins including the low- and high-mobility proteins (HMG-1, -2, -14, and -17). The second source of these proteins consisted of proteins soluble in a 10% TCA cut of this 0.35 *M* NaCl supernatant (after removing the low-mobility proteins with a 2% TCA cut). This 10% TCA supernatant represents purified HMG-14 and -17.

Experiments were performed in which different weight ratios of core histones per milligram of purified DNA were reconstituted as described in Section II. The resulting reconstituted chromatin was digested with micrococcal nuclease to determine the weight ratio yielding optimum reconstitution. Optimum reconstitution was determined by the weight ratio resulting in the highest number of separate nucleosome bands (monomer, dimer, trimer, etc.). The results showed that 1000 μg of histone per milligram of DNA produced the optimum reconstituted material for all DNA samples.

These reconstitution experiments were all carried out using urea and salt to solubilize the otherwise highly insoluble nonhistone proteins associated with DNA-P. Experiments on DNA/histone reconstitution from high salt or high salt-urea has been examined, and revealed no difference between the two methods.[3,74,76,89] The fidelity of chromatin reconstitution still remains questionable, however.

Experiments in which the purified HMG-14 and -17 proteins and 0.35 *M* NaCl proteins were reconstituted to core histone/DNA complexes were done at 200 μg of protein per milligram of DNA. DNase I digestion kinetics of the purified HMG-14 and -17 reconstitutes showed essentially no digestion, even after 60 min at 22°C. The DNase I digestion kinetics of various reconstituted DNA-P samples appear in Figure 7. The results from the DNA-P/HMG-14 and -17 reconstitutes suggested that this material precipitated out of solution and was therefore not susceptible to nuclease digestion. Figure 7 also shows DNA-P to be digested more rapidly by DNase I than DNA-P reconstituted with core histones. Reconstitution of the 0.35 *M* NaCl-soluble proteins with DNA/core histone complexes resulted in significant digestion by DNase I compared to the results obtained with purified HMG-14 and -17 as seen in the case of DNA-P in Figure 7. The rate of digestion was, however, slower than that of the reconstituted materials not containing any HMG proteins. Once it was determined that the 0.35 *M* NaCl-soluble proteins resulted in less aggregation than the purified HMG-14 and -17 proteins, the 0.35 *M* NaCl fraction was used exclusively as a source of HMG-14 and -17 for the reconstitution experiments.

D. Dot Hybridization Experiments

The probe used to assay for the presence of β-globin genes in the various DNA samples was the plasmid pHB1001 containing a cDNA copy of chicken β-globin mRNA (obtained from Dr. W. A. Salser, UCLA). The ovalbumin probe was pOV230 (cDNA clone) and was obtained from Dr. B. W. O'Malley (Baylor College of Medicine). The clones were amplified and purified as described in Section II. The plasmids were nick translated and used as in Section II.

It was of interest to determine the extent to which DNA-P from chicken erythrocytes was enriched in β-globin gene sequences (pHB1001) using the dot hybridization technique as described in Section II. Figure 8 shows the results of such an experiment in which DNA-P and -T were spotted in different amounts on the filter followed by hybridization. The data clearly demonstrate globin gene enrichment in DNA-P compared to total unfractionated DNA (DNA-T). From the scan of the developed film (containing increasing amounts of DNA

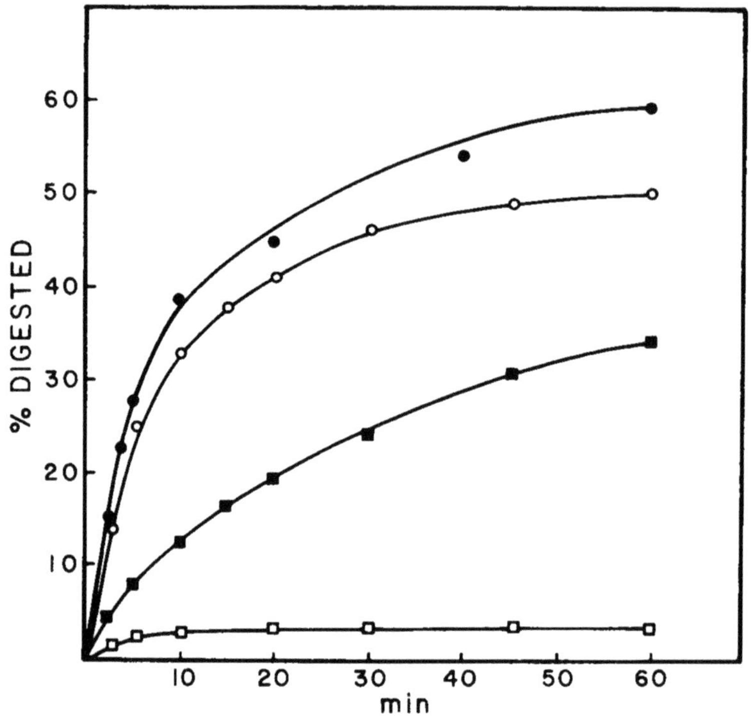

FIGURE 7. The kinetics of digestion with DNase I of variously reconstituted DNA-P samples; (●) native, protein-bound DNA-P, (○) protein-bound DNA-P reconstituted with equivalent amounts of core histones, (■) protein-bound DNA-P/core histones reconstituted with 0.2 mg/mg DNA of 0.35 *M* NaCl-soluble chromosomal proteins, (□) protein-bound DNA-P/core histones reconstituted with 0.2 mg/mg DNA of purified HMG-14 and -17. All samples were isolated from chicken erythrocyte nuclei as described in Section II.

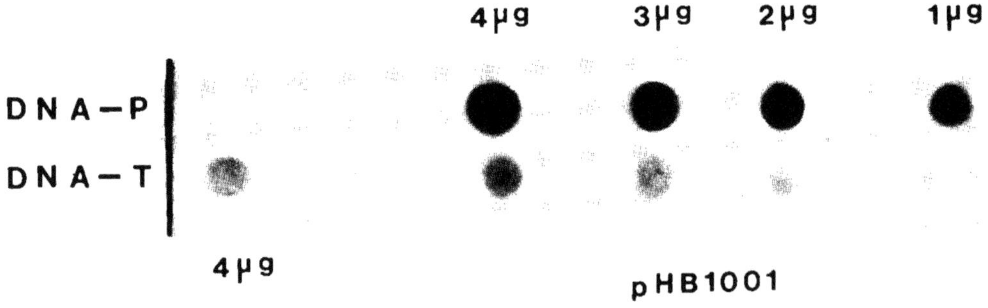

FIGURE 8. Enrichment of β-globin gene sequences in DNA tightly bound with nonhistones (DNA-P) as measured by dot hybridization with nick translated pHB1001. DNA-T shows the relative abundance of this gene in total unfractionated DNA. DNA-P and DNA-T were isolated from chicken erythrocytes as described in Section II. Dot to left is a duplicate of the 4-μg dot of total unfractionated chicken DNA.

from 1 to 4 μg per dot) and integrating the areas under each scan, it was possible to estimate the relative concentration of specific DNA sequences found in common between unfractionated DNA and DNA tightly bound to nonhistones (Table 3). The table shows that the β-globin gene concentration in DNA-P relative to DNA-T varied from 15x (at the 4 μg DNA dots) to 30x (at the 2 μg DNA dots) with an average enrichment of 23-fold. This

Table 3
RELATIVE INTENSITIES OF PROTEIN-FREE AND RNA-FREE DNA-P AND DNA-T HYBRIDIZED TO pHB1001[a]

Sample	Concentration (μg DNA per dot) 1	2	3	4
DNA-P	0.24	0.57	0.76	1.00
DNA-T	0.01	0.019	0.033	0.066
Fold enrichment	24	30	21	15

[a] Data from Figure 8.

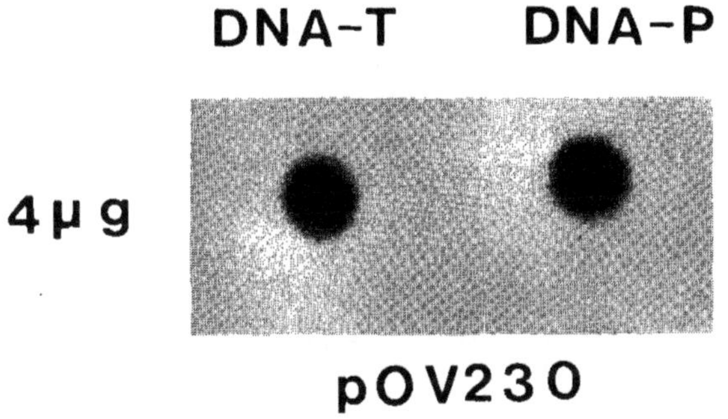

FIGURE 9. Abundance of ovalbumin gene sequences in DNA-T and DNA-P isolated from chicken erythrocytes as measured by dot hybridization with nick translated pOV230.

confirms earlier liquid hybridization data from this laboratory. The intensity of the various DNA-P dots is directly proportional to the amount of DNA applied (Table 3), confirming the accuracy of the technique. Hybridization of erythrocyte DNA-P with pOV230 (cDNA clone of chicken ovalbumin, an unexpressed gene in erythrocytes) showed no significant differences in the distribution of this gene between total unfractionated DNA and erythrocyte DNA-P (Figure 9).

It has been shown that the globin gene in the transcriptionally inactive erythrocytes retains its susceptibility to DNase I digestion.[84] Further dot hybridization experiments support these results as shown in Figure 10 and Table 4. This data demonstrate that when native erythrocyte nuclei were digested with DNase I to 15% acid solubility, most of the β-globin gene sequences have been digested (approximately 80% by intensity). Nuclei from which HMG-14 and -17 and many nonhistone proteins have been eluted with 0.35 *M* NaCl lose their sensitivity to DNase I as seen by only an 8% decrease in intensity (Figure 10 and Table 4). This observation confirms the work of Weisbrod and Weintraub[87] and Gazit et al.[26]

Evidence of preferential digestion of DNA-P is suggested when one compares the DNase I sensitivity of protein-free DNA-P reconstituted with core histones both in the presence and absence of the 0.35 *M* NaCl-soluble nonhistone proteins (containing HMG-14 and -17) relative to the DNase I sensitivity of similarly reconstituted protein-bound DNA-P (containing the tightly bound nonhistones). The results appear in Figure 11 and Table 5. The increase in intensity caused by the addition of the 0.35 *M* NaCl-soluble nonhistone proteins was due to partial aggregation of the sample (also seen in Figure 7). The significant difference between the intensities obtained from the protein-free and protein-bound DNA-P reconstitutes (Figure 11 and Table 5) when digested with DNase I is very suggestive of preferential DNase I

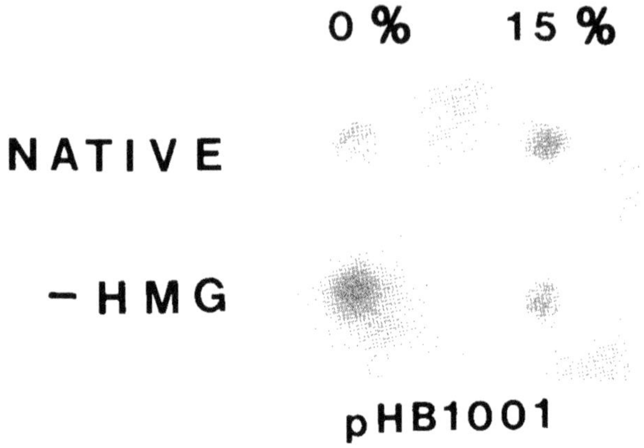

FIGURE 10. Effects of DNase I digestion of chicken erythrocyte nuclei on the distribution of globin gene sequences either in native or 0.35 *M* NaCl-extracted nuclei (-HMG); (%) indicates % DNA digested under the conditions described in Section II.

Table 4
RELATIVE INTENSITIES OF pHB1001 HYBRIDIZED TO DNA ISOLATED FROM DNaseI DIGESTED AND UNDIGESTED ERYTHROCYTE NUCLEI BEFORE AND AFTER EXTRACTION WITH 0.35 *M* NaCl[a]

Sample	% Digested	pHB1001
Native nuclei	0	1.00
	15	0.214
0.35 *M* NaCl-extracted nuclei	0	1.00
	15	0.92

[a] Data from Figure 10.

sensitivity. The results demonstrate that the protein-bound DNA-P was depleted twofold more than the protein-free DNA-P in β-globin sequences. The only difference between the two samples was the presence of the 2 *M* NaCl-resistant proteins in the sample more sensitive to DNase I. This suggests that these tightly bound nonhistone proteins do confer further sensitivity and specificity to digestion by DNase I beyond that imparted by the HMG-14 and -17 nonhistone proteins.

IV. DISCUSSION

The experiments presented here show that the DNA sequences associated with the tightly bound nonhistone proteins in DNA-P are enriched 23-fold in β-globin sequences when compared to total unfractionated DNA as demonstrated by the dot hybridization experiments shown in Figure 8 and Table 3. The similar experiment probing for the ovalbumin gene demonstrated no difference between DNA-P and total unfractionated DNA (Figure 9). This

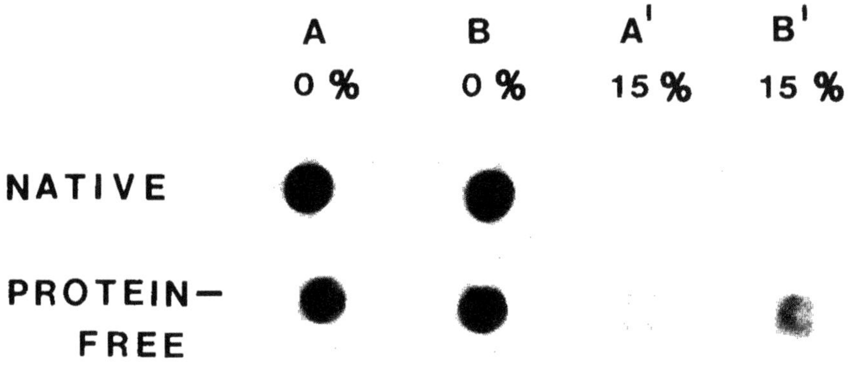

FIGURE 11. Abundance of β-globin gene sequences following digestion of the various DNA-P reconstitutes (as described in the text) with DNase I. "Native" designates protein-bound DNA-P reconstitutes. "Protein-free" designates DNA purified from protein-bound DNA-P then reconstituted with chromosomal proteins as described in Section II. (A) Native or protein-free DNA-P reconstituted with core histones; (B) native or protein-free DNA-P reconstituted with core histones and 0.5 *M* NaCl-soluble chromosomal proteins; (') designates the digested samples.

Table 5
RELATIVE INTENSITIES OF VARIOUS RECONSTITUTES HYBRIDIZED TO pHB1001[a]

Reconstitute	% Digested	Amt. added of 0.35 *M* NaCl-soluble proteins: 0	Amt. added of 0.35 *M* NaCl-soluble proteins: 0.2 mg/mg DNA
		Relative intensity	
Protein-bound DNA-P/core histones (1:1)	0	1.00	0.91
	15	0.19	0.34
Protein-free DNA-P/core histones (1:1)	0	1.00	1.00
	15	0.53	0.68

[a] Data from Figure 11.

significant enrichment in tissue-specific gene sequences has been replicated several times using the identical 2 *M* NaCl extraction fractionation procedure to prepare DNA-P (using liquid hybridization techniques). Bekhor and Mirell[8] found chicken reticulocyte DNA-P to be enriched 20-fold in the β-globin sequences while the nonerythroid liver DNA-P revealed no enrichment. Gates and Bekhor[25] reported that chicken liver DNA-P hybridized to 25% of a cDNA probe complementary to total chicken liver polyadenylated messenger RNA. Further work by Norman and Bekhor[53] found that human placenta DNA-P was enriched 22-fold in active human placental lactogen gene sequences. It is particularly noteworthy that the enrichment factors for the tissue-specific gene sequences probed in our laboratory average 22 (see References 8 and 53 in addition to this chapter). This enrichment is a result of specific DNA sequences being concentrated in DNA-P and is not a result of RNA coprecipitating with DNA-P since the DNA samples were routinely treated with NaOH prior to hybridization. It is also well established that mature chicken erythrocytes no longer transcribe globin mRNA.[43]

Others have utilized the 2 *M* NaCl extraction fractionation method introduced by Bekhor and Mirell[8] to further support the active gene enrichment results found for our DNA-P fraction. Rakowicz-Szulczyńska and Horst[62] isolated DNA-P from mouse spleen cells which represented 25% of the total DNA and 40% of which was DNase II/Mg^{2+} soluble. In addition, the template activity (using *E. coli* RNA polymerase) of the DNA-P was increased six- to sevenfold compared to that of chromatin. Kuo[36] used a modification of our procedure in which he digested the oviduct DNA-P fraction with restriction endonuclease EcoRI and collected the DNA-P remaining on glass fiber filters. The yield of DNA-P obtained was 0.5% of the total DNA which is within the range of values found for chicken erythrocytes in this work. Kuo[36] reported a threefold enrichment in ovalbumin gene sequences in DNA-P isolated from laying hen oviducts, but found no enrichment in β-globin gene sequences in the same fraction. These experiments further demonstrate the tissue-specific nature of the DNA-P gene sequences.

The analysis of the nonhistone proteins tightly associated with DNA-P from reticulocytes and erythrocytes revealed that tissue-specific differences in nonhistone protein composition exist. Figures 2C and 3 show that reticulocyte DNA-P consists primarily of three nonhistone proteins called G1 (molecular weight 80,000), G2 (molecular weight 50,000), and G3 (molecular weight 58,000), while the erythrocyte DNA-P protein composition was much more heterogeneous. The erythrocyte DNA-P nonhistones did suggest that one protein focusing at pI 5 with a molecular weight of 80,000 may be identical to the reticulocyte G1 protein. All of the DNA-P nonhistone proteins from both cell types exhibited molecular weights greater than 45,000 daltons.

These results are quite similar to those of others using the DNA-P fraction. Norman and Bekhor[53] found five major proteins (molecular weight 180,000, 150,000, 136,000, 120,000, and 58,000) in human placenta DNA-P. Kuo[36] showed that several major nonhistone proteins from chicken oviduct DNA-P had molecular weights ranging from 45 to 67,000 daltons. Perle et al.[56] reported that during the differentiation of chick embryo precartilage mesenchyme into cartilage, the DNA-P fraction loses two tightly bound nonhistone proteins of molecular weight 35,500 and 125,000. Pumo et al.[60] isolated DNA-P from chicken reticulocytes and determined by gel electrophoresis and microcomplement fixation that these tightly bound nonhistone proteins were very similar to the tissue- and species-specific chicken nuclear antigens found in reticulocyte chromatin. Wang[83] identified androgen acceptors in DNA-P prepared from rat prostate glands.

The amino acid analysis of the reticulocyte nonhistone proteins G1, G2, and G3 reveals that the acidic to basic amino acid ratios increase in the order G1 $<$ G2 $<$ G3 (Table 2). The pIs of these three nonhistone proteins also increase in the order G1 $<$ G2 $<$ G3 as shown in Figure 2. This apparent contradiction may be explained by blocking of the non-peptidyl carboxyl groups, perhaps by methylation or amination. These amino acid modifications would increase the basic nature of the three nonhistones, while methylation would also result in increased hydrophobic properties. Table 2 shows the three proteins contain a high proportion of hydrophobic residues (up to 21% for nonhistone protein G1). This suggests that these tightly bound nonhistone proteins may interact with DNA hydrophobically.

The electron micrograph of the DNA-P complexes confirms that the proteins are indeed tightly bound to the DNA (Figure 4). These nonhistone proteins appear to associate with DNA at 50- to 100-bp intervals and intervals smaller than 50 to 100 bp, while some proteins bind to DNA only infrequently. The nonhistone proteins display a nonrandom distribution on DNA. This periodic binding suggests that these nonhistone proteins play an important structural and functional role in chromatin.

The nonrandom distribution of nonhistone proteins on DNA-P was further confirmed by the DNA reassociation studies (Figures 5 and 6). It is evident that DNA-P prepared from liver is significantly depleted in reticulocyte DNA-P sequences (20%) from the results in Figure 5. Furthermore, both liver and reticulocyte DNA-P show a complete absence of repetitive sequences. This lack of repetitive sequences was also found for human placenta

DNA-P by Norman and Bekhor.[53] Comparing reticulocyte and erythrocyte DNA-P to each other reveals that approximately 95% of the sequences are homologous (Figure 6). Once again, the total lack of repetitive sequences should be noted. It is clear that specific gene sequence enrichment cannot be determined from these particular results since the various DNA-P fractions contain many homologous sequences.

The reconstitution experiments using TCA purified HMG-14 and -17 showed that the DNA-P/HMG-14 and -17 reconstituted material was insoluble in aqueous solutions, and was therefore inaccessible for study (Figure 7). Using the 0.35 *M* NaCl-soluble proteins as a source of HMG-14 and -17 resulted in reconstituted material which was much more soluble in low salt (Figure 7).

Digestion of native erythrocyte nuclei with DNase I to 15% results in a preferential 80% loss in β-globin sequences compared to HMG-14 and -17-depleted nuclei, which lost only 8% of these sequences (Figure 10 and Table 4). These results confirm the work of Weintraub and Groudine,[84] Weisbrod and Weintraub,[87] and Gazit et al.[26] Extraction of nuclei with 0.35 *M* NaCl solubilizes the HMG proteins and the preferential susceptibility to DNase I digestion is eliminated. Upon reconstitution of this 0.35 *M* NaCl-soluble protein fraction back to depleted nuclei, the preferential sensitivity to DNase I digestion is regained (data not shown).

The role of the 2 *M* NaCl-resistant nonhistone proteins in conferring DNase I sensitivity to reconstituted DNA/chromosomal protein complexes was investigated in Figure 11 and Table 5. Weisbrod and Weintraub[88] and Weisbrod[85] have found that HMG-14 and -17 alone do not confer binding specificity to active gene sequences. It is possible that the tightly bound DNA-P nonhistone proteins in some unknown manner help determine this specificity. The results in Figure 11 and Table 5 show that both in the presence and absence of the 0.35 *M* NaCl proteins (containing HMG-14 and -17 proteins) the DNA-P which is tightly associated with the 2 *M* NaCl-resistant nonhistone proteins is depleted twofold more in β-globin gene sequences than the corresponding protein-free DNA-P. This increase in DNase I sensitivity and specificity is due entirely to the presence of the tightly bound nonhistone proteins. Thus, it is tempting to postulate that the unknown factor which is required for the specific binding of HMG-14 and -17 to active gene sequences[85,88] is in part, or totally, represented by the 2 *M* NaCl-resistant nonhistone proteins.

The data presented here suggest that (1) the tightly bound nonhistone proteins may play a significant role in the expression of tissue-specific genes; (2) the tightly bound nonhistone proteins confer preferential DNase I sensitivity to DNA-P sequences; and (3) the DNA which is tightly associated with the 2 *M* NaCl-resistant proteins in DNA-P is very significantly enriched in tissue-specific gene sequences.

ACKNOWLEDGMENTS

We wish to thank Dr. C. C. Liew for the protein analysis and Dr. J. Griffith for the electron microscope examination. We also thank Dr. W. Salser and Dr. B. W. O'Malley for their generous donations of pHB1001 and pOV230, respectively. Finally, the authors wish to thank Leslie Hentzen for her skillful editorial assistance.

REFERENCES

1. **Adolph, K. W., Cheng, S. M., Paulson, J. R., and Laemmli, U. K.,** Isolation of protein scaffold from mitotic HeLa cell chromosomes, *Proc. Natl. Acad. Sci. U.S.A.*, 74, 4937, 1977.
2. **Agutter, P. S. and Richardson, J. C. W.,** Nuclear non-chromatin proteinaceous structures: their role in the organization and function of the interphase nucleus, *J. Cell Sci.*, 44, 395, 1980.
3. **Allan, J., Fey, S. J., Cowling, G. J., Gould, H. J., and Maryanka, D.,** Pathway-dependent reconstitution of chromatin structure from separated constituents, *J. Biol. Chem.*, 254, 11061, 1979.
4. **Barrack, E. R. and Coffey, D. S.,** The specific binding of estrogens and androgens to the nuclear matrix of sex hormone responsive tissues, *J. Biol. Chem.*, 255, 7265, 1980.
5. **Basler, J., Hastie, N. D., Pietras, D., Matsui, S. I., Sandberg, A. A., and Berezney, R.,** Hybridization of nuclear matrix attached deoxyribonucleic acid fragments, *Biochemistry*, 20, 6921, 1981.
6. **Bekhor, I., Anne, L., Kim, J., Lapeyre, J. N., and Stambaugh, R.,** Organ discrimination through organ-specific nonhistone chromosomal proteins, *Arch. Biochem. Biophys.*, 161, 1, 1974.
7. **Bekhor, I., Lapeyre, J. N., and Kim, J.,** Fractionation of nonhistone chromosomal proteins isolated from rabbit liver and submandibular salivary glands, *Arch. Biochem. Biophys.*, 161, 1, 1974.
8. **Bekhor, I. and Mirell, C. J.,** Simple isolation of DNA hydrophobically complexed with presumed gene regulatory proteins (M3), *Biochemistry*, 18, 609, 1979.
9. **Bekhor, I. and Samal, B.,** Nonhistone chromosomal protein interaction with DNA/histone complexes, *Arch. Biochem. Biophys.*, 179, 537, 1977.
10. **Bellard, M., Kuo, M. T., Detzen, G., and Chambon, P.,** Differential nuclease sensitivity of the ovalbumin and beta-globin chromatin regions in erythrocytes and oviduct cells of laying hen, *Nucleic Acids Res.*, 8, 2737, 1980.
11. **Berezney, R. and Coffey, D. S.,** Identification of a nuclear protein matrix, *Biochem. Biophys. Res. Commun.*, 60, 1410, 1974.
12. **Bloom, K. S. and Anderson, J. N.,** Conformation of ovalbumin and globin genes in chromatin during differential gene expression, *J. Biol. Chem.*, 254, 10532, 1979.
13. **Buckler-White, A. J., Humphrey, G. W., and Pigiet, V.,** Association of polyoma T antigen and DNA with the nuclear matrix from lytically infected 3T6 cells, *Cell*, 22, 37, 1980.
14. **Burton, K.,** A study of the conditions and mechanism of the diphenylamine reaction for the colorimetric estimation of deoxyribonucleic acid, *Biochem. J.*, 62, 315, 1955.
15. **Chiu, J.-F., Wang, S., Fujitani, H., and Hnilica, L.,** DNA-binding chromosomal non-histone proteins. Isolation, characterization, and tissue specificity, *Biochemistry*, 14, 4552, 1975.
16. **Davis, A. W. and Philips, D. R.,** A defined molecular-weight distribution of DNA after extensive sonication, *Biochem. J.*, 173, 179, 1978.
17. **Dimitriadis, G. J. and Tata, J. R.,** Subnuclear frctionation by mild micrococcal-nuclease treatment of nuclei of different transcriptional activities causes a partition of expressed and non-expressed genes, *Biochem. J.*, 187, 467, 1980.
18. **Dische, Z.,** Color reactions of nucleic acid components, in *The Nucleic Acids*, Vol. 1, Chargaff, E. and Davidson, J. N., Eds., Academic Press, New York, 1955, 285.
19. **Elgin, S. C. R. and Weintraub, H.,** Chromosomal proteins and chromatin structure, *Ann. Rev. Biochem.*, 44, 725, 1975.
20. **Felber, B. K., Gerber-Huber, S., Meier, C., May, F. E. B., Westley, B., Weber, R., and Ryffel, G. U.,** Quantitation of DNaseI sensitivity in Xenopus chromatin containing active and inactive globin, albumin and vitellogenin genes, *Nucleic Acids Res.*, 9, 2455, 1981.
21. **Felsenfeld, G.,** Chromatin, *Nature (London)*, 271, 115, 1978.
22. **Garel, A. and Axel, R.,** The structure of the transcriptionally active ovalbumin genes in chromatin, *Cold Spring Harbor Symp. Quant. Biol.*, 42, 701, 1978.
23. **Garel, A., Zolan, M., and Axel, R.,** Genes transcribed at diverse rates have a similar conformation in chromatin, *Proc. Natl. Acad. Sci. U.S.A.*, 74, 4867, 1977.
24. **Gates, D. M. and Bekhor, I.,** DNA sequence selection by tightly-bound nonhistone chromosomal proteins, *Nucleic Acids Res.*, 6, 1617, 1979.
25. **Gates, D. M. and Bekhor, I.,** Distribution of active gene sequences: a subset associated with tightly bound chromosomal proteins, *Science*, 207, 661, 1980.
26. **Gazit, B., Panet, A., and Cedar, H.,** Reconstitution of a deoxyribonucleaseI-sensitive structure on active genes, *Proc. Natl. Acad. Sci. U.S.A.*, 77, 1787, 1980.
27. **Goodwin, G. H., Walker, J. M., and Johns, E. W.,** The HMG nonhistone chromosomal proteins, *Cell Nucleus*, 6, 181, 1978.
28. **Griffith, J. D. and Christiansen, G.,** Electron microscope visualization of chromatin and other DNA-protein complexes, *Ann. Rev. Biophys. Bioeng.*, 7, 19, 1978.

29. **Groudine, M., Eisenmann, R., and Weintraub, H.,** Chromatin structure of endogenous retroviral genes and activation by an inhibitor of DNA methylation, *Nature (London)*, 292, 311, 1981.
30. **Groudine, M. and Weintraub, H.,** Activation of globin genes during chicken development, *Cell*, 24, 393, 1981.
31. **Hardy, K., Chiu, J.-F., Beyer, A. L., and Hnilica, L. S.,** Immunological properties of fractionated avian erythroid nuclei, *J. Biol. Chem.*, 253, 5825, 1978.
32. **Helling, R. B., Goodman, H. M., and Boyer, H. W.,** Analysis of endonuclease R EcoR1 fragments of DNA from lambdoid bacteriophages and other viruses by agarose-gel electrophoresis, *J. Virol.*, 14, 1235, 1974.
33. **Jagodzinski, L. L., Chilton, J. C., and Sevall, J. S.,** DNA-binding nonhistone proteins: DNA site association, *Nucleic Acids Res.*, 5, 1487, 1978.
34. **Jeppeson, P. G. N. and Bankier, A. T.,** A partial characterization of DNA fragments protected from nuclease degradation in histone depleted metaphase chromosomes of the Chinese hamster, *Nucleic Acids Res.*, 7, 49, 1979.
35. **Klyzsejko-Stefanowicz, L., Chiu, J.-F., Tsai, Y. H., and Hnilica, L. S.,** Acceptor proteins in rat androgenic tissue chromatin, *Proc. Natl. Acad. Sci. U.S.A.*, 73, 1954, 1976.
36. **Kuo, M. T.,** Distribution of tightly bound proteins in the chicken ovalbumin gene region, *Biochemistry*, 21, 321, 1982.
37. **Kuo, M. T., Mandel, J. L., and Chambon, P.,** DNA methylation: correlation with DNaseI sensitivity of chicken ovalbumin and conalbumin chromatin, *Nucleic Acids Res.*, 7, 2105, 1979.
38. **Laemmli, U. K.,** Cleavage of structural proteins during the assembly of the head of bacteriophage T4, *Nature (London)*, 227, 680, 1970.
39. **Lapeyre, J. N. and Bekhor, I.,** Chromosomal protein interactions in chromatin and with DNA, *J. Mol. Biol.*, 104, 25, 1974.
40. **Levy-Wilson, B., Connor, W., and Dixon, G. H.,** A subset of trout testis nucleosomes enriched in transcribed DNA sequences contains high mobility group proteins as major structural components, *J. Biol. Chem.*, 254, 609, 1979.
41. **Levy-Wilson, B., Kuehl, L. R., and Dixon, G. H.,** The release of high mobility group protein H6 and protamine gene sequences upon selective DNaseI degradation of trout testis chromatin, *Nucleic Acids Res.*, 8, 2859, 1980.
42. **Lewin, B., Ed.,** *Gene Expression 2*, 2nd ed., John Wiley & Sons, New York, 1980.
43. **Linder, S., Zuckerman, S. M., and Ringertz, N. R.,** Reactivation of chicken erythrocyte nuclei in heterokaryons results in expression of adult chicken globin genes, *Proc. Natl. Acad. Sci. U.S.A.*, 78, 6286, 1981.
44. **Long, B. H., Huang, C. Y., and Pogo, A. O.,** Isolation and characterization of nuclear matrix in erythroleukemia cells: chromatin and hnRNA interactions with the nuclear matrix, *Cell*, 18, 1079, 1979.
45. **Lowry, O. H., Rosebrough, N. J., Farr, A. L., and Randall, R. J.,** Protein measurement with the folin phenol reagent, *J. Biol. Chem.*, 193, 265, 1951.
46. **Mardian, J. K. W., Paton, A. E., Bunick, G. J., and Olins, D. E.,** Nucleosome cores have two specific binding sites for nonhistone chromosomal proteins HMG 14 and 17, *Science*, 209, 1534, 1980.
47. **Mathis, D. J., Oudet, P., Wasylyk, B., and Chambon, P.,** Effect of histone acetylation on structure and in vitro transcription of chromatin, *Nucleic Acids Res.*, 5, 3523, 1978.
48. **McGhee, J. D. and Felsenfeld, G.,** Nucleosome structure, *Ann. Rev. Biochem.*, 49, 1115, 1980.
49. **Morioka, K., Tanaka, K., and Ono, T.,** Comparison of nuclear proteins of differentiation-induced and uninduced Friend erythroleukemia cells, *J. Biochem.*, 89, 1633, 1981.
50. **Mullinger, A. M. and Johnson, R. T.,** Packing DNA into chromosomes, *J. Cell Sci.*, 46, 61, 1980.
51. **Naveh-Many, T. and Cedar, H.,** Active gene sequences are undermethylated, *Proc. Natl. Acad. Sci. U.S.A.*, 78, 4246, 1981.
52. **Nelkin, B. D., Pardoll, D. M., and Vogelstein, B.,** Localization of SV40 genes within supercoiled loop domains, *Nucleic Acids Res.*, 8, 5623, 1980.
53. **Norman, G. L. and Bekhor, I.,** Enrichment of selected active human gene sequences in the placental deoxyribonucleic acid fraction associated with tightly bound nonhistone chromosomal proteins, *Biochemistry*, 20, 3568, 1981.
54. **Pardoll, D. M. and Vogelstein, B.,** Sequence analysis of nuclear matrix associated DNA from rat liver, *Exp. Cell Res.*, 128, 466, 1980.
55. **Pardoll, D. M., Vogelstein, B., and Coffey, D. S.,** A fixed site of DNA replication in eucaryotic cells, *Cell*, 19, 527, 1980.
56. **Perle, M. A., Leonard, C. M., and Newman, S. A.,** Developmentally regulated nonhistone proteins: evidence for deoxyribonucleic acid binding role and localization near deoxyribonucleaseI sensitive domains of precartilage cell chromatin, *Biochemistry*, 21, 2379, 1982.

57. **Peterson, J. L. and McConkey, E. H.,** Non-histone chromosomal proteins from HeLa cells, *J. Biol. Chem.*, 251, 548, 1976.
58. **Philips, I. R., Shephard, E. A., Stein, J. L., and Stein, G. S.,** Role of nonhistone chromosomal proteins in selective gene expression, in *Eukaryotic Gene Regulation*, Vol. 2, Kolodny, G. M., Ed., CRC Press, Boca Raton, Fla., 1980, 113.
59. **Pumo, D. E., Seward, E. A., Childs, S. J., and Chiu, J.-F.,** Erythroid specific nuclear antigen and globin gene binding protein, *Biochem. Biophys. Res. Commun.*, 97, 487, 1980.
60. **Pumo, D. E., Wierzbicki, R., Sainten, A., and Chiu, J.-F.,** Immunospecificity of nuclear antigens in chicken erythroid cells, *Mol. Cell Biochem.*, 32, 49, 1980.
60a. **Quick, P., Orchard, P. J., and Duerre, J. A.,** Carboxyl methylation of nonhistone chromosomal proteins, *Biochemistry*, 20, 4724, 1981.
61. **Rabbani, A., Goodwin, G. H., and Johns, E. W.,** HMG nonhistone chromosomal proteins from chicken erythrocytes, *Biochem. Biophys. Res. Commun.*, 81, 351, 1978.
62. **Rakowicz-Szulczyńska, E. M. and Horst, A.,** Nonhistone chromosomal proteins in fractionated chromatin before and after DNaseII treatment, *Biochem. Biophys. Acta*, 653, 69, 1981.
63. **Razin, S. V., Chernokhovostov, V. V., Roodyn, A. V., Zbarsky, I. B., and Razin, G. P.,** Proteins tightly bound to DNA in the regions of DNA attachment to the skeletal structures of interphase nuclei and metaphase chromosomes, *Cell*, 27, 65, 1981.
64. **Robinson, S. I., Nelkin, B. D., and Vogelstein, B.,** The ovalbumin gene is associated with the nuclear matrix of chicken oviduct cells, *Cell*, 28, 99, 1982.
65. **Ross, D. A., Yen, R. W., and Chae, C. B.,** Association of globin ribonucleic acid and its precursors with the chicken erythroblast nuclear matrix, *Biochemistry*, 21, 764, 1982.
66. **Samal, B. and Bekhor, I.,** Nonhistone chromosomal protein interaction with DNA/histone complexes, *Arch. Biochem. Biophys.*, 179, 527, 1977.
67. **Sandeen, G., Wood, W. I., and Felsenfeld, G.,** The interactions of HMG 14 and 17 with nucleosomes, *Nucleic Acids Res.*, 8, 3757, 1980.
68. **Sealy, L. and Chalkley, R.,** DNA associated with hyperacetylated histone is preferentially digested by DNaseI, *Nucleic Acids Res.*, 5, 1863, 1978.
69. **Sevall, J. S., Cockburn, A., Sevage, M., and Bonner, J.,** DNA-protein interactions of the rat liver nonhistone chromosomal proteins, *Biochemistry*, 14, 782, 1975.
70. **Shaper, J. H., Pardoll, D. M., Kaufmann, S. H., Barrack, E. R., Vogelstein, B., and Coffey, D. S.,** The relationship of the nuclear matrix to cellular structure and function, in *Advances in Enzyme Regulation*, Vol. 17, Weber, G., Ed., Pergamon Press, Oxford, 1978, 213.
71. **Sheehan, D. M. and Olins, D. E.,** The binding of nuclear non-histone protein to DNA, *Biochem. Biophys. Acta*, 353, 438, 1974.
72. **Shelton, K. and Allfrey, V. G.,** Selective synthesis of a nuclear acidic protein in liver cells stimulated by cortisol, *Nature (London)*, 228, 132, 1970.
73. **Stalder, J., Groudine, M., Dodgson, J. B., Engel, J. D., and Weintraub, H.,** Hb switching in chickens, *Cell*, 19, 973, 1980.
74. **Stein, G. S., Mans, R. J., Gabbay, E. J., Stein, J. L., Davis, J., and Adawadkar, P. D.,** Evidence for fidelity of chromatin reconstitution, *Biochemistry*, 14, 1859, 1975.
75. **Suria, D. and Liew, C. C.,** Isolation and analysis of non-histone chromatin proteins from rat-liver nuclei by three different methods, *Can. J. Biochem.*, 52, 1143, 1974.
76. **Tatchell, K. and Van Holde, K. E.,** Reconstitution of chromatin core particles, *Biochemistry*, 16, 5295, 1977.
77. **Teng, C. S., Teng, C. T., and Allfrey, V. G.,** Studies on nuclear acidic proteins, *J. Biol. Chem.*, 246, 3597, 1971.
78. **Thomas, T. L. and Patel, G. L.,** Optimal conditions and specificity of interaction of a distinct class of nonhistone chromosomal proteins with DNA, *Biochemistry*, 15, 1481, 1976.
79. **Vidali, G., Boffa, L. C., Littau, V. C., Allfrey, K. M., and Allfrey, V. G.,** Changes in nuclear acidic protein complement of red blood cells during embryonic development, *J. Biol. Chem.*, 248, 4065, 1973.
80. **Vogelstein, B., Pardoll, D. M., and Coffey, D. S.,** Supercoiled loops and eukaryotic DNA replication, *Cell*, 22, 79, 1980.
81. **Wahl, G. M., Stern, M., and Stark, G. R.,** Efficient transfer of large DNA fragments from agarose gels to diazobenzyloxy-methylated paper and rapid hybridization by using dextran sulfate, *Proc. Natl. Acad. Sci. U.S.A.*, 76, 3683, 1979.
82. **Wakabayashi, K. and Hnilica, L. S.,** The immunospecificity of nonhistone protein complexes with DNA, *Nature (London) New Biol.*, 242, 153, 1973.
83. **Wang, T. Y.,** The role of nonhistone chromosomal proteins in the interaction of prostate chromatin with androgen-receptor complex, *Biochem. Biophys. Acta*, 518, 81, 1978.

84. **Weintraub, H. and Groudine, M.,** Chromosomal subunits in active genes have an altered conformation, *Science*, 193, 848, 1976.
85. **Weisbrod, S. T.,** Properties of active nucleosomes as revealed by HMG 14 and 17 chromatography, *Nucleic Acids Res.*, 10, 2017, 1982.
86. **Weisbrod, S., Groudine, M., and Weintraub, H.,** Interaction of HMG 14 and 17 with actively transcribed genes, *Cell*, 19, 289, 1980.
87. **Weisbrod, S., and Weintraub, H.,** Isolation of a subclass of nuclear proteins responsible for conferring a DNaseI-sensitive structure on globin chromatin, *Proc. Natl. Acad. Sci. U.S.A.*, 76, 630, 1979.
88. **Weisbrod, S. and Weintraub, H.,** Isolation of actively transcribed nucleosomes using immobilized HMG 14 and 17 and an analysis of alpha-globin chromatin, *Cell*, 23, 391, 1981.
89. **Woodcock, C. L. F.,** Reconstitution of chromatin subunits, *Science*, 195, 1350, 1977.
90. **Wu, F. C., Elgin, S. C. R., and Hood, L. E.,** Nonhistone chromosomal proteins of rat tissues. A comparative study by gel electrophoresis, *Biochemistry*, 12, 2792, 1973.
91. **Yeoman, L. C., Taylor, C. W., Jordan, J. J., and Busch, H.,** Two-dimensional polyacrylamide gel electrophoresis of chromatin proteins of normal rat liver and Novikoff hepatoma ascites cells, *Biochem. Biophys. Res. Commun.*, 53, 1067, 1973.

Chapter 5

CARBOXYL METHYLATION-DEMETHYLATION OF NONHISTONE CHROMOSOMAL PROTEINS

John A. Duerre, Michael D. Traynor, and Keith C. Leatherbury

TABLE OF CONTENTS

I. INTRODUCTION

In 1965, Axelrod and Daly[1] reported the presence of a methanol-forming enzyme in bovine pituitary gland. Subsequent reports by Liss et al.[22-24] and Kim and Paik[17,18] described a protein-carboxyl-O-methyltransferase (EC 2.1.1.24) which catalyzed the transfer of the methyl group from S-adenosylmethionine (AdoMet) to the carboxyl side chain of glutamate and/or aspartate in various proteins (Scheme 1).

This reaction results in the formation of protein-methyl esters which are quite labile, undergoing spontaneous hydrolysis to form methanol at temperatures and pH valves within the physiological range.[36] For some time it was thought that this instability accounted for the formation of methanol. However, in 1979 Gagnon[10] found that various tissues from the rat contained a protein methylesterase which would readily hydrolyze protein-methyl esters to produce methanol.

Over the past few years a specific carboxyl methylation-demethylation system has been described in bacteria. Methyl groups are transferred enzymatically from AdoMet to glutamate residues of specific methyl receptor proteins in the envelope during the chemotaxic response.[2] The glutamyl esters formed are subsequently hydrolyzed via a specific methylesterase to regenerate the receptor proteins and yield methanol.[35]

The major problem in establishing a specific function for carboxyl methylation in eukaryotic cells has been the identification of the specific substrate(s). Of all the compounds tested in vitro, the polypeptide hormones from the post-anterior lobe of the pituitary gland appear to be the best methyl acceptors.[5,16] In addition, significant carboxyl methyl groups are incorporated into γ-globulin,[15] ovalbumin,[5] gelatin,[10] calmodulin,[31] chromaffin granule membranes,[12] erythrocyte membrane proteins,[15] neurophysin,[8] and acetylcholine receptors.[19] Failure to detect significant amounts of any of these carboxyl methylated proteins in vivo has deterred significantly from delineating the function of this particular posttranslational modification.

Recent studies in our laboratory indicate that chromatin contains measurable amounts of protein-methyl esters.[30] The instability of these protein-methyl esters may be attributed to the presence of a methylesterase in the nucleus. The rapid reversible alteration of charge resulting from methylation-demethylation of specific amino acid side chains may be one mechanism by which posttranslational control of cellular and/or organ function could occur.

II. DISTRIBUTION OF PROTEIN CARBOXYL METHYLTRANSFERASE IN RAT TISSUE

The levels of protein-carboxyl methyltransferase (PCM) in the various tissues from 46-day-old rats that we obtained were quite similar to the values reported by Diliberto and Axelrod.[5] Significant levels of enzyme were detectable in the cytosolic fractions from all organs examined with the highest activity found in the testes, pituitary, and brain (Table 1). In the absence of added substrate, measurable activity was noted in all fractions with the pituitary having the highest level of endogenous substrates. After dialysis, the endogenous activity was reduced some 50 to 70% indicating that polypeptides or small molecular weight proteins constituted the majority of endogenous substrates found in the cytoplasm (Figure 1).

The majority of the PCM has been found associated with the cytosolic fraction.[5,6,27] In bovine adrenal medulla 97% of the enzyme was found in the cytosol. While some 40% of the enzyme in brain is associated with the particulate fraction,[5] it appears that measurable levels of the enzyme also are present in the nucleoplasm. This may have resulted from cytoplasmic contamination; however, the levels of enzyme associated with the brain or thymus nucleoplasm were as high or higher than that in the cytoplasm from the same organ

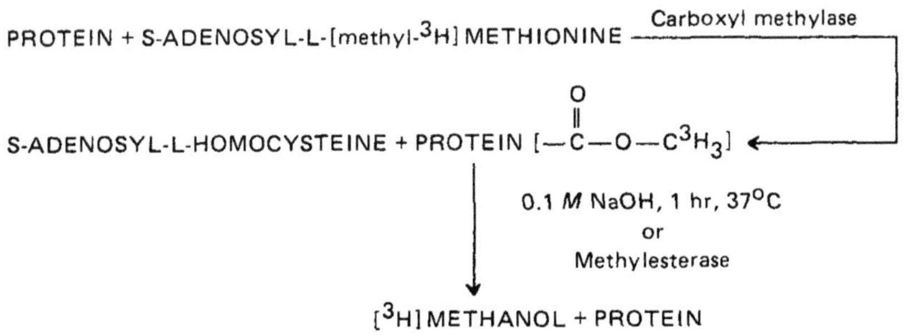

Scheme 1

Table 1
DISTRIBUTION OF CARBOXYL METHYLTRANSFERASE IN VARIOUS TISSUES FROM 46-DAY-OLD RATS

	Cytoplasm		Nucleoplasm	
Organ	**None**	**ACTH**[a]	**None**	**ACTH**[a]
Cerebrum	0.4	1.23	0.42	1.0
Cerebellum	0.37	1.8	0.35	1.25
Pituitary	0.67	0.90	—	—
Testes	0.22	0.83	—	—
Adrenal	0.24	0.63	0.16	0.22
Heart	0.13	0.70	—	—
Liver	0.23	0.31	0.17	0.13
Thymus	0.13	1.46	0.22	2.56
Spleen	0.10	0.33	—	—
RBCs	0.13	0.41	—	—
Muscle	0.08	0.50	—	—
Kidney	0.14	0.31	0.06	None

Note: Reaction mixtures contained 67 m*M* Na acetate buffer, pH 6.5, 0.5 mg ACTH, 10 μ*M* *S*-adenosyl-L-[methyl-^{3}H]methionine (1.0 Ci/mmol), and either 0.7 mg cytoplasmic or nucleoplasmic proteins. After 6 min at 37°C the reaction was terminated with the addition of 10% Cl_3CCO_2H. After washing the precipitate three times with 5% Cl_3CCO_2H, 0.5 mℓ of 0.1 *M* NaOH was added and the sample heated for 1 hr at 37°C. The radioactive methanol liberated was extracted with organic solvents and radioactivity determined as previously outlined.[30] Results are expressed as picomoles of alkaline labile ^{3}H-methyl liberated as [^{3}H]methanol per milligram per minute. Dash signifies that assay was not performed due to technical difficulties in obtaining sufficient amounts of clean nuclei. Values represent the mean of 30 animals.

[a] Endogenous activity has been subtracted.

(Table 1). This was particularly apparent when we examined the level of PCM in the tissues from young rats. The concentration of enzyme in the cytosol and nucleoplasm from the liver, thymus, and kidney was elevated in the young rat to a level equal to or greater than that found in the brain (Table 2). Unlike neuronal tissue, the level of the enzyme in these

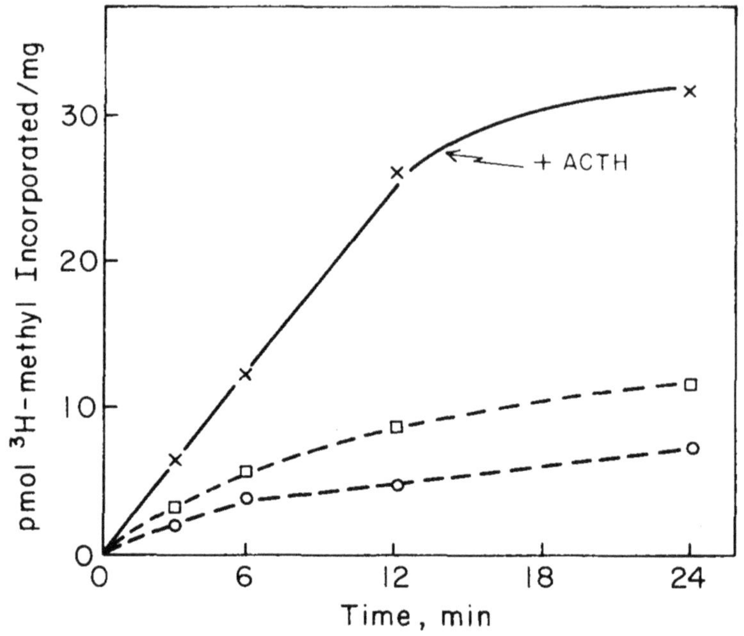

FIGURE 1. Protein-carboxyl methyltransferase activity in cytosol from 20-day-old rat cerebrum. Endogenous reaction was performed with dialyzed (○---○) and undialyzed cytosol (□---□).

Table 2
CARBOXYL METHYLTRANSFERASE LEVELS IN THE CYTOPLASM (C) AND NUCLEOPLASM (N) FROM VARIOUS ORGANS OF DIFFERENT AGE RATS

	Age in days							
	5		20		150		350	
Tissue	**C**	**N**	**C**	**N**	**C**	**N**	**C**	**N**
Liver	0.83	0.43	0.93	0.35	0.25	0.08	0.25	0.08
Cerebrum	1.1	1.2	1.35	1.6	1.85	1.0	2.25	1.1
Cerebellum	0.9	1.35	1.2	1.9	2.7	0.75	2.85	0.3
Thymus	—	—	1.85	2.95	0.85	0.45	0.25	0.4
Kidney	1.0	0.18	0.95	0.1	0.3	0.05	0.25	0.05

Note: Activity is expressed as picomoles of alkaline labile ^{3}H-methyl per milligram of protein per minute.

tissues decreased some four- to fivefold by day 150. A similar decrease in the level of nucleoplasmic enzyme was observed in the brain, whereas the concentration of the enzyme in the cytoplasm increased gradually reaching a level some threefold greater in the adult animal. The marked increase in PCM during neuronal development has led Gagnon et al.[12] to suggest that the enzyme is involved in neurosecretion. This particular explanation might suffice for the brain; however, the level of the enzyme is so high in other organs during development that it could not possibly result from neuronal development. The enzyme must play some more general function. O'Dea et al.[27] have found that human monocytes display

Table 3
SUBSTRATE SPECIFICITY OF CARBOXYL METHYLTRANSFERASE FROM 20-DAY-OLD CEREBRUM

	Cytoplasm[a]		Nucleoplasm[a]	
Substrate	pmol/mg/min	% Prolactin	pmol/mg/min	% Prolactin
Prolactin (0.5 mg)	1.53 ± 0.01		1.53 ± 0.1	
ACTH (0.5 mg)	1.14 ± 0.2	75	1.00 ± 0.6	65
Ovalbumin (1.0 mg)	1.00 ± 0.05	65	0.85 ± 0.05	56
LH (0.5 mg)	0.56 ± 0.1	37	0.70 ± 0.1	46
γ-Globulin (1.0 mg)	0.33 ± 0.03	22	0.34 ± 0.12	22
Gelatin (1.0 mg)	0.19 ± 0.03	12	0.25 ± 0.08	16

[a] Activity from endogenous control with cytoplasm (0.26 pmol/mg/min) or nucleoplasm (0.11 pmol/mg/min) was subtracted.

a dramatic increase in both cellular PCM and endogenous methyl-acceptor proteins concomitant with morphological changes that occur when this cell differentiates into a macrophage. In contrast, human pulmonary alveolar macrophages, a fully differentiated cell, do not display any such increase in PCM while methyl-acceptor proteins increased markedly. It may well be that growth and differentiation are in some way linked to carboxyl methylation of specific proteins. This would explain why PCM decreased in the liver, kidney, and thymus after the organs were fully developed. However, to date we have no logical explanation as to why the enzyme increased in the brain with age.

III. SUBSTRATE SPECIFICITY OF PROTEIN-CARBOXYL METHYLTRANSFERASE

The PCM in the cerebrum cytosol and nucleoplasm was assayed against a wide variety of proteins. Of the proteins tested, the pituitary hormones and ovalbumin were the best substrates (Table 3). Significant activity also was obtained using γ-globulin and gelatin, whereas little or no activity was detectable with rat brain histones, protamine, serum albumin, myosin, ribonuclease, deoxyribonuclease, histone methyltransferase, cytochrome *c*, rat brain nuclear membranes, or erythrocyte membranes from humans or rats. Several of these proteins have been reported to serve as methyl acceptors for the enzyme from a variety of tissues.[29] Similar substrate patterns were observed with nucleoplasmic and cytoplasmic enzyme from the cerebellum, kidney, and liver.[38] Consequently, the enzyme from the nucleoplasm and cytosol does not appear to differ, nor does the enzyme demonstrate any organ specificity.

IV. PROTEIN METHYLESTERASES IN RAT TISSUE

When adrenocorticotropin (ACTH) was incubated with cerebrum cytosol the substrate appeared to be saturated with methyl groups within 20 min (Figure 1). At saturation ACTH contained about 0.15 mmol ^{3}H-methyl per mol. This represented less than one methyl group per 6600 ACTH molecules. This could be attributed to several factors including product stability, enzyme stability, purity of substrate, availability of sites on the substrate, or presence of protein methylesterases (PME). Early investigators attributed such low levels of methylation to product stability.[29] However, the carboxyl methylated ACTH has been found to be quite stable at pH values of 5.0 to 6.5.[16] Some 95% of the ^{3}H-methyl groups were retained for up to 1 hr at 37°C at pH values from 5.0 to 7.0, whereas 85 to 90% were labile at pH 11. Hence, it appears that stability is not a factor under the experimental

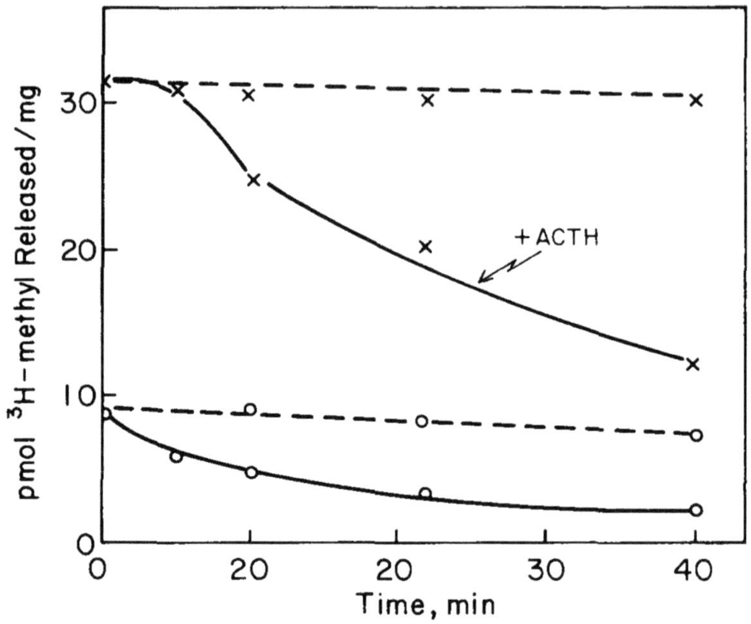

FIGURE 2. Demethylation of endogenous substrates and ACTH by protein methylesterase from rat cerebrum cytosol. After incubating the reaction mixtures with (x—x) and without (O——O) ACTH for 24 min the PCM was inhibited with the addition of AdoHcy (0.2 m*M*). Incubation was continued for the designated time periods and the reaction was terminated by the addition of 10% Cl_3CCO_2H. Controls (---) were treated exactly as described above, except PME was denatured wih acid prior to incubation at pH 6.5.

conditions employed (Figure 1). It was also doubtful that the level of activity obtained was related to enzyme stability since the PCM was quite stable at temperatures up to 41°C for 2 hr. Failure to achieve higher levels of methylation by increasing the concentration of enzyme might lead one to suspect that an impurity in the ACTH preparation was indeed the actual methyl acceptor; however, Paik and Kim[29] have identified the glutamyl residue at position 28 of the ACTH molecule as the major methylation site. Hence, it would appear that the extracts contain an extremely active protein methylesterase. Gagnon[10,11] has reported the presence of PME in various tissues from the rat. The enzyme is equally distributed between the cytosol and various organelles. As with PCM the level of the enzyme varies from organ to organ as well as within cells from a particular organ.[11] The ratio of PME to PCM has been reported to be close to 1 in the kidney, while all other tissues range from 0.01 to 0.1.[11] When the methylation reaction in rat brain cytosol was blocked with *S*-adenosylhomocysteine (AdoHcy), a marked decrease in the level of methylation of endogenous substrates and ACTH was observed (Figure 2). Similar demethylation patterns were observed with ovalbumin and gelatin (data not presented). PME activity observed in the cerebrum cytosol was some fourfold higher than that previously reported.[11] However, we are not certain as to the age of the animals that were used in that particular study. Variance in tissue concentrations of PME as well as PCM during development and aging could well account for any such discrepancy. Similarly, variances in the ratio of PME to PCM could alter the level of endogenous methyl acceptors with a particular preparation. Such variance can be readily demonstrated when tissue extracts are preincubated prior to measuring the level of endogenous substrate activity. When brain extracts were so treated the endogenous substrate activity increased from 0.35 to 1.4 pmol ^{3}H-methyl incorporated per milligram per minute.

Table 4
INCORPORATION OF ^{3}H-METHYL GROUPS INTO NUCLEAR PROTEINS

Fraction	Total ^{3}H-Methyl incorporated (pmol/mg)	Carboxyl methyl	
		pmol/mg	% Total
Nucleoplasm	25	0.4	1
NHCP (0.35 *M* salt soluble)	27	0.9	3
Histones H2A and H2B	29	1.2	4
NHCP (firmly bound)	99	76	77
Histones H1, H3, and H4	26	0.2	8

Note: Liver nuclei (10 mg DNA) from 20-day-old rats were incubated with *S*-adenosyl-L-[methyl-^{3}H]methionine for 90 min and the proteins fractionated as outlined by Quick et al.[30]

V. CARBOXYL METHYLATION OF NHCP IN NUCLEI

When rat liver nuclei were incubated with [methyl-^{3}H] AdoMet measurable amounts of ^{3}H-methyl were incorporated into all protein fractions (Table 4). Most of the radioactivity can be accounted for as N-methylated amino acid residues, or as contaminating methylated phospholipids. The presence of N-methylated histones as well as other proteins have been well established;[7,28] however, contrary to a previous report[9] N-methylated amino acids could not be detected in the firmly bound NHC proteins (for definitions refer to Figure 5). Treatment of these proteins with phospholipase C, ribonuclease, deoxyribonuclease, or organic solvents failed to dissociate the ^{3}H-methyl groups from this fraction. However, some 75 to 80% of the counts could be dissociated from these proteins under mild alkaline conditions (Table 4). Microdistillation revealed that the radioactive product was volatile at 65°C. The boiling point of methanol is 64.5°C. Recovery of the alkaline labile product could be achieved by partition chromatography or microdiffusion using a Conway cell.[30] When these samples were analyzed by gas chromatography the only radioactive compound detected co-chromatographed with authentic [^{14}C]methanol. Identity of the product was verified by reacting the product with 3,5-dinitrobenzoyl chloride.[25] In the resultant 3,5-dinitrobenzoate (mp 107°C), 80% of the radioactivity was recovered.

The protein-methyl esters isolated from chromatin were found to be quite stable at pH 8 if kept below 10°C. This allowed for electrophoretic separation of the carboxyl methylated NHC proteins. The radioautographic patterns obtained with the carboxyl methylated NHC proteins from brain, liver, and thymus were quite distinct (Figure 3). Brain exhibited 6-methylated NHC proteins with the most intensely labeled components having molecular weights of 35,000 and 66,000. The liver NHC proteins contained four methylated components with the band at 66,000 being the most heavily labeled, while the thymus did not contain any highly methylated components. The carboxyl methylated NHC proteins appear to be tissue-specific. Except for the band at 66,000, none of the methylated proteins in these tissues were similar.

VI. ORIGIN OF THE METHYL-ACCEPTOR PROTEINS

Gagnon et al.[12] have reported that the methyl-acceptor proteins in adrenal chromaffin granules were mainly associated with the membrane. Erythrocyte membranes[15,26] as well as lymphocytes[32] have been found to contain methyl-acceptor proteins. It was possible that the NHC proteins could have been contaminated with trace quantities of highly radiolabeled

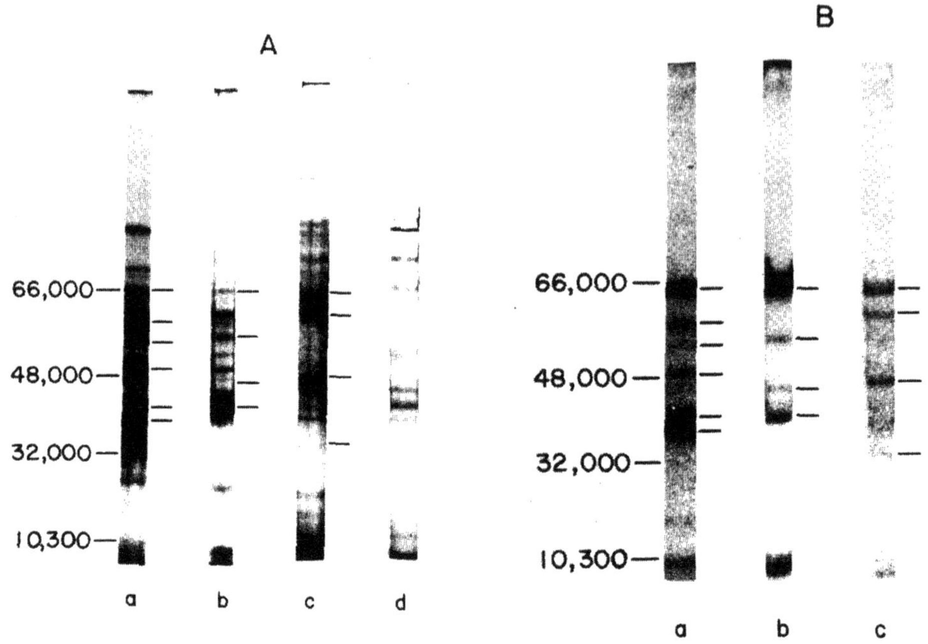

FIGURE 3. Electrophoresis (A) and radioautography (B) of carboxyl methyated NHC proteins. The NHC proteins from brain (a), liver (b), and thymus nuclei (c) were subjected to slab gel acrylamide electrophoresis in SDS.[30] (d) Denotes liver nucleoplasmic proteins. Nuclei were isolated from the various tissues by methods previously described,[30] and the firmly bound NHC proteins were separated from histones and DNA by fractionation on hydroxylapatite as previously described.[30] The NHC proteins from peak 2 (eluted off hydroxylapatite with 0.45 *M* NaCl, 5 *M* urea, and 0.05 *M* sodium phosphate, pH 6.8) was subjected to slab gel electrophoresis as described.[30] For reference on hydroxylapatite chromatography: peak 1 represents histones H2A and H2B (22.5% of total chromatin proteins); peak 2, firmly bound NHC proteins (11.25%); peak 3, histones H1, H3, and H4 (7.5%); peak 4, other NHC proteins (6.25%). The remaining 52.5% chromosomal proteins constitute the loosely bound proteins (Table 1, Reference 30).

proteins from one or more of these sources. However, red blood cell ghosts prepared from rat RBCs were extremely poor methyl acceptors. Nuclear membranes isolated by the method of Kashnig and Kasper[14] did not contain high levels of protein-methyl esters. After incubating nuclei with [methyl-^{3}H]AdoMet for 30 min the membrane proteins contained about 7 pmol alkaline labile ^{3}H-methyl per milligram compared to 70 to 80 pmol ^{3}H-methyl per milligram in the NHC proteins. Consequently, these proteins must be present within the nucleus.

Considerable amounts of protein can be solubilized from rat brain or liver nuclei with dilute buffers or physiological saline. These proteins are essentially devoid of alkaline labile methyl groups (Table 4). However, this fraction contained proteins with molecular weights similar to the firmly bound carboxyl methylated NHC proteins (Figure 3). The presence of identical molecular weights in the nucleoplasm and on chromatin has been documented by Cantino et al.[4] and Kostraba et al.[20]

When the nucleoplasmic fraction was incubated with AdoMet several of the proteins served as methyl acceptors (Table 1). This observation led us to propose that methylation of specific carboxyl side chains would alter the charge and allow these proteins to bind to DNA.[30] The rapid reversal of this reaction via a protein methylesterase could account for the presence of methyl-acceptor proteins in the nucleoplasm in the absence of protein synthesis. The presence of PME in the nuclei could be readily demonstrated upon inhibition of the PCM with AdoHcy. The majority of the PME was bound to the chromatin, since removal

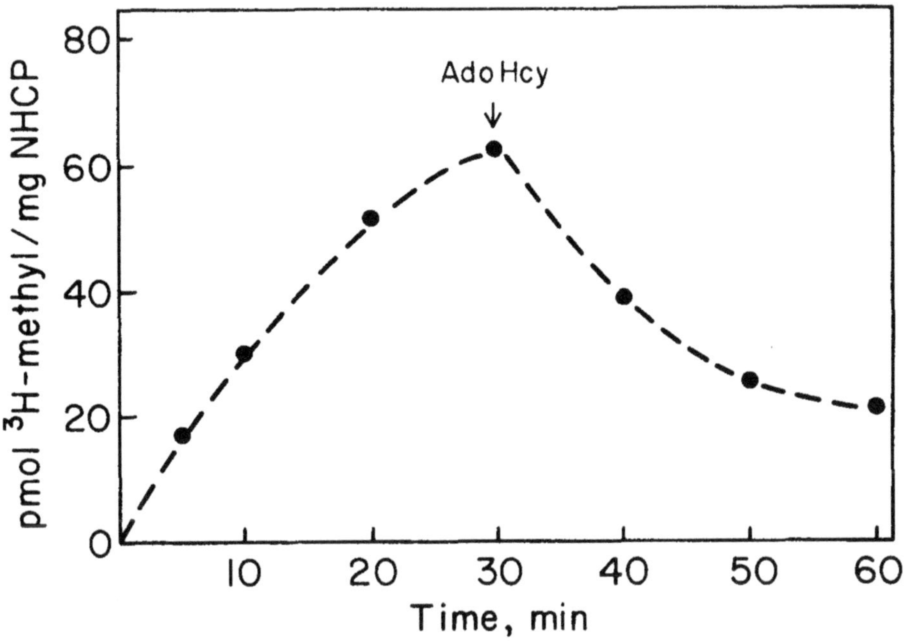

FIGURE 4. Carboxyl methylation-demethylation of NHC proteins. Rat liver NHC proteins were isolated after nuclei had been incubated with [methyl-^{3}H]AdoMet (Table 4). After 30 min PCM was blocked with AdoHcy (0.2 m*M*).

of the nucleoplasmic proteins did not alter the rate of demethylation. The chromosomal-bound PME yielded some 3 pmol [^{3}H]methanol per milligram of protein per minute (Figure 4). This activity was some tenfold higher than the endogenous activity observed with either cytoplasm of nucleoplasm (data not presented).

The high concentration of PME associated with chromatin may also account for the low levels of methylation observed upon incubating exogenous methyl-acceptor proteins with nuclei. For example, when ACTH was incubated with nuclei little or no carboxyl methylation was observed in the presence of AdoMet (Table 5). Furthermore, the addition of cytosol had only marginal effects on the methylation of ACTH in the presence of chromatin (Table 5). Paik and Kim[29] reported a similar reduction in carboxyl methylation upon the addition of particulate fraction to liver cytosol. They attributed the marked reduction in activity to the presence of a heat labile particulate-bound inhibitor. It is just as probable that this reduction was due to the presence of a protein methylesterase.

It is attractive to think that carboxyl methylation might allow for the activation of certain hormones. Once activated these proteins could bind to chromatin either directly or in conjunction with some receptor protein. To date we have found no evidence to support this theory. To the contrary, when ACTH, luteinizing hormone (LH), or prolactin was incubated with nuclei, no significant increase in the level of methylation of the NHC proteins occurred (Table 5). Although the further addition of cytoplasm allowed for the detection of trace amounts of [methyl-^{3}H]ACTH in the soluble fraction, no increase in chromosomal-bound methylated proteins was observed.

Various investigators have demonstrated that NHC proteins are synthesized in the cytoplasm and are then transferred to the nucleus.[3,33,37] Interestingly, the amount of carboxyl methylated proteins bound to chromatin increased some two- to threefold upon addition of cytoplasmic proteins (Table 5). That cytosol contained methyl-acceptor proteins that bind to chromatin in the presence of AdoMet was verified by utilizing ^{14}C-cytoplasmic proteins.

Table 5
CARBOXYL METHYLATION OF NHC PROTEINS IN PRESENCE OR ABSENCE OF ACTH AND CYTOPLASMIC PROTEINS

	Soluble and loosely bound NHC proteins		Firmly bound NHC proteins	
Additions	pmol/[^{3}H]methyl ester	% Control	pmol/[^{3}H]methyl ester	% Control
Control	0.58		0.74	
ACTH (0.5 mg)	0.57	100	0.59	80
Cytoplasm (1.7 mg)	1.5	260	1.83	250
Cytoplasm + ACTH	2.9	500	1.65	220

Note: Reaction mixture contained rat brain nuclei (1.5 mg DNA), 0.32 *M* sucrose, Ka phosphate buffer, pH 6.9, 0.4 m*M* paramethyl sulfonyl fluoride (PMSF), 1.0 m*M* $MgCl_2$, and 10 μ*M* [methyl-^{3}H]AdoMet (1.0 Ci/mmol). Tubes were incubated for 6 min at 37°C and the reaction terminated with the addition of AdoHcy (0.5 m*M*). The chromatin was washed three times with phosphate buffer, pH 6.0, and once with 0.34 *M* Na citrate, pH 6.0. The soluble proteins in these washes were precipitated with Cl_3CCO_2H (10%) and washed three times with Cl_3CCO_2H prior to releasing [^{3}H]methanol with alkali.

When radiolabeled brain cytosolic proteins were incubated with nuclei, little or no radioactive protein was associated with the chromatin. However, upon the further addition of AdoMet significant amounts of ^{14}C-protein were firmly bound to chromatin.

VII. IN VIVO METHYLATION OF NHC PROTEINS

When 6-day-old rat pups were given [methyl-^{3}H]methionine by i.p. injection maximum incorporation of the isotope occurred within 8 hr. Of all the liver fractions, the firmly bound NHC proteins attained the highest specific radioactivity (Figure 5). The level being some twofold greater than the microsomal fraction and 4 to 5 times greater than the other cellular fractions. The firmly bound NHC proteins also have the shortest half-lives of all the protein fractions. The average half-life of these proteins from the liver was about 15 hr compared to half-lives of 3 to 5 days for the microsomal, nucleoplasmic, and loosely associated NHC proteins (0.34 *M* salt soluble). Half-lives of less than 24 hr for liver NHC proteins have been reported by Legraverend and Glazer.[21]

Of all the cellular fractions the only proteins that contained significant amounts of alkaline labile methyl groups were the firmly bound NHC proteins. These proteins were completely devoid of alkaline labile methyl groups within 15 hr post-injection (Figure 5). The half-lives of the protein-methyl esters in this fraction were found to be markedly less than the polypeptide chains in all tissues examined (Table 6). This would suggest either that the carboxyl methylated NHC proteins have much shorter half-lives than the majority of proteins in this fraction or the carboxyl methyl groups turn over independently of the polypeptide chains. The small amount of these proteins associated with chromatin and the instability of the carboxyl methyl groups make it extremely difficult to subfractionate and identify the various proteins with this fraction. However, both of these possibilities give support to the hypothesis that carboxyl methylated NHC proteins have certain properties expected of gene-regulatory proteins.[34]

Even in the firmly bound NHC proteins the level of carboxyl methylation attained in vivo was quite tenuous. The amount of ^{3}H-methyl isolated as [^{3}H]methanol represented less than

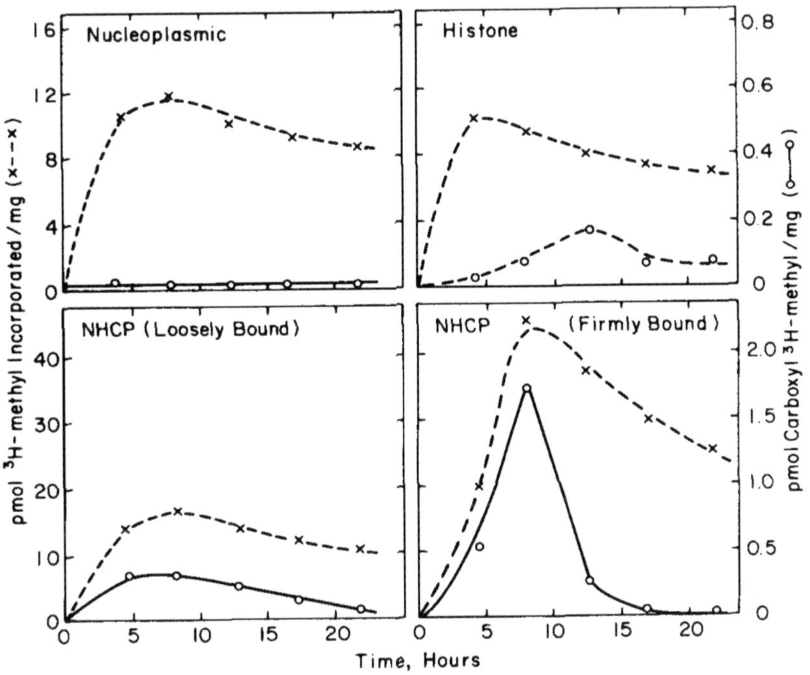

FIGURE 5. In vivo incorporation of [methyl-^{3}H]methionine into rat liver nuclear proteins. Forty 6-day-old rat pups were given 9 μCi/g of L-[methyl-^{3}H]methionine and nuclear proteins were analyzed for alkaline labile methyl groups at various times. Loosely bound NHC proteins refer to those chromosomal proteins which dissociate from chromatin in 0.34 *M* sodium citrate buffer, pH 6.0. Firmly bound NHC proteins refer to those chromosomal proteins which remain bound to chromatin after extraction with 0.34 *M* sodium citrate buffer, pH 6.0.[30]

Table 6
AVERAGE BIOLOGICAL HALF-LIVES OF [^{3}H]METHYL ESTERS AND NHC PROTEINS IN VARIOUS ORGANS FROM 6-DAY-OLD RATS

	Turnover rate (hr)	
Organ	**[^{3}H]Methyl esters**	**NHC proteins**
Cerebellum	0.7	50
Cerebrum	1.3	74
Kidney	7.2	72
Liver	1.5	15
Thymus	0.9	28

Note: The turnover rates of the methyl esters were determined between 3 and 24 hr and the turnover rates of the NHC proteins were calculated between 7 and 120 hr. The biological half-lives were calculated from the regression coefficients by the following equation: $T_{1/2} = - \log 2 \div$ regression coefficient. Activity is expressed as pmol/mg.

5% of the total isotopic label incorporated into these proteins (Figure 5). The reason why we attain much higher levels of methylation with nuclei remains unanswered. Perhaps the time required for distribution of the isotope throughout the animal allows for most of the carboxyl methylated proteins to turn over.

VIII. CONCLUDING REMARKS

It is attractive to suppose that methylation-demethylation of specific proteins could activate or inactivate genes by the binding to and release of these proteins from chromatin. Gene-regulatory proteins are expected to possess certain characteristics including a high rate of turnover, tissue specificity, and the ability to repress or stimulate RNA polymerase activity.[34] The carboxyl methylated NHC proteins appear to be tissue-specific and turn over very rapidly. Legraverend and Glazer[21] have isolated a 66,000-molecular weight NHC protein that stimulates RNA synthesis. Interestingly, the only carboxyl methylated NHC protein common to all tissues examined has an identical molecular weights. Carboxyl methylation could alter the charge on such proteins allowing them to bind at specific sites, thus altering RNA polymerase activity. Alternately, demethylation via a chromosomal bound protein methylesterase could release such proteins from the DNA.

REFERENCES

1. **Axelrod, J. and Daly, J.,** Pituitary gland. Enzymatic formation of methanol from S-adenosylmethionine, *Science*, 150, 892, 1965.
2. **Black, R. A., Hobson, A. C., and Adler, J.,** in *Biochemistry of S-Adenosylmethionine and Related Compounds*, Usdin, E., Borchardt, R., and Creveling, C. R., Eds., Macmillan, London, 1982, 91.
3. **Borun, T. W. and Stein, G. S.,** The synthesis of acidic chromosomal proteins during the cell cycle of HeLa S-3 cells. II. The kinetics of residual protein synthesis and transport, *J. Cell Biol.*, 52, 308, 1972.
4. **Cantino, J. J., Busch, H., Daskal, Y., and Yeoman, L. C.,** Subcellular localization of DNA-binding protein BA by immunofluorescence and immunoelectron microscopy, *J. Cell Biol.*, 83, 462, 1979.
5. **Diliberto, E. J., Jr. and Axelrod, J.,** Regional and subcellular distribution of protein carboxymethylase in brain and other tissues, *J. Neurochem.*, 26, 1159, 1976.
6. **Diliberto, E. J., Jr., Viveros, O. H., and Axelrod, J.,** Subcellular distribution of protein carboxymethylase and its endogenous substrates in the adrenal medulla: possible role in excitation-secretion coupling, *Proc. Natl. Acad. Sci. U.S.A.*, 73, 4050, 1976.
7. **Duerre, J. A. and Quick, D. P.,** in *Transmethylation*, Usdin, E., Borchardt, R., and Creveling, C. R., Eds., Elsevier/North Holland, Amsterdam, 1979, 583.
8. **Edgar, D. H. and Hope, D. B.,** Protein-carboxyl methyltransferase of the bovine posterior pituitary gland: neurophysin as a potential endogenous substrate, *J. Neurochem.*, 27, 949, 1976.
9. **Friedman, M., Shull, K. H., and Farber, E.,** Highly selective in vivo ethylation of rat liver nuclear protein by ethionine, *Biochem. Biophys. Res. Commun.*, 34, 857, 1968.
10. **Gagnon, C.,** Presence of a protein methylesterase in mammalian tissues, *Biochem. Biophys. Res. Commun.*, 88, 847, 1979.
11. **Gagnon, C.,** in *Biochemistry of S-Adenosylmethionine and Related Compounds*, Usdin, E., Borchardt, R., and Creveling, C. R., Eds., Macmillan, London, 1982, 55.
12. **Gagnon, C., Viveros, O. H., Diliberto, E. J., Jr., and Axelrod, J.,** Enzymatic methylation of carboxyl groups of chromaffin granule membrane proteins, *J. Biol. Chem.*, 253, 3778, 1978.
13. **Gilad, G. M., Gagnon, C., and Kopin, I. J.,** Protein carboxymethylase activity in the rat superior cervical ganglion during development and after axonal injury, *Brain Res.*, 183, 393, 1980.
14. **Kashnig, D. M. and Kasper, C. B.,** Isolation, morphology, and composition of the nuclear membrane from rat liver, *J. Biol. Chem.*, 244, 3786, 1969.
15. **Kim, S., Galleti, P., and Paik, W. K.,** In vivo carboxyl methylation of human erythrocyte membrane protein, *J. Biol. Chem.*, 255, 338, 1980.

16. **Kim, S. and Li, C. H.,** Enzymatic methyl esterification of pituitary polypeptides, *Int. J. Peptide Protein Res.*, 13, 282, 1979.
17. **Kim, S. and Paik, W. K.,** Purification and properties of protein methylase II., *J. Biol. Chem.*, 245, 1806, 1970.
18. **Kim, S. and Paik, W. K.,** Studies on the structural requirements of substrate protein for protein methylase II., *Biochemistry*, 10, 3141, 1971.
19. **Kloog, Y., Flynn, D., Hoffman, A. R., and Axelrod, J.,** Enzymatic carboxymethylation of the nicotinic acetylcholine receptor, *Biochem. Biophys. Res. Commun.*, 97, 1474, 1980.
20. **Kostraba, N. C., Montagna, R. A., and Wang, T. Y.,** Study of the loosely bound nonhistone chromatin proteins. Stimulation of deoxyribonucleic acid templated ribonucleic acid synthesis by a specific deoxyribonucleic acid binding phosphoprotein fraction, *J. Biol. Chem.*, 250, 1548, 1975.
21. **Legraverend, M. and Glazer, R. I.,** Characterization of a nonhistone chromosomal protein which stimulates RNA polymerase II, *Biochim. Biophys. Acta*, 607, 92, 1980.
22. **Liss, M. and Edelstein, L. M.,** Evidence for the enzymatic methylation of crystaline ovalbumin preparation, *Biochem. Biophys. Res. Commun.*, 26, 497, 1967.
23. **Liss, M., Maxam, A. M., and Cuprak, L. J.,** Methylation of protein by calf spleen methylase. A new protein methylation reaction, *J. Biol. Chem.*, 244, 1617, 1969.
24. **Morin, A. M. and Liss, M.,** Evidence for a methylated protein intermediate in pituitary methanol formation, *Biochem. Biophys. Res. Commun.*, 52, 373, 1973.
25. **Neish, A. C.,** in *Meth. Enzymol.*, 3, 255, 1957.
26. **O'Dea, R. F., Viveros, O. H., Acheson, A., Gorman, C., and Axelrod, J.,** Protein carboxymethylase and methyl-receptor protein in human platelets and erythrocytes, *Biochem. Pharmacol.*, 27, 679, 1978.
27. **O'Dea, R. F., Viveros, O. H., and Diliberto, E. J.,** Protein carboxymethylation: role in the regulation of cell functions, *Anal. Pharmacol.*, 30, 1163, 1978.
28. **Paik, W. K. and Kim, S.,** Solubilization and partial purification of protein methylases from calf thymus nuclei, *J. Biol. Chem.*, 245, 6010, 1970.
29. **Paik, W. K. and Kim, S.,** in *Protein Methylation*, John Wiley & Sons, New York, 1980.
30. **Quick, D. P., Orchard, P. J., and Duerre, J. A.,** Carboxyl methylation of nonhistone chromosomal proteins, *Biochemistry*, 20, 4724, 1981.
31. **Sitaramayya, A., Wright, L. S., and Siegel, F. L.,** Enzymatic methylation of calmodulin in rat brain cytosol, *J. Biol. Chem.*, 255, 8894, 1980.
32. **Springer, M. S., Goy, M. E., and Adler, J.,** Protein methylation in behavioural control mechanism and signal transduction, *Nature (London)*, 280, 279, 1979.
33. **Stein, G. and Baserga, R.,** Cytoplasmic synthesis of acidic chromosomal proteins, *Biochem. Biophys. Res. Commun.*, 44, 218, 1971.
34. **Stein, G. S., Stein, J. L., Klensmith, L. J., Lancing, R. L., Park, W. D., and Thompson, J. A.,** Nonhistone chromosomal proteins: their role in the regulation of histone gene expression, *Biochem. Soc. Symp.*, 42, 137, 1977.
35. **Stock, J. B. and Koshland, D. E., Jr.,** A protein methylesterase involved in bacterial sensing, *Proc. Natl. Acad. Sci. U.S.A.*, 75, 3659, 1978.
36. **Strittmatter, W. J., Gagnon, C., and Axelrod, J.,** Beta adrenergic stimulation of protein carboxymethylation and amylase secretion in rat parotid gland, *J. Pharmacol. Exp. Ther.*, 207, 419, 1978.
37. **Yamaizumi, M., Uchida, T., Okada, Y., Furusawa, M., and Mitsui, H.,** Rapid transfer of nonhistone chromosomal protein to the nucleus of living cells, *Nature (London)*, 273, 782, 1978.
38. **Duerre, J. A.,** Unpublished data.

Chapter 6

IDENTIFICATION OF NUCLEAR ANTIGENS IN HUMAN CELLS BY IMMUNOFLUORESCENCE, IMMUNOELECTRON MICROSCOPY, AND IMMUNOBIOCHEMICAL METHODS USING MONOCLONAL ANTIBODIES

Alan L. Epstein and Charles V. Clevenger

TABLE OF CONTENTS

I. INTRODUCTION

The recent discovery of the nucleosome as the fundamental structural component of eukaryotic chromatin[1] has greatly added to our knowledge of DNA-histone interactions. Likewise, the discovery of discrete multienzyme complexes involved in DNA replication[2] and heterogeneous ribonucleoprotein particles[3-6] associated with newly transcribed mRNA have shown that nonhistone chromosomal (NHC) proteins and small nuclear RNA are compartmentalized within the nucleus to efficiently execute functional processes. Recent investigations by Berezney and Coffey,[7] Adolph et al.,[8] and Long et al.[9] have suggested the presence of a nuclear matrix within eukaryotic nuclei composed of core NHC proteins which are highly insoluble under dissociating conditions. Although conventional techniques of fine structural analysis such as transmission electron microscopy (TEM) have yielded significant information on the structure of the nucleus, the biochemical studies mentioned above have provided important new data on the organization of the nucleus and on the existence of nuclear components designed to perform specific functions. High resolution two-dimensional gel electrophoresis of nuclear fractions have revealed the presence of a large number of NHC proteins in eukaryotic cells.[10-12] The basic mechanisms of gene function including the selective activation and inactivation of chromosomal segments for transcription, and other nuclear functions such as chromosome replication, processing and transportation of mRNA, and the condensation and movement of chromosomes during cell division, have as yet not been elucidated. The complexity of chromatin revealed by two-dimensional gel electrophoresis underscores the importance of understanding the structure of the nucleus so that subclasses of NHC proteins may be defined with respect to their natural location and function within the cell.

Recently, our laboratory has been studying the usefulness of monoclonal antibodies in the identification of nuclear antigens in normal and malignant human cells. Because of the remarkable specificity of these reagents, we have found that monoclonal antibodies recognize cell cycle specific antigens which unlike malignant cells, appear to be tightly regulated in normally dividing cells. In addition, several different patterns of immunofluorescence were noted and were used to identify antibodies directed against different intracellular antigens including those associated with the cytoskeleton. Immunoelectron microscopic analysis using colloidal gold-linked second antibody demonstrated that monoclonal antibodies are able to identify the location of nuclear antigens at the ultrastructural level. These studies strongly suggest that monoclonal antibodies may be a promising new tool to investigate the structural and functional organization of the eukaryotic nucleus. With these reagents, the molecular mechanisms associated with complex events such as blastogenesis, differentiation, hormonal stimulation, and malignant transformation may be studied in relation to normal cell growth and division.

II. MATERIALS AND METHODS

A. Cell Lines and Cultures

Eight human malignant lymphoma and leukemia cell lines were used in these experiments as the source of nuclear antigens and for the antibody screening procedures. These include the EBV-positive nonproducer Raji[13] and producer AG876[13] African Burkitt's lymphoma cell lines; the T-cell acute lymphoblastic leukemia CEM cell line;[15] the IgE secreting multiple myeloma U-266 cell line;[16] the erythroleukemia K562 cell line;[17] and the histiocytic type SU-DHL-1[18] and U-937[19] and B-cell type SU-DHL-4[20] diffuse histiocytic lymphoma cell lines. In addition to these cultures, normal peripheral blood lymphocytes pooled from several individuals and separated by the ficoll-hypaque technique[21] were used alone and after four days of stimulation with 5μg/mℓ of Pokeweed mitogen (Sigma). In order to help characterize and screen the monoclonal antibodies, HeLa cells and a normal diploid human fibroblast cell strain established from a skin punch biopsy were used in the immunofluorescence and immunoelectron microscopy experiments. All of these cultures were grown in RPMI-1640 medium (GIBCO) supplemented with 15% fetal calf serum, 100 units /mℓ Penicillin-G, and 100 μg/mℓ streptomycin sulfate. In order to produce large quantities of cells, 3-ℓ hanging bar spinner flasks (Bellco) were used after seeding with approximately 4 × 10^8 cells. The cells were harvested 4 to 5 days later, washed twice in phosphate buffered saline (PBS), resuspended in 10 mℓ of buffer A (150 m*M* NaCl, 20 m*M* Tris-HCl, pH 8.0, 1 m*M* EDTA, 1 m*M* 2-mercaptoethanol, 0.5 m*M* phenylmethylsulfonylfluoride, and 1% aprotinin) containing 10% glycerol, and frozen at −85°C for future use. Generally, 3 ℓ of cells yielded 2.5 mℓ of packed cells per harvest.

B. Preparation of Nuclear Extracts

Nuclear extracts from the human malignant lymphoma and leukemia cell lines and peripheral blood lymphocytes were prepared as described by Lydersen and Pettijohn.[22] Briefly, 2.5 mℓ of packed cells were thawed and washed once with Ca/pipes buffer (0.01 *M* $CaCl_2$, 2 × 10^{-3} *M* piperazine-N, N′-bis (2-ethanesulfonic acid)) in a 50-mℓ centrifuge tube. All procedures were performed at 4°C and centrifugation was at 1000 × *g* for 10 min. The pellet was then resuspended in 40 mℓ of Ca/pipes buffer and thoroughly homogenized using a motor-driven Teflon® pestle to disrupt the swollen cells. The nuclei were then pelleted leaving a cloudy supernatant as the cytoplasmic fraction and resuspended in Ca/pipes buffer containing 1% NP-40 (Gallard-Schlesinger). The nuclei were then rehomogenized to remove nuclear membranes and checked by phase-contrast microscopy to be free of contaminating cytoplasmic and membranous debris. Nuclei were then washed twice in Ca/pipes buffer to remove the detergents, resuspended in 8 mℓ of phosphate-buffered saline, and sonicated three times for 15 sec at 30-sec intervals in order to produce a more homogeneous suspension. The nuclear preparations were then frozen in 1-mℓ aliquots at −85°C for use in the immunization procedures. Using this method, Lydersen and Pettijohn[22] have demonstrated by electrophoretic analysis that nuclear extracts are free of major cytoplasmic proteins.

C. Immunization Protocol

One-milliliter aliquots of the nuclear preparations from each of the eight human lymphoma and leukemia cell lines and the peripheral blood lymphocytes were thawed, resonicated to reduce viscosity, and emulsified in 1.5 mℓ of complete Freund's adjuvant (Difco) using two glass syringes and a 20-gauge microemulsifying needle (Bolab). Three 8-week-old Balb/c female mice per sample were injected s.c. at multiple sites using a 22-gauge needle and glass syringe. The mice were reinoculated 2 weeks later as above except the nuclear extracts were prepared in incomplete adjuvant. The mice received a third inoculation of antigen 10 days later, this time without adjuvant and by i.p. injection. The mice were sacrificed by cervical dislocation 4 days later and the spleens removed using aseptic technique for the hybridoma experiments.

D. Generation of Hybridomas

Spleen cells were fused with 8-azaguanine resistant mouse myeloma NS-1 cells at a ratio of 5:1, respectively, using 40% polyethylene glycol (PEG) molecular weight 1540 (J. T. Baker) [23]. After fusion, the cells were cultured in selective HAT medium containing 10^{-4} *M* hypoxanthine, 4 × 10^{-7} *M* aminopterin, and 1.5 × 10^{-5} *M* thymidine (Sigma) in a 96-well microtiter plate (Costar) at a concentration of 2 × 10^5 cells per well. The medium was changed every 3 days and after the first week the aminopterin was discontinued. Culture supernatants from wells with active cell growth were tested by indirect immunofluorescence for antibody activity. To help harvest the supernatants, a Titertek Multichannel Pipette®

(Flow Laboratories) was used to transfer the supernatants to other 96-well plates. In order to prevent cross-contamination from well to well, a new sterile pipette tip was used for each supernatant fluid. Positive cultures were transferred to a 24-well cluster plate (Costar) for expansion and an aliquot of cells was cloned on 0.5% Noble agar (Difco) containing RPMI-1640 medium, 20% fetal calf serum, and antibiotics. Agar plates were prepared in 60 × 15 mm tissue culture petri dishes the day before cloning and were stored at 4°C until use. In order to assure the growth of an adequate number of colonies, plates were seeded with 500 or 1000 cells in a volume of 50 μℓ. The cells were then spread on the surface of the agar with a glass rod bent at 90° and the plates were incubated undisturbed for 10 to 12 days in a well-humidified 5% CO_2 incubator at 37°C. At this time, a maximum of 24 colonies per hybridoma were picked with Pasteur pipettes and transferred to a 96-well plate for continued growth. Supernatants of these cultures were then retested 4 to 5 days later when the media turned acidic. Positive cultures were then expanded slowly and when actively growing in tissue culture flasks, they were frozen in liquid nitrogen for safe storage in RPMI-1640 medium containing 20% fetal calf serum, antibiotics, and 10% dimethylsulfoxide (DMSO) at a concentration of 5 × 10^6 cells per milliliter.

E. Immunofluorescence and Antibody Screening Procedures

Hybridoma supernatants were examined by indirect immunofluorescence using cell preparations of the human lymphoma and leukemia cell lines and peripheral blood lymphocytes. In order to screen large numbers of supernatants in the required time, washed cells at a concentration of 5 × 10^6 cells per milliliter in PBS were pipetted dropwise on Teflon®-coated printed microscope slides (Cel-Line Associates, Inc., Newfield, N. J.) containing 10 5-mm wells per slide. After the cells were allowed to settle to the surface of the glass, the overlying fluid was quickly removed by aspiration and the cells were dried to the slide by a gentle stream of warm air. The slides were then immediately fixed in 2% formaldehyde (ultrapure, E. M. grade, Polysciences #4018) in PBS for 15 min at room temperature. After fixation, the slides were rinsed in PBS and placed in acetone at −20°C for 3 min to make the cells permeable. After a final rinse in PBS to remove the acetone, the slides were stored in PBS at 4°C indefinitely in covered Coplan jars. Alternatively, the cells were fixed directly in acetone, methanol to acetone (1:1), or 100% ethanol for 5 min at −20°C, rinsed in PBS, and stored as described above. In addition to the lymphoid cultures, HeLa cells and normal adult skin fibroblasts were used a substrates for the indirect immunofluorescence studies. Since the HeLa cells and human fibroblasts grew as monolayer cultures, they were plated directly onto the printed microscope slides after trypsinization and allowed to attach and grow overnight at 37°C in complete medium. The following day, the slides were briefly rinsed in PBS to remove the medium and the cells were fixed as described above. In general, the slides were not allowed to air dry either during or after the fixation procedures in order to maintain the cellular integrity and antigenicity of intracellular molecules. For photographic analysis, viable cell preparations obtained from ficoll-hypaque gradient separation were cytocentrifuged directly onto slides at 1250 rpm for 10 min. This procedure flattened the lymphoid cells and greatly improved the visibility of intranuclear and cytoplasmic antigens. Slides prepared in this manner were fixed in the same way directly after cytocentrifugation.

In order to screen the hybridoma supernatants by indirect immunofluorescence, 35 μℓ of each supernatant was pipetted on a well of the printed microscope slides using a different tip for each supernatant. After 60 min of incubation at 37°C in a humidified chamber, the slides were rinsed three times with PBS at room temperature, and again incubated for 30 min at 37°C with 20 μℓ of a 1 to 20 dilution of fluorescein-conjugated goat anti-mouse IgG antibody which was $F(ab^1)_2$ fragment specific (Cappel Laboratories, Malverne, Pa.). The slides were then rinsed three times with PBS, counterstained with Evans Blue for 5 min at room temperature using a freshly prepared solution containing 50 μℓ of a 1% stock solution

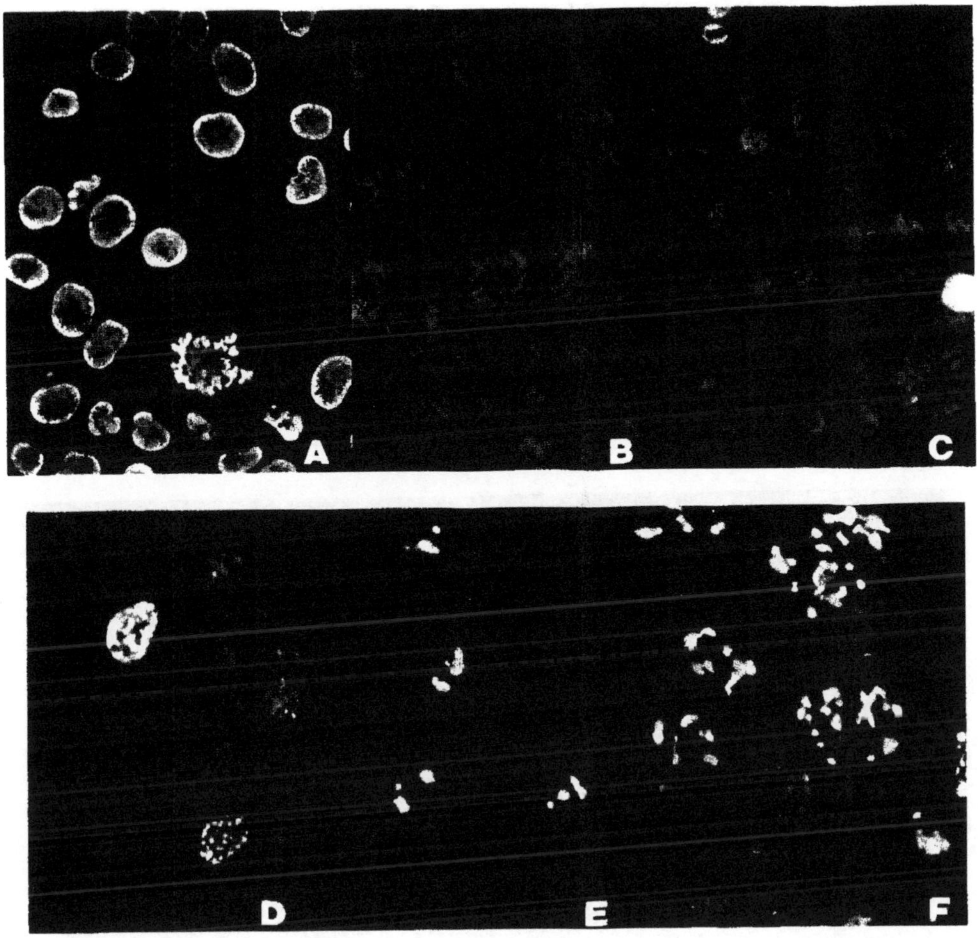

FIGURE 1. Indirect immunofluorescence staining patterns of monoclonal antibodies reactive with nuclear antigens. (A) Homogeneous nuclear staining with prominent rim pattern by N-CS244-7 on Raji cells. Mitotic chromosomes are brightly stained. (B) Homogeneous nuclear staining pattern by N235-1 on Raji cells. During mitosis, the chromosomes are negatively stained and the antigen migrates into the cytoplasm causing a red to orange color shift in the Evans Blue counterstain. (C) Finely granular nuclear staining pattern by N-Cy58-15 on HeLa cells. Note the presence of negatively stained nuclei in cells with cytoplasmic staining around the external periphery of the nucleus. (D) Speckled nuclear staining pattern by N780-3 on human fibroblasts. Note the presence of negatively stained nuclei which suggests that this antibody is directed against a cell cycle-specific nuclear antigen. (E) Homogeneous nucleolar staining pattern by Nu125-10 on human fibroblasts. (F) Homogeneous nuclear and nucleolar staining pattern by N-Nu364-5 on SU-DHL-1 cells. All preparations were fixed with formalin-acetone, counterstained with Evans Blue, and photomicrographed at × 500.

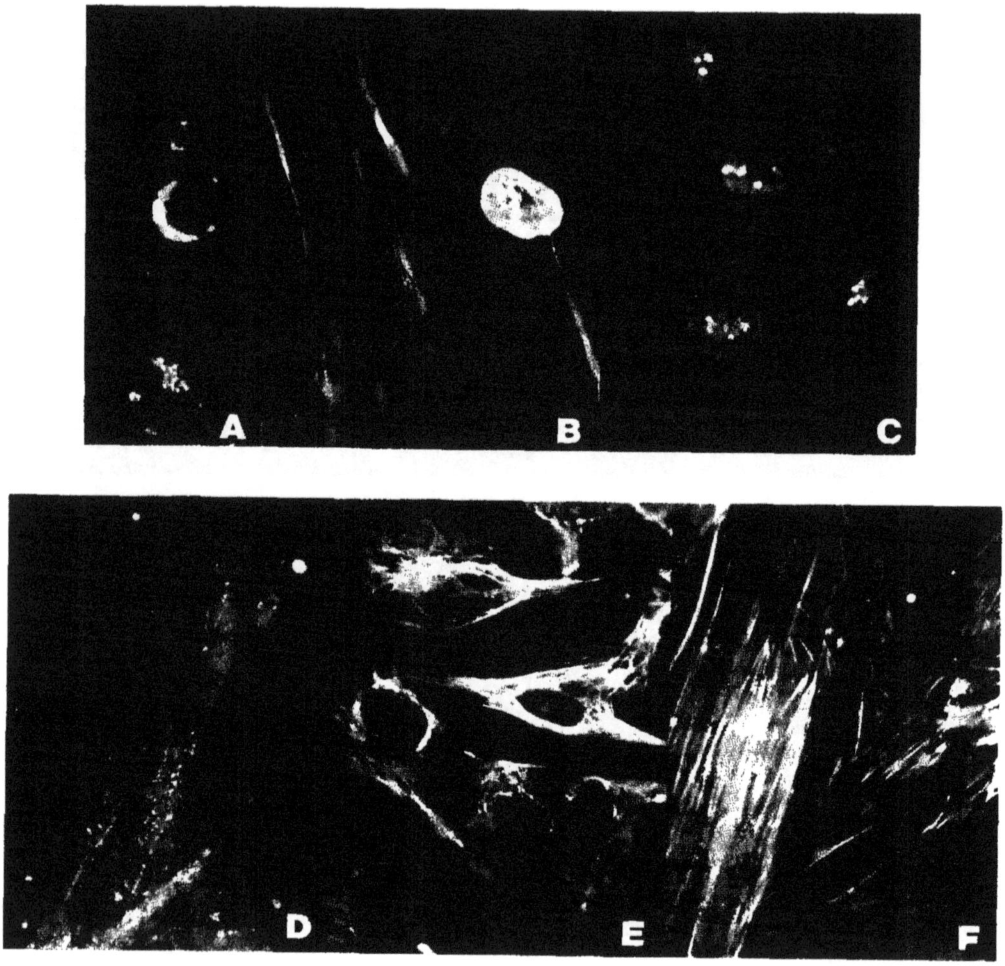

FIGURE 2. Indirect immunofluorescence staining patterns of monoclonal antibodies reactive with nuclear and cytoplasmic antigens. (A) Perinuclear staining pattern by pN731-3 on Raji cells. (B) Cell cycle-specific homogeneous nuclear staining pattern by N-CS244-7 on human fibroblasts. Note that of the two cells shown here, only one nucleus is positively stained while both cells have brightly positive cytoplasmic filaments. (C) Nuclear, nucleolar, and cytoplasmic filamentous staining pattern by N-CS899-1 on human fibroblasts. (D) Speckled nuclear and cytoplasmic staining pattern by N-Cy652-2 on human fibroblasts. (E) Intermediate filament staining pattern by CS443-4 on human fibroblasts. (F) Actin bundle staining pattern by CS420-8 on human fibroblasts. All preparations were fixed with formalin-acetone, counterstained with Evans Blue, and photomicrographed at × 500.

of Evans Blue in 80 mℓ of PBS, rinsed a final time in PBS, and coverslipped using a 1:1 solution of glycerol to PBS. The slides were then examined by epi-fluorescence microscopy using a Leitz Orthoplan® microscope with Ploemopak® 2.1 fluorescence illuminator, HBO 100 mercury lamp, and 50x water immersion objective. Microphotographs were taken with a Leitz Vario Orthomat® automatic camera system using Kodak Ektachrome® fast color 35-mm daylight film (ASA 400). Since many of the monoclonal antibodies produced a rapidly diminishing fluorescent reaction, exposure times optimally were less than 5 sec. Finally, to maximize the film exposure, the Ektachrome® film was routinely pushed one stop to ASA 800 during development.

F. Immunoelectron Microscopy

Cells were prepared using a modified post-embedding procedure first set forth by Moriarty et al.[24,25] and Childs and Ellison.[26] Confluent cultures of Raji and HeLa cells were harvested and washed twice with PBS. They were then fixed with 1% glutaraldehyde in 0.1 *M* phosphate buffer, pH 7.4 for 45 min. The cells were dehydrated in ethanol and embedded in Araldite 6005 (Electron Microscopy Services). The embedded cells were cut using glass knives and light gold sections were mounted on 200-mesh nickel grids.

For the immunogold staining, the grids were etched by flotation on 10% hydrogen peroxide for 10 min. After rinsing in distilled water, nonspecific sites of protein adsorption were blocked with a 1/100 dilution of normal goat serum for 3 min. The grids were then incubated for 48 hr in hybridoma supernatants diluted 1:10 to 1:100 in PBS containing 1 mg/mℓ BSA (bovine serum albumin) and 0.02% sodium azide. After incubation, the grids were rinsed in PBS, blocked again with a 1:100 dilution of normal goat serum, and floated for 10 min on a 1:2 dilution of goat anti-mouse immunoglobulin linked to 20 nm colloidal gold particles (GAM20, Janssen Pharmaceuticals, Beerse, Belgium). The grids were then rinsed in PBS and the immunogold stain fixed with 4% OsO_4 for 10 min. Nuclear morphology was then enhanced by staining with 5% uranyl acetate for 10 min and lead citrate made according to the methods of Venable and Coggeshall[27] for 5 min. The sections were then examined using a JEOL JEM-100S electron microscope at 60 kV and photographed with Kodak® 4489 EM film.

G. Serological Characterization of Antibody Isotype

Hybridoma supernatants from 4-day cultures were concentrated 10 to 20 times in B15 minicon concentrators (Amicon) and tested in double diffusion Ouchterlony plates against rabbit anti-mouse immunoglobulin heavy chain subclass antisera (Miles). The precipitin bands were read after 2 to 3 days of incubation in a well-humidified 37°C incubator.

H. Immunoprecipitation and SDS-Polyacrylamide Gel Electrophoresis

Lymphoma cells were metabolically labeled in methionine and leucine-free RPMI-1640 medium containing 5% fetal calf serum, antibiotics, 30 μCi/mℓ (^{35}S)-methionine, and 10 μCi/mℓ (^{14}C)-leucine. After labeling, the cells were washed twice with PBS and resuspended in lysis buffer at 4°C. The cells were sonicated three times for 15-sec intervals, ultracentrifuged at 30,000 rpm for 30 min, and divided into 1-mℓ aliquots in 12 × 75 mm plastic tubes (Falcon #2052). The tubes were incubated overnight at 4°C with constant shaking following the addition of 1 mℓ hybridoma supernatant or 30 μℓ of ascites fluid. The antigen-antibody complexes were removed from solution by the addition by Protein-A Sepharose beads (Pharmacia) which were pretreated consecutively with 10 mg/mℓ BSA in PBS and cold Raji lysate to prevent the nonspecific binding of radiolabeled proteins. If the monoclonal antibodies were of the IgM heavy chain subclass, 10 μℓ of rabbit anti-mouse IgM specific antiserum per sample were premixed with the Protein-A Sepharose beads. The immunoprecipitates were then washed twice in buffer containing 0.15 *M* NaCl, 0.1 *M* Tris-HCl, pH

8.0 and three times in buffer containing 0.25 *M* LiCl, 0.1 *M* Tris-HCl, pH 8.0, then switched to a new set of tubes before being denatured in 65 μℓ of sample buffer (0.17 *M* Tris-HCl, pH 6.8, 11% glycerol, 0.003% bromophenol blue, 3% SDS, and 5% 2-mercaptoethanol).

The samples were then analyzed by SDS-PAGE (polyacrylamide gel electrophoresis) using a 10% separating gel and a 4.2% stacking gel as described by Laemmli.[28] After electrophoresis, the gels were processed for fluorography according to the method of Chamberlain,[29] dried under vacuum, and then exposed to Cronex 4 X-ray film (Dupont) at −70°C prior to development.

I. Immunoblot Analysis

For this analysis, 2.5×10^7 lymphoma cells were washed twice with PBS and dissolved in 2 mℓ of SDS sample buffer. Repeated vortexing and sonication were required to completely dissolve the cells and reduce the viscosity of the lysate. After boiling for 90 sec, 30-μℓ aliquots were loaded onto a 10% polyacrylamide gel and electrophoresed for 3.5 hr with 20 mA of current. Molecular weight standards (Biorad®) were routinely run with each gel. The proteins were then transferred electrophoretically to nitrocellulose paper (Hahy paper, Millipore) using an Electroblot apparatus (E-C Apparatus Corp.) set at 0.2 A for 10 hr. The chamber buffer contained 24 m*M* Tris-HCl, pH 8.3, 192 m*M* glycine, 20% methanol, and 0.1% SDS and was used a maximum of three times. After completing the transfer, the section of the nitrocellulose paper which contained the molecular weight markers was cut and stained for 1 hr with a 0.1% Amido Black solution containing methanol, dH_2O, and glacial acetic acid (5:5:1). The remaining piece of nitrocellulose paper containing four lanes of cellular proteins was rinsed three times in PBS containing 0.5% NP-40 to remove excess SDS, and then blocked for 3 hr in freshly prepared 10X Denhardt's solution (0.2% ficoll-400, 0.2% BSA and 0.2% polyvinyl pyrrolidone-360) to prevent the nonspecific binding of antibody to the nitrocellulose paper. The four lanes were then carefully separated with a scalpel blade and each strip incubated in plastic heat-sealable Scotchpak® pouches (4 × 6 in.) containing 10 mℓ of hybridoma supernatants and 0.02% sodium azide for 1 to 3 days at room temperature with constant mixing. The strips were then rinsed in 8 × 8 in. glass dishes (no more than two per dish) for 4 hr using several changes of PBS and then incubated overnight as above with 1×10^6 cpm of (^{125}I)-goat anti-mouse antibody (New England Nuclear) diluted in 10 mℓ of PBS containing 1 mg /mℓ BSA and 0.02% sodium azide. The following day, the strips were washed in PBS for 6 hr (3 times per hour) as above. The nitrocellulose strips were finally placed on Saran® wrap, blotted dry with filter paper, and sandwiched with a second piece of Saran® wrap. The enclosed strips were placed between an intensifying screen (Cronex Hi-Plus, DuPont) and Cronex 4 X-ray film (DuPont) in an 8 × 10 in. X-ray holder (Kodak®) and exposed for 1 to 3 days at −70°C. The molecular weights of the radioactive bands were determined by placing the Amido Black stained marker nitrocellulose strip in juxtaposition to the developed X-ray film.

III. RESULTS

A. Production of Monoclonal Antibodies

Supernatant fluids from wells with actively growing cells were screened for the presence of antibody by indirect immunofluorescence. The screening procedures were performed on fixed cell preparations made from the same cell used to immunize the mice. After 3 weeks of incubation, approximately 60% of the wells were acidic and showed active cellular proliferation by phase-contrast microscopy. From these wells, approximately half of the supernatants were found to give a positive staining reaction by indirect immunofluorescence. Those wells giving the brightest reaction for each distinctive pattern of immunofluorescence were then selected for cloning and subsequent culture. For each hybridoma, 24 clones were

Table 1
MONOCLONAL ANTIBODIES REACTIVE WITH NUCLEAR AND NUCLEOLAR ANTIGENS[a]

Monoclonal antibodies	Isotype	Fixation	Description of immunofluorescence pattern
Nuclear (N)			
73-7	IgG_1	A	Speckled pattern, lymphocyte-specific
144-1	IgM	F-A	Speckled pattern
181-8	IgG_1	F-A	Densely speckled pattern
235-1	IgG_1	F-A	Homogeneous pattern, human-specific
343-5	IgG_1	F-A	Speckled pattern
671-3	IgM	F-A	Faintly speckled pattern
780-3	IgM	F-A	Coarse granular, G_2 and M cell cycle-specific
818-18		IgG_1	F-A homogeneous pattern, myeloid-specific
825-18		IgG_1	F-A homogeneous pattern, myeloid-specific
856-4	IgG_1	F-A	Homogeneous pattern, myeloid-specific
888-5	IgG_1	F-A	Homogeneous pattern, myeloid-specific
983-5	IgG_3	F-A	Speckled pattern
Nuclear and nucleolar (N-Nu)			
364-5	IgG_1	F-A	Homogeneous nuclear and nucleolar
794-3	IgM	E	Speckled nuclear and nucleolar
887-8	IgM	F-A	Speckled nuclear and nucleolar, spindle and telephase midbody positively stained
Nucleolar (Nu)			
116-2	IgG_1	F-A	Speckled nucleolar
125-10		IgG_1	F-A homogeneous nucleolar
160-4	IgG_1	F-A	Homogeneous nucleolar
162-4	IgG_1	F-A	Homogeneous nucleolar
164-4	IgG_1	F-A	Homogeneous nucleolar

[a] See footnotes to Table 3.

picked and tested for antibody production by indirect immunofluorescence. From a total of 45 hybridomas cloned, 28 were found to have ⩾ 20/24 positive clones, 10 were shown to have only ⩽ 3/24 positive clones, and 7 were negative and thus lost to further study. For each hybridoma, three clones producing the brightest immunofluorescence staining were then selected for passage in culture and frozen in liquid nitrogen for safe storage. In a few instances, supernatants from clones picked from the same plate gave different staining reactions indicating that the original microtiter plate well was seeded with more than one surviving antibody-producing hybridoma cell.

B. Immunofluorescence Staining Patterns

As shown in Tables 1 to 3, several distinct patterns of immunofluorescence were obtained including nuclear (homogeneous, speckled, granular), nucleolar, perinuclear, cytoplasmic, cytoskeletal, and combinations of the above. In Figures 1 and 2 (see color plate), representative patterns of immunofluorescence obtained by selected monoclonal antibodies are shown.

C. Effects of Fixation on the Detection of Nuclear Antigens by Immunofluorescence

Since it was necessary to perform the initial screening procedures on a large number of supernatants in a short period of time, it was only possible to use one fixation procedure in these experiments. Because the choice of the fixation procedure was felt to greatly affect the outcome of the results, a number of different procedures, including formalin-acetone (F-A fixation), acetone, methanol, ethanol, methanol to acetic acid (3:1), acetone to methanol (1:1), acetone to ethanol (1:1), and periodate-lysine-paraformaldehyde were tested on ran-

Table 2
MONOCLONAL ANTIBODIES REACTIVE WITH NUCLEAR AND CYTOPLASMIC ANTIGENS[a]

Monoclonal antibodies	Isotype	Fixation	Description of immunofluorescence pattern
Perinuclear (pN)			
102-1	IgM	F-A	Finely speckled
106-1	IgM	F-A	Finely speckled, sparce
142-3	IgM	F-A	Speckled
155-19	IgM	F-A	Finely speckled
706-6	IgG_1	F-A	Speckled, lymphoid-specific
731-3	IgM	F-A	Densely speckled
Nuclear and cytoplasmic (N-Cy)			
58-15	IgM	F-A	Densely speckled cytoplasmic, coarse granular nuclear
129-11	IgM	F-A	Densely speckled cytoplasmic, coarse granular nuclear
130-5	IgM	F-A	Speckled cytoplasmic and nuclear
231-7	IgM	F-A	Densely speckled cytoplasmic and nuclear
318-2	IgM	F-A	Speckled cytoplasmic and nuclear, sparse
503-1	IgG_{2b}	A	Densely speckled cytoplasmic and nuclear, lymphocyte-specific
652-2	IgM	F-A	Densely speckled cytoplasmic and nuclear
752-10	IGM	F-A	Homogeneous cytoplasmic and nuclear, cell cycle-specific
761-6	IgM	F-A	Densely speckled cytoplasmic and nuclear, cell cycle-specific
841-19	IgM	F-A	Densely speckled cytoplasmic and nuclear, cell cycle-specific
859-4	IgM	F-A	Densely speckled cytoplasmic, coarsely granular nuclear
882-17	IgG_1	F-A	Densely speckled cytoplasmic and homogeneous nuclear, myeloid-specific
Nuclear membrane (NM)			
539-9	IgG_1	F-A	Bright ring pattern, lymphocyte-specific

[a] See footnotes to Table 3.

domly selected hybridoma supernatants. The results showed that the F-A fixation method enabled the identification of over 90% of the antibody-producing hybridomas. When tested against a large number of positive hybridoma clones, it also gave the best results with regard to the brightness of immunofluorescence and the clarity of antigen localization. From these results, this procedure was chosen to be used in these experiments. In some instances, however, the F-A fixation method was found to destroy the antigenicity of a few intracellular molecules, notably those identified by monoclonal antibodies N671-3, N73-7, and N-Nu794-3. Except for antibody N-Nu794-3 which required ethanol fixation, these hybridoma clones required the use of cold acetone fixation in order to give a positive immunofluorescence pattern. A complete listing of the optimal fixation procedures for each hybridoma clone is shown in Tables 1 to 3. In general, the immunofluorescence techniques used in these experiments were an excellent method of screening, since negative supernatants gave essentially no background immunofluorescence staining. Because of this, rare antigens or antigens associated with minute subcellular structures could be identified. As an example of the sensitivity of this technique, monoclonal antibody N671-3 gave a minutely speckled nuclear staining pattern which was only apparent after careful scrutiny under high magnification. Finally, the use of the Evans Blue counterstain was found to greatly enhance the

Table 3
MONOCLONAL ANTIBODIES REACTIVE WITH NUCLEAR AND CYTOSKELETAL ANTIGENS[a]

Monoclonal antibodies	Isotype	Fixation	Description of immunofluorescence pattern
Nuclear and cytoskeletal (N-CS)			
244-7	IgM	F-A, E	Intermediate filaments, nuclear rim pattern, chromosomes positive
314-4	IgM	F-A, A	Speckled cytoplasmic and nuclear
502-3	IgM	F-A	Speckled nuclear and cytoplasmic filaments
702-4	IgM	F-A	Speckled intermediate filaments, speckled nuclear
891-5	IgM	F-A	Intermediate filaments, homogeneous nuclear, cell cycle-specific
898-9	IgM	F-A	Intermediate filaments, homogeneous nuclear, cell cycle-specific
899-4	IgM	F-A	Speckled intermediate filaments, coarsely granular nuclear
Cytoskeletal (CS)			
244-6	IgM	F-A, E	Intermediate filaments
372-6	IgM	F-A, E	Intermediate filaments
420-8	IgM	F-A	Actin bundles
443-4	IgM	F-A	Intermediate filaments
691-9	IgM	F-A	Intermediate filaments

Note: Abbreviations: F-A, formalin-acetone fixation; A, acetone; E, ethanol.

[a] In order to designate the hybridoma cultures in an informative manner, the number of the original microtiter plate well (1 to 96) was used in combination with a number prefix to indicate the source of the antigen. The designation of these prefixes are none, AG876; 1, Raji; 2, CEM; 3, SU-DHL-1; 4, K562; 5, SU-DHL-4; 6, U-266; 7, Pokeweed mitogen-stimulated peripheral blood lymphocytes; 8, adult peripheral blood lymphocytes; and 9, U-937.

immunofluorescence staining of antinuclear antibodies since its red color localized to the RNA of the cytoplasm and nucleolus and was complementary to the apple-green color of the second fluorescent probe.

D. Specificity of Monoclonal Antibodies to Nuclear Antigens and Their Relationship to the Cell Cycle

Once the hybridoma cells were cloned and sufficiently tested, they were screened against several of the human malignant lymphoma cell lines, human fibroblasts, HeLa cells, and mouse and hamster cell lines in order to determine their specificities. With one exception, all of the monoclonal antibodies directed against nuclear, nucleolar, and perinuclear antigens were found to be human specific (data not shown) and several of these antigens were lymphoid or myeloid specific. In contrast, all of the monoclonal antibodies directed against nuclear-cytoplasmic, nuclear-cytoskeletal, cytoplasmic, and cytoskeletal antigens with one exception were not human-specific. These preliminary results suggest that some of these antibodies may have interesting specificities that correlate with the state of differentiation of the cell. In addition, it is clear that the mouse is capable of responding to self-antigens especially if they are associated with subcellular structures such as the cytoskeleton. Continued screening may reveal hitherto unrecognized specificities which may make these antibodies useful in the identification of cellular subtypes in both malignant and nonmalignant tissues.

Table 4
IMMUNOFLUORESCENCE STAINING OF POKEWEED MITOGEN-STIMULATED HUMAN PERIPHERAL BLOOD LYMPHOCYTES BY MONOCLONAL ANTIBODIES

Monoclonal antibody	Days after pokeweed mitogen stimulation				
	0	1	2	3	4
N235-1	+[a]	++	+++	+++	+++
N780-3	+	++	+++	+++	+++
pN731-3	–	–	+++	+++	++
CS691-9	+++	+++	+++	+++	+++

[a] Intensity of immunofluorescence staining.

Four monoclonal antibodies were also studied for their reactivity in Pokeweed mitogen-stimulated normal peripheral blood lymphocytes. In these experiments, aliquots of ficoll-hypaque separated peripheral blood lymphocytes were taken at different times after the addition of mitogen. Since the cells traverse the cell cycle in partial synchrony, this system was used to identify cell cycle-associated antigens. As shown in Table 4, the nuclear antigens detected by monoclonal antibodies N235-1 and N780-3 were present in only small amounts in resting lymphocytes. After 2 days of mitogen stimulation, these antigens were greatly increased and continued to be expressed at the same level through day 4. The perinuclear antigens recognized by monoclonal antibody pN731-3 were not expressed at all until 48 hr after mitogen stimulation and decreased in quantity by 96 hr. In contrast, the cytoskeletal antigen identified by monoclonal antibody CS691-9 was present at all times and apparently was unaffected by blastogenesis. These preliminary studies strongly suggest that monoclonal antibodies can identify intracellular cell cycle-associated antigens which appear to be tightly regulated in proliferating cells.

E. Immunoelectron Microscopic Analysis of Monoclonal Antibody Reactivity

Immunocytochemical staining techniques using post-imbedding methods with colloidal gold reagents enabled the ultrastructural localization of monoclonal antibody binding in the cell nucleus. Using this technique, 5/7 monoclonal antibodies were found to have excellent binding specificities on multiple cell lines and produced low background staining. Negative controls using colloidal gold-linked second antibody alone were uniformly nonreactive in these preparations. The binding characteristics of two monoclonal antibodies, N-Cy58-15 and N-CS244-7, are shown in Figures 3 and 4, respectively. Electron micrographs of Raji nuclei stained with monoclonal antibody N-Cy58-15 revealed a significant staining reaction over euchromatic regions while N-CS244-7 gave an opposite staining reaction and bound to heterochromatic regions. Other monoclonal antibodies are presently being studied using these techniques.

F. Relationship of Immunoglobulin Subclass to Antigen

After reviewing the isotypes of the various hybridoma clones, two interesting relationships became apparent. First, there was a surprisingly large number of hybridoma clones of the IgM immunoglobulin subclass. This is especially true for those antibodies directed against antigens associated with the cytoskeleton and cytoplasm. Since in all of these experiments the mice were immunized repeatedly with identical aliquots of antigen over a 6-week period,

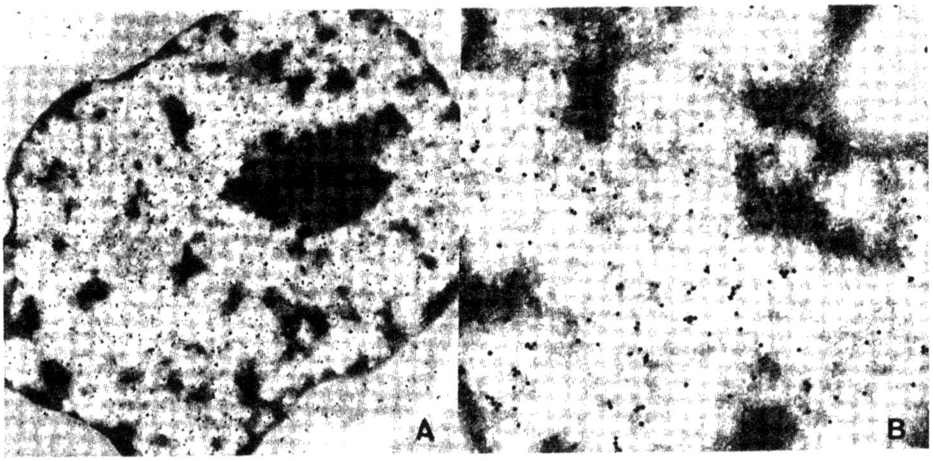

FIGURE 3. Immunogold staining pattern by monoclonal antibody N-Cy58-15 using 20-nm gold conjugated second antibody. (A) Electron micrograph of Raji cells showing a preferential reaction over euchromatic regions of the cell nucleus (Magnification × 11,000); (B) higher magnification of same preparation. (Magnification × 34,000.)

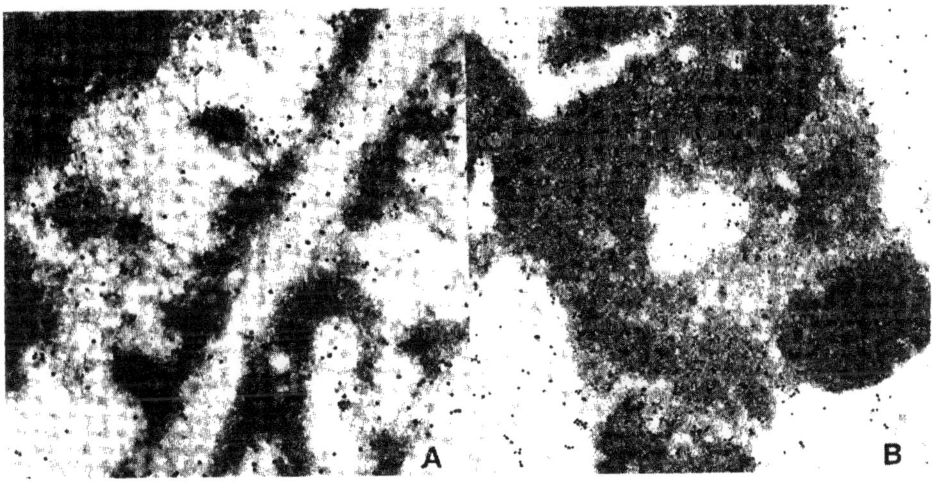

FIGURE 4. Immunogold staining pattern of monoclonal antibody N-CS244-7 using 20-nm gold conjugated second antibody. (A) Electron micrograph of Raji cell nucleus with cytoplasmic invagination showing a preferential staining pattern over heterochromatic regions. (Magnification × 30,000); (B) dense immunogold staining of a Raji cell mitotic figure. (Magnification × 25,000.)

it was expected that the majority of hybridomas would be of the IgG heavy chain subclass. Secondly, when hybridoma clones derived from different mice were found to be directed against the same antigen as shown by immunofluorescence and immunoprecipitation data, they were found to have the same immunoglobulin isotype. For example, hybridoma clones N235-1, N531-11, N12-14, N571-8, and N77-1 which were found to recognize the human-specific nuclear antigen p87;72, all had the IgG_1 immunoglobulin subclass even though they were the product of different fusions using nuclear extracts from different cell lines. These findings seem to indicate that certain antigens favor the production of specific subclasses of antibody.

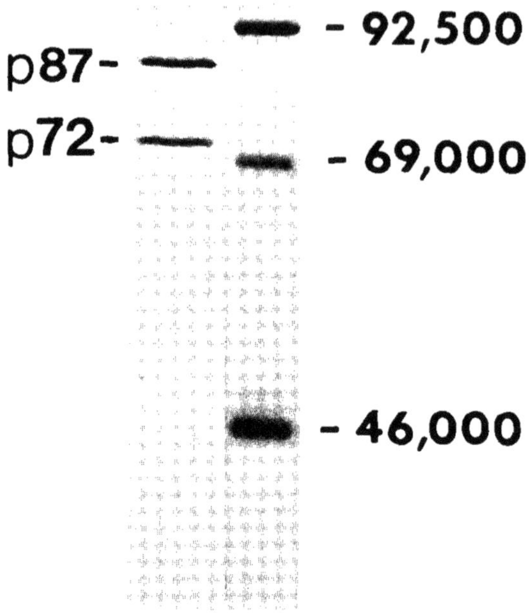

FIGURE 5. Immunoprecipitation of ^{35}S-methionine labeled nuclear antigen p87,72 by monoclonal antibody N235-1. High salt (1 *M* NaCl) was required to solubilize this antigen from Raji cells.

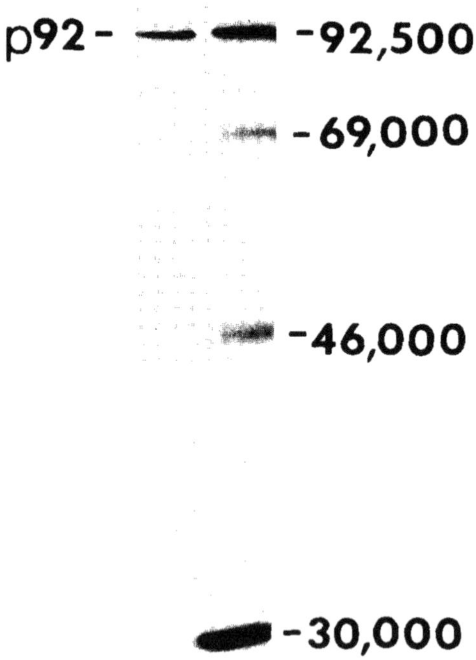

FIGURE 6. Immunoprecipitation of ^{35}S-methionine labeled nuclear antigen p92 from Raji cells by monoclonal antibody N-Cy231-7.

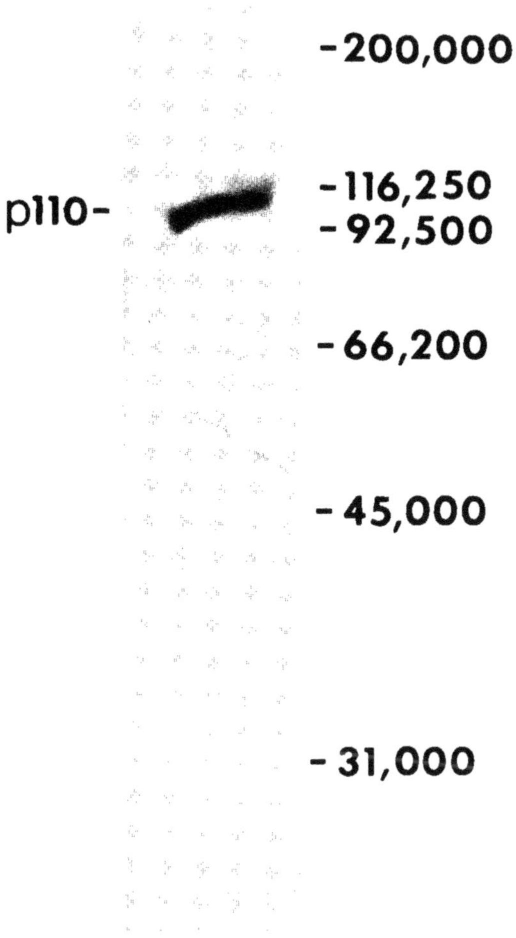

FIGURE 7. Immunoblot of nuclear and nucleolar antigen p110 from Raji cells using monoclonal antibody N-Nu364-5.

G. Immunobiochemical Identification of Nuclear Antigens Using Immunoprecipitation and Immunoblot Analysis

Immunoprecipitation and immunoblot procedures were performed in order to identify the molecular weight and chemical nature of the nuclear antigens recognized by specific monoclonal antibodies. Both procedures yielded limited result, however, since the majority of the monoclonal antibodies tested were negative. In Figures 5 to 7, representative immunoprecipitation and immunoblot autoradiograms are presented for three monoclonal antibodies. Monoclonal antibody N235-1 was found to immunoprecipitate two polypeptides with molecular weights of 87,000 and 72,000 daltons (Figure 5). In order to successfully immunoprecipitate p87;72, 1 *M* NaCl was required in the lysis buffer. In Figure 6, the immunoprecipitation results with monoclonal antibody N-Cy231-7 are shown. Unlike N235-1, low salt lysis buffer was used in the immunoprecipitation of a 92,000-dalton polypeptide by monoclonal antibody N-Cy231-7. Using the immunoblot procedure, monoclonal antibody N-Nu364-5 was found to recognize a 110,000-dalton polypeptide from Raji cells. These results show that both procedures are helpful in the immunobiochemical identification of nuclear antigens using monoclonal antibodies.

IV. DISCUSSION

The results of this investigation have shown that monoclonal antibodies can be used to effectively identify previously unrecognized nuclear and cytoskeletal antigens present in normal and malignant human lymphoid cells. Indirect immunofluorescence studies on fixed cellular preparations have revealed a number of distinctive patterns of staining demonstrating the intracellular location of these antigens. Previous attempts to define and characterize nuclear proteins by fractionation and chromatographic procedures have been greatly hampered by the physical properties of NHC proteins and by the lack of functional assays. In addition, the presence of chromatin-bound proteases and the contamination of nuclear preparations by extrachromosomal proteins during isolation procedures have been major obstacles in studies aimed at purifying and defining nuclear-specific proteins.[30] Because of these difficulties, a number of laboratories have used immunobiochemical techniques to study the antigenic composition of the eukaryotic nucleus. By injecting animals with whole chromatin, dehistonized chromatin, or purified fractions of NHC proteins, investigators have produced antisera against tissue-,[31-35] species-,[22,36-38] and tumor-specific[39,40] antigens. More recently, Saumweber et al.[41] and Howard et al.[42] have successfully produced monoclonal antibodies against nonhistone chromosomal proteins of Drosophila. Using indirect immunofluorescence techniques on salivary gland polytene chromosomes, Howard et al.[42] were able to identify sites of antibody binding which show that a specific antigen is associated with genetically active loci. These investigations represent a significant advance since they demonstrate the remarkable specificity of monoclonal antibodies in the *in situ* identification of chromosomal proteins.

Using somatic cell hybridization techniques, our laboratory has produced a series of monoclonal antibodies to nuclear antigens in human lymphoid cells. In these experiments, screening procedures for the detection of antinuclear monoclonal antibodies required sufficient sensitivity to identify rare antigens and had to be rapid enough to detect the presence of positive supernatants from hundreds of samples in a short period of time. Indirect immunofluorescence on fixed cell preparations was found to meet these demanding criteria and also had the advantage of yielding important information concerning the intracellular location of the antigens. Of critical importance was the selection of a standard fixation method which was able to stabilize the majority of intracellular antigens and produce superior morphological preservation. Technically, the routine screening of the hybridoma supernatant fluids was facilitated by the use of multiwell slides. In addition, the Evans Blue counterstain significantly enhanced the immunofluorescence staining patterns in the cells. Theoretically, the only limitation of the indirect immunofluorescence procedure is the inherent lability of a given antigen to know fixation procedures. More generally, however, the production of monoclonal antibodies to nuclear proteins is dependent upon the antigenicity of the immunogen, and highly conserved proteins may prove to be poorly antigenic by conventional immunization procedures.

As illustrated in Figures 1 and 2, it was possible to recognize several distinct immunofluorescence staining patterns produced by the hybridoma clones. For those antibodies directed against intranuclear antigens, several interesting characteristics were noted in our experiments. First, most of the intranuclear antigens gave a granular pattern of staining which in certain instances was localized to specific regions of the nucleus. These results suggest the presence of discrete antigen complexes in the nucleoplasm of the cells, a finding which is in agreement with the investigations of Prem veer Reddy and Pardee[2] who have reported on the existence of multienzyme complexes in the nucleus of eukaryotic cells. Conceivably, monoclonal antibodies with the same immunofluorescence staining pattern may be directed against different antigens in the same protein complex. Second, some of the nuclear antigens identified by the monoclonal antibodies were only expressed in a fraction of the cells, indicating their association with specific phases of the cell cycle. Interestingly, when these antigens were studied in the malignant lymphoma cell lines, they were found to be present in all the cells. It appears that there is an abnormal expression of these antigens in malignant cells, and that the regulation of nuclear antigens during different phases of the cell cycle may be blocked after malignant transformation. Third, during mitosis some of the intranuclear antigens dissociated from the condensing chromatin and dispersed into the surrounding cytoplasm. After the nucleus reformed, these antigens once again were localized in the nucleoplasm. This pattern of staining has been observed for a number of other proteins including terminal transferase[43] and the S-II antigen associated with RNA polymerase II[44] and may be a characteristic of nuclear proteins which mediate gene replication and transcription. Fourth, several monoclonal antibodies identified antigens which were found to translocate from the cytoplasm to the nucleus apparently during specific phases of the cell cycle. One of the antigens, identified by monoclonal antibody N-CS244-7, was associated with cytoplasmic intermediate filaments in human epithelial cells and appeared in the nucleus during the G_2 and M phases of the cell cycle. These findings indicate that the compartmentalization of intracellular proteins may be an important process in the regulation of normal cellular events. Likewise, small nuclear RNA molecules have been shown to shuttle from the nucleus to cytoplasm and have been postulated to be involved in RNA transcription, cellular "reprogramming",[45-47] and RNA splicing.[48,49] It would be of interest to determine whether these small nuclear RNAs are a component of the antigens identified by the monoclonal antibodies mentioned above.

Based upon the results of this investigation and a review of the literature, we have found that the immunofluorescence staining patterns of nuclear antigens can be classified into 11 categories. As shown in Table 5, prototypic monoclonal antibodies are listed which identify antigens at specific intracellular sites with respect to the cell cycle. In Table 6, other known nuclear antigens are listed and categorized according to the classification schema presented in Table 5.

In general, the detection of nuclear antigens by indirect immunofluorescence microscopy has been best described by investigators studying human autoimmune diseases. Monoclonal antibodies to several of the autoimmune antigens have been produced using MLR and NZB mice which display lupus erythematosus-like diseases.[70-73] Unlike the majority of the intranuclear antigens identified by our studies, antigens recognized by autoimmune patient sera thus far have been found to be composed of highly conserved molecules and therefore are not human-specific. In addition to autoimmune patients, cancer patients have been reported to produce antinuclear antibodies as reported by Klein et al.[74] and Steiner et al.[75] who detected an acute myelogenous leukemia-associated coarsely granular nuclear antigen by anticomplement immunofluorescence techniques. It will be of interest to determine whether additional types of staining patterns exist and whether a classification scheme based upon the intracellular location and distribution of a given antigen will be useful in future investigations on the structure and function of the eukaryotic nucleus.

The utility of colloidal gold-linked reagents in the ultrastructural localization of intracellular antigens has previously been demonstrated by several investigators.[76,77] For studies involving nuclear fine structure, colloidal gold offers several advantages over conventional reagents, such as horseradish peroxidase and ferritin. The electron-dense nature of colloidal gold allows easy visualization of antibody binding sites in electron-dense regions of the cell nucleus. In addition, the availability of different sizes of colloidal gold particles facilitates double labeling experiments. In our laboratory, post-embedding procedures were used to identify the binding sites of monoclonal antibodies in the nucleus. Our results demonstrated the ability of these reagents to identify specific nuclear substructures such as euchromatin and heterochromatin. The immunogold staining patterns were found to correlate well with indirect immunofluorescence results but revealed a hitherto unrecognized specificity of bind-

Table 5
CLASSIFICATION SCHEMA FOR NUCLEAR ANTIGENS

Type	Prototypic antibody	Interphase nucleus	Nucleolus	Mitotic chromosomes	Mitotic spindle	Mitotic poles	Telophase midbody	Interphase cytoplasm	Mitotic cytoplasms	Cytoskeleton
I	Anti-histone	+	−	+	−	−	−	−	−	−
II	235-1	+	−	−	−	−	−	−	+	−
III	Anti-NuMa	+	−	−	−	+	−	−	−	−
IV	364-5	+	+	−	−	−	−	−	+	−
V	899-1	+	+	+	−	−	−	−	−	+
VI	877-8	+	+	−	+	−	+	−	−	−
VII	125-10	−	+	−	−	−	−	−	+	−
VIII	Anti-C23	−	+	+	−	−	−	−	−	−
IX	244-7	+ Cell cycle	−	+	−	−	−	−	−	+
X	58-15	+ Cell cycle	−	−	−	−	−	+	+	−
XI	231-7	+ Cell cycle	−	−	−	−	−	+ Cell cycle	−	−

Table 6
IMMUNOFLUORESCENCE STAINING PATTERNS OF NUCLEAR ANTIGENS IN MAMMALIAN CELLS

Type	
I.	Interphase nucleus and metaphase chromosomes
	Histones [50-52]
	Single- and double-stranded DNA[35,53]
	Scl-70 antigen[54]
	Scleroderma-associated centromere antigen[55]
	Epstein-Barr virus nuclear-associated antigen[56]
	Some HMG proteins[57]
	Mitotic scaffold proteins[32,58]
II.	Interphase nucleus, antigen liberated into cytoplasm during mitosis
	Nuclear laminar pore proteins[58,60]
	Terminal transferase[43]
	Ribonucleoprotein complexes[60,62]
	Sm antigen of systemic lupus erythematosus[62]
	SS-B antigen of Sjogren's syndrome[63]
	SV-40 T antigen[64]
	S-II antigen associated with RNA polymerase II[44]
III.	Interphase nucleus, antigen associates with poles of mitotic cells
	NuMA[22]
VII.	Nucleolus
	Autoimmune antigens[65-67]
	Tumor-associated nucleolar antigen 54/6.3[39,68]
X.	Interphase nucleus and cytoplasm
	Some HMG proteins[69]
	3T6 Chromatin antigens[37]

ing to nuclear structures. This technique has great potential, therefore, as a method of identifying specific antigenic domains in the eukaryotic interphase and mitotic nucleus. Further application of these techniques to other electron microscopic methods should yield valuable information about the ultrastructural localization of nuclear antigens.

Besides detecting nuclear-associated antigens, our studies produced a number of monoclonal antibodies to cytoskeletal structures which apparently were present in the chromatin preparations used to immunize the mice. Previous workers such as Douvas et al.[30] have been able to identify by polyacrylamide disc gel electrophoresis the presence of major cytoskeletal proteins in rat liver chromatin preparations, and Osborn and Weber[78] have shown that detergent-resistant cytoskeletal structures of tissue culture cells retain the nucleus. These studies emphasize the intimate special relationship of the cytoskeleton and the nucleus and suggest the presence of common antigens. Because of the importance of the microtubule-microfilament cytoskeletal system in cell locomotion, shape, and structure and in its postulated role in malignant transformation,[79] specific reagents which can identify cytoskeletal-associated antigens may be especially useful in future biochemical studies. In addition, the presence or absence of specific cytoskeletal antigens may help to define the cell of origin of human tumors including those of hematopoietic derivation.

ACKNOWLEDGMENTS

The authors are especially grateful for the continuous support and the interest of Dr. David Pettijohn and Dr. Eng Tan during the course of this work. This research was partly performed at the Eleanor Roosevelt Institute for Cancer Research, University of Colorado Medical Center, Denver, under the auspices of Dr. Theodore Puck when Alan Epstein was a Hubert H. Humphrey Distinguished Research fellow of the Damon Runyon-Walter Winchell Cancer

Fund. This work was supported in part by grants from the Milheim Foundation for Cancer Research, the Leukemia Research Foundation, the Chicago Community Trust-Searle Funds, and by grants PCM-8140431 from the National Science Foundation and RO1-CA 3062 from the National Institutes of Health. Alan Epstein is a Searle Scholar and the recipient of a Junior Faculty Research Award from the American Cancer Society. Charles Clevenger is a medical and graduate student supported in part by the Medical Student Short-Term Research Training Grant 1 P35A107224 from the National Institutes of Health.

NOTE ADDED IN PROOF

Since the submission of this chapter, four additional manuscripts have been published or are in press which utilize the monoclonal antibodies described here.[80-83] These publications from a laboratory further demonstrate the importance of monoclonal antibodies as useful tools in the study of nuclear substructures with respect to cellular differentiation, cell growth, and malignant transformation.

REFERENCES

1. **Laskey, R. A. and Earnshaw, W. C.,** Nucleosome assembly, *Nature (London)*, 286, 763, 1980.
2. **Prem veer Reddy, G. and Pardee, A. B.,** Multienzyme complex for metabolic channeling in mammalian DNA replication, *Proc. Natl. Acad. Sci.* U.S.A., 77, 3312, 1980.
3. **Beyer, A. L., Christensen, M. E., Walker, B. W., and LeStourgeon, W. M.,** Identification and characterization of the packaging proteins of core 40S hnRNP particles, *Cell*, 11, 127, 1977.
4. **Beyer, A. L., Miller, O. L., Jr., and McKnight, S. L.,** Ribonucleoprotein structure in nascent hnRNA is nonrandom and sequence dependent, *Cell*, 20, 75, 1980.
5. **Karn, J., Vidali, G., Boffa, L. C., and Allfrey, V. G.,** Characterization of the non-histone nuclear proteins associated with rapidly labeled heterogeneous nuclear RNA, *J. Biol. Chem.*, 252, 7307, 1977.
6. **Suria, D. and Liew, C. C.,** Characterization of proteins associated with nuclear ribonucleoprotein particles by two-dimensional polyacrylamide gel electrophoresis, *Can. J. Biochem.*, 57, 32, 1979.
7. **Berezney, R. and Coffey, D. S.,** Nuclear matrix: isolation and characterization of the framework structure from rat liver nuclei, *J. Cell Biol.*, 73, 616, 1977.
8. **Adolph, K. W., Cheng, S. M., Paulson, J. R., and Laemmli, U. K.,** Isolation of a protein scaffold from mitotic HeLa cell chromosomes, *Proc. Natl. Acad. Sci.* U.S.A., 74, 4937, 1977.
9. **Long, B. H., Huang, C. Y., and Pogo, A. O.,** Isolation and characterizatin of the nuclear matrix in friend erythroleukemia cells: chromatin and hnRNA interactions with the nuclear matrix, *Cell*, 18, 1079, 1979.
10. **MacGillivray, A. and Rickwood, D.,** The heterogeneity of mouse-chromatin nonhistone proteins as evidenced by two-dimensional polyacrylamide-gel electrophoresis and ion-exchange chromatography, *Eur. J. Biochem.*, 41, 181, 1974.
11. **Peters, K. E. and Comings, D. E.,** Two-dimensional gel electrophoresis of rat liver nuclear washes, nuclear matrix, and hnRNA proteins, *J. Cell Biol.*, 86, 135, 1980.
12. **Peterson, J. L. and McConkey, E. H.,** Non-histone chromosomal proteins from HeLa cells: a survey by high resolution, two-dimensional electrophoresis, *J. Biol. Chem.*, 251, 548, 1976.
13. **Pulvertaft, R. J. V.,** Cytology of Burkitt's tumor (African lymphoma), *Lancet*, 1, 238, 1964.
14. **Pizzo, P. A., Magrath, I. T., Chattopadhyay, S. K., Biggar, R. J., and Gerber, P.,** A new tumor-derived transforming strain of Epstein-Barr virus, *Nature*(London), 272, 629, 1978.
15. **Foley, G. E., Lazarus, H., Farber, S., Uzman, B. G., Boone, B. A., and McCarthy, R. E.,** Continuous culture of human lymphoblasts from peripheral blood of a child with acute leukemia, *Cancer*, 18, 522, 1965.
16. **Nilsson, K., Bennich, H., Johansson, S. G. O., and Ponten, J.,** Established immunoglobulin producing myeloma (IgE) and lymphoblastoid cell lines from an IgE myeloma patient, *Clin. Exp. Immunol.*, 7, 477, 1970.

17. **Lozzio, C. B. and Lozzio, B. B.,** Human chronic myelogenous leukemia cell line with positive Philadelphia chromosome, *Blood*, 45, 321, 1975.
18. **Epstein, A. L. and Kaplan, H. S.,** Biology of the human malignant lymphomas. I. Establishment in continuous cell culture and heterotransplantation of diffuse histiocytic lymphomas, *Cancer*, 34, 1851, 1974.
19. **Sundstrom, C. and Nilsson, K.,** Establishment and characterization of a human histiocytic lymphoma cell line (U-937), *Int. J. Cancer*, 17, 565, 1976.
20. **Epstein, A. L., Levy, R., Kim, H., Henle, W., Henle, G., and Kaplan, H. S.,** Biology of the malignant lymphomas. IV. Functional characterization of ten diffuse histiocytic lymphoma cell lines, *Cancer*, 42, 2379, 1978.
21. **Thorsby, E. and Bratlie, A.,** A rapid method for preparation of pure lymphocyte suspensions, in *Histocompatibility Testing*, Terasaki, P., Ed., Munksgaard, Copenhagen, 1970, 655.
22. **Lydersen, B. K. and Pettijohn, D. E.,** Human specific nuclear protein that associates with the polar region of the mitotic apparatus: distribution in a human/hamster hybrid cell, *Cell*, 22, 489, 1980.
23. **Oi, V. T. and Herzenberg, L. A.,** Immunoglobulin-producing hybid cell lines, in *Selected Methods in Cellular Immunology*, Mishell, B. B. and Shiigi, S. M., Eds., W. H. Freeman, San Francisco, 1980, 351.
24. **Moriarty, G. C. and Halmi, N. S.,** Electron microscopic study of the adrenocorticotropin-producing cell with the use of unlabeled antibody and the soluble peroxidase-anti-peroxidase complex, *J. Histochem. Cytochem.*, 20, 590, 1972.
25. **Moriarty, G. C.,** Adenohypophysis ultrastructural cytochemistry, *J. Histochem. Cytochem.*, 21, 855, 1973.
26. **Childs, G. V. and Ellison, D. G.,** Critique of contributions of immunochemistry to our understanding of pituitary cell structure and function, *Histochem. J.*, 12, 405, 1980.
27. **Venable, J. H. and Coggeshall, R.,** A simplified lead citrate stain for use in electron microscopy, *J. Cell Biol.*, 25, 407, 1965.
28. **Laemmli, U. K.,** Cleavage of structural proteins during the assembly of the head of bacteriophage T4, *Nature (London)*, 227, 680, 1970.
29. **Chamberlain, J. P.,** Fluorographic detection of radioactivity in polyacrylamide gels with the water soluble fluor, sodium salicylate, *Anal. Biochem.*, 98, 132, 1979.
30. **Douvas, A. S., Harrington, C. A., and Bonner, J.,** Major nonhistone proteins of rat liver chromatin: preliminary identification of myosin, actin, tubulin, and tropomyosin, *Proc. Natl. Acad. Sci.*, U.S.A., 72, 3902, 1975.
31. **Briggs, R. C., Campbell, A., Chiu, J.-F., Hnilica, L. S., Lincoln, G., Stein, J., and Stein, G.,** Specificity of DNA-associated nuclear antigens in HeLa cells and distribution during the cell cycle, *Cancer Res.*, 39, 3683, 1979.
32. **Campbell, A., Briggs, R. C., Bird, R. E., and Hnilica, L. S.,** Cell specific antiserum to chromosome scaffold proteins, *Nucleic Acids Res.*, 6, 205, 1979.
33. **Chytil, F. and Spelsberg, T. C.,** Tissue differences in antigenic properties of non-histone protein-DNA complexes, *Nature (London) New Biol.*, 233, 215, 1971.
34. **Glass, W. F., II, Briggs, R. C., and Hnilica, L. S.,** Identification of tissue-specific nuclear antigens transferred to nitrocellulose from polyacrylamide gels, *Science*, 211, 70, 1980.
35. **Wakabayashi, K. and Hnilica, L. S.,** The immunospecificity of nonhistone protein complexes with DNA, *Nature (London) New Biol.*, 242, 153, 1973.
36. **Alevy, Y. G. and Fleischman, J. B.,** Immunospecific isolation of a human chromatin fraction from mouse-human hybrid cells, *Mol. Immunol.*, 17, 275, 1980.
37. **Okita, K. and Zardi, L.,** Immunofluorescent study of chromatin proteins in cultured fibroblasts, *Exp. Cell Res.*, 86, 59, 1974.
38. **Tsutsui, Y., Suzuki, I., and Iwai, K.,** Immunofluorescent study of nonhistone protein-DNA complexes in cultured cells and lymphocytes, *Exp. Cell Res.*, 101, 202, 1976.
39. **Busch, H., Gyorkey, F., Busch, R. K., Davis, F. M., Gyorkey, P., and Smetana, K.,** A nucleolar antigen found in a broad range of human malignant tumor specimens, *Cancer Res.*, 39, 3024, 1979.
40. **Fujitana, H., Chiu, J.-F., and Hnilica, L. S.,** Purification of nuclear antigens in Novikoff hepatoma, *Proc. Natl. Acad. Sci.*, U.S.A., 75, 1943, 1978.
41. **Saumweber, H., Symmons, P., Kabish, R., Will, H., and Bonhoeffer, F.,** Monoclonal antibodies against chromosomal proteins of Drosophila melanogaster, *Chromosoma (Berlin)*, 80, 253, 1980.
42. **Howard, G. C., Abmayr, S. M., Shinefeld, L. A., Sato, V. L., and Elgin, S. C. R.,** Monoclonal antibodies against a specific nonhistone chromosomal protein of Drosophila associated with active genes, *J. Cell Biol.*, 88, 219, 1981.
43. **Bollum, F. J.,** Deoxynucleotide-polymerizing enzymes in mammalian cells: immunofluorescence, in *Antiviral Mechanisms in the Control of Neoplasia*, Chandra, P., Ed., Plenum Press, New York, 1979, 587.
44. **Sekimizu, K., Kobayashi, N., Mizuno, D., and Natori, S.,** Purification of a factor from Ehrlich ascites tumor cells specifically stimulating RNA polymerase II, *Biochemistry*, 15, 5064, 1976.

45. **Goldstein, L. and Ko, C.,** Electrophoretic characterization of shuttling and nonshuttling small nuclear RNAs, *Cell,* 2, 259, 1974.
46. **Goldstein, L.,** Role for small nuclear RNAs in "programming" chromosomal information? *Nature (London),* 261, 519, 1976.
47. **Zieve, G. and Penman, S.,** Small RNA species of the HeLa cell: metabolism and subcellular localization, *Cell,* 8, 19, 1976.
48. **Lerner, M. R., Boyle, J. A., Mount, S. M., Wolin, S. L., and Steitz, J. A.,** Are snRNPs involved in splicing?, *Nature (London),* 283, 220, 1980.
49. **Yang, V. W., Lerner, M. R., Steitz, J. A., and Flint, S. J.,** A small nuclear ribonucleoprotein in required for splicing of adenoviral early RNA sequences, *Proc. Natl. Acad. Sci.,* U.S.A., 78, 1371, 1981.
50. **Bustin, M., Yamasaki, H., Goldblatt, D., Shani, M., Huberman, E., and Sachs, L.,** Histone distribution in chromosomes revealed by anti-histone sera, *Exp. Cell Res.,* 97, 440, 1976.
51. **Pothier, L., Gallagher, J. F., Wright, C. E., and Libby, P. R.,** Histones in fixed cytological preparations of Chinese hamster chromosomes demonstrated by immunofluorescence, *Nature (London),* 255, 350, 1975.
52. **Tan, E. M., Robinson, J., and Robitaille, P.,** Studies on antibodies to histones by immunofluorescence, *Scand. J. Immunol.,* 5, 811, 1976.
53. **Arana, R. and Seligmann, M.,** Antibodies to native and denatured deoxyribonucleic acid in systemic lupus erythematosus, *J. Clin. Invest.,* 46, 1867, 1967.
54. **Douvas, A. S., Achten, M., and Tan, E. M.,** Identification of a nuclear protein (Scl-70) as a unique target of human antinuclear antibodies in scleroderma, *J. Biol. Chem.,* 254, 10514, 1979.
55. **Moroi, Y., Peebles, C., Fritzler, M. J., Steigerwald, J., and Tan, E. M.,** Autoantibody to centromere (kinetochore) in scleroderma sera, *Proc. Natl. Acad. Sci.* U.S.A., 77, 1627, 1980.
56. **Reedman, B. and Klein, G.,** Cellular localization of an Epstein-Barr virus (EBV)-associated complement-fixing antigen in producer and non-producer lymphoblastoid cell lines, *Int. J. Cancer,* 11, 499, 1973.
57. **Smith, B., Robertson, D., Birbeck, M., Goodwin, G., and Johns, E.,** in *Exp. Cell Res.,* 115, 420, 1978.
58. **Laemmli, U., Cheng, S., Adolph, K., Paulson, J., Brown, J., and Baumbach, W.,** The role of non-histone proteins, *Cold Spring Harbor Symp. Quant. Biol.,* 42, 351, 1977.
59. **Gerace, L., Blum, A., and Blobel, G.,** Immunocytochemical localization of the major polypeptides of the nuclear pore complex-lamina fraction, *J. Cell Biol.,* 79, 546, 1978.
60. **Gerace, L. and Blobel, G.,** The nuclear envelop lamina is reversibly depolymerized during mitosis, *Cell,* 19, 277, 1980.
61. **Jones, R. E., Okamura, C. S., and Martin, T. E.,** Immunofluorescent localization of the proteins of nuclear ribonucleoprotein complexes, *J. Cell Biol.,* 86, 235, 1980.
62. **Northway, J. D. and Tan, E. M.,** Differentiation of antinuclear antibodies giving speckled staining patterns in immunofluorescence, *Clin. Immunol. Immunopathol.,* 1, 140, 1972.
63. **Alspaugh, M. A., Talal, N., and Tan, E. M.,** Differentiation and characterization of auto-antibodies and their antigens in Sjogren's syndrome, *Arthritis Rheum.,* 19, 216, 1976.
64. **Gurney, E. G., Harrison, R. O., and Fenno, J.,** Monoclonal antibodies against Simian virus 40 T antigens: evidence for distinct subclasses of large T antigen and for similarities among non-viral T antigens, *J. Virol.,* 34, 752, 1980.
65. **Ritchie, R. F.,** Antinucleolar antibodies: their frequency and diagnostic association, *New Engl. J. Med.,* 282, 1174, 1970.
66. **Pinnas, J. L., Northway, J. D., and Tan, E. M.,** Antinucleolar antibodies in human sera, *J. Immunol.,* 111, 996, 1973.
67. **Smetana, K., Busch, R. K., Hermansky, F., and Busch, H.,** Nucleolar immunofluorescence in human hematological malignancies, *Life Sci.,* 25, 227, 1979.
68. **Chan, P. K., Feyerabend, A., Busch, R. K., and Busch, H.,** Identification and partial purification of human tumor nucleolar antigen 54/6.3, *Cancer Res.,* 40, 3194, 1980.
69. **Bustin, M. and Neihart, N. K.,** Antibodies against chromosomal HMG proteins stain the cytoplasm of mammalian cells, *Cell,* 16, 181, 1979.
70. **Andrezejewski, C., Jr., Stollar, B. D., Lalor, T. M., and Schwartz, R. S.,** Hybridoma autoantibodies to DNA, *J. Immunol.,* 124, 1499, 1980.
71. **Lee, J. S., Lewis, J. R., Morgan, A. R., Mosmann, T. R., and Singh, B.,** Monoclonal antibodies showing sequence specificity in their interaction with single-stranded DNAs, *Nucleic Acids Res.,* 9, 1707, 1981.
72. **Lerner, E. A., Lerner, M. R., Janeway, C. A., Jr., and Steitz, J. A.,** Monoclonal antibodies to nucleic acid-containing cellular constituents: probes for molecular biology and autoimmune disease, *Proc. Natl. Acad. Sci.* U.S.A., 78, 2737, 1981.
73. **Billings, P. B., Allen, R. W., Jensen, F. C., and Hoch, S. D.,** Anti-RNP monoclonal antibodies derived from a mouse strain with lupus-like autoimmunity, *J. Immunol.,* 128, 1176, 1982.

Chapter 7

NONHISTONE CHROMOSOMAL PROTEINS OF THE MITOTIC APPARATUS: PROPOSED ROLE IN NUCLEAR REASSEMBLY

David E. Pettijohn

TABLE OF CONTENTS

There is a group of nonhistone chromosomal (NHC) proteins which during mitosis becomes part of the mitotic apparatus. The first example of this class was a large nonhistone chromosomal protein named NuMA (for *Nu*clear *M*itotic *A*pparatus protein). The purpose of this chapter is to review experimental approaches which have been developed to study these proteins, to describe some of their properties, and to discuss ideas for their function. The description will focus on NuMA protein because it has been studied the most, but will also briefly review findings with other probably related NHC proteins.

I. PROPERTIES OF NuMA PROTEIN

A. Identification and Purification of NuMA Protein

The NuMA protein was initially recognized in a study of total human NHC proteins intended to identify chromosomal proteins limited to human cells.[1] NHC proteins from human cell lines were compared by one- and two-dimensional gel electrophoresis with the NHC proteins from other species. Among those NHC proteins which had no electrophoretic homologues in nonhuman lines, NuMA had the lowest electrophoretic mobility. Its apparent molecular weight in dissociating solvents containing sodium dodecyl sulfate (SDS) estimated by comparison with protein standards was 300,000.

Figure 1 shows a typical one-dimensional analysis of the large NHC proteins of HeLa cells, a comparison with Chinese hamster large NHC proteins and with those of certain human/hamster hybrids, some of which do and some of which do not have protein bands of the mobility of NuMA. The hybrid cells which can synthesize NuMA seem to be limited to those having certain human chromosomes. It can be seen that the hamster NHC proteins have only a very minor component with an electrophoretic mobility near that of NuMA protein and this minor protein is actually of slightly higher mobility than NuMA. Other nonprimates similarly lack a protein of this size.

Because of its unusually large size, purification of the NuMA protein by molecular sizing or by sedimentation methods was relatively easy.[3] The protein was difficult to solubilize from chromatin, so that sonication of isolated nuclei and solubilization in solutions of SDS or other denaturing solvents were required. Therefore, the protein has so far been purified only in a denatured form. Purification has proceeded to the point where analysis by gel electrophoresis indicates only trace amounts of contaminants in bands other than that of NuMA (Figure 2).

B. NuMA Specific Antibodies and Their Use in Defining Species Distribution of NuMA Antigens

Purified NuMA protein was used to immunize mice and to prepare both polyclonal and monoclonal antibodies. Careful analysis of both types of antibody using immunoprecipitation techniques[3] as well as western blot analysis[4] has shown that antibody preparations are specific to antigenic determinants on NuMA protein. These antibodies have been used to further study the species specificity of NuMA and to elaborate other properties of this protein.

Radioimmunoassay using the anti-NuMA sera showed that NuMA antigens which were readily detectable in human chromatin were not present in hamster chromatin (Figure 3). Other immune analyses using indirect immunofluorescence microscopy, western blot analysis, and immunoprecipitation demonstrated that NuMA antigens were present in any primate cell line but in no other mammalian or eukaryotic cell tested (Table 1). Initial studies of NuMA in African green monkey cells had indicated that NuMA antigens might be absent in this primate,[3] but subsequent analyses using monoclonal antibodies readily detected the antigens. Moreover, the protein could be recognized electrophoretically as a 300,000-dalton component of all primate cells that have been tested, but of no other mammal. Thus, the physical and immunological studies are in agreement. The available evidence suggests that

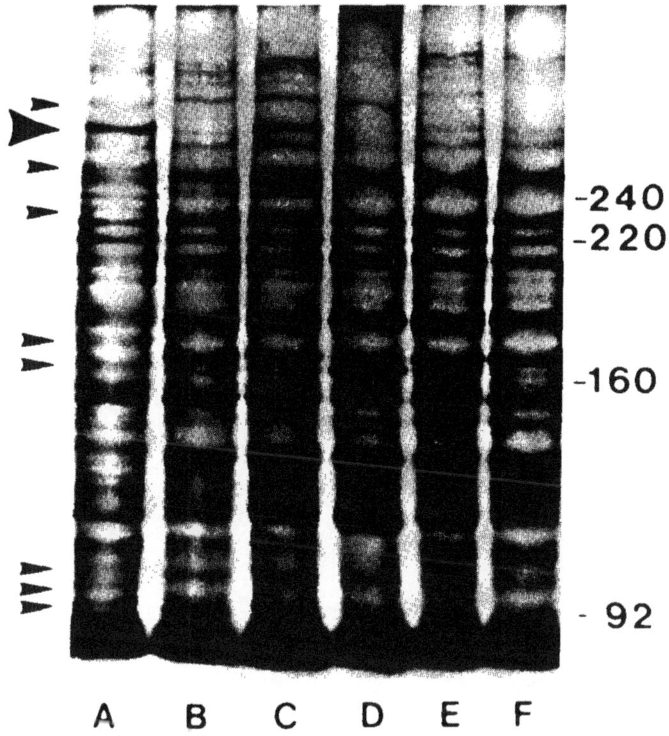

FIGURE 1. Analysis by polyacrylamide gel electrophoresis (PAGE) of high-molecular-weight NHC proteins from different cell lines. NHC ^{35}S-labeled proteins solubilized from isolated chromatin with 0.5% sodium dodecyl sulfate (SDS) were prepared as described and analyzed by electrophoresis on 5% polyacrylamide gels.[1] The proteins were prepared from (A) HeLa; (B) Chinese hamster, CHO, and from several human-CHO hybrids with defined[3] compositions of human chromosomes; (C) CP29; (D) CP27; (E) CP31; and (F) CP32. Duration of electrophoresis was sufficiently long that proteins of molecular weight > 90,000 passed through the gel. The large arrow on the left indicates the 300,000-molecular weight NuMA protein. Other smaller arrows indicate other bands present in HeLa nuclei but not in CHO. The numbers on the right indicate the positions of unlabeled markers on the same gel detected by Coomassie blue stain. The bands shown above were detected by autoradiography. (From Lydersen, B. K., Kao, F.-T., and Pettijohn, D. E., *J. Biol. Chem.*, 255, 3002, 1980. With permission.)

the NuMA protein has an unusually recent evolutionary history. Considerations of the significance of these changes will be delayed until later in this chapter.

C. Intracellular Localization of NuMA Protein

The anti-NuMA antibodies were also used to define the location of NuMA protein in cells. Earlier electrophoretic analysis had indicated that purified nuclei contained the detectable 300,000-molecular weight NuMA protein and the protein could not be found in cytoplasmic fractions of human cells. The nuclear location was confirmed using indirect immunofluorescence microscopy (Figure 4). In mitotic cells, using the same assays with anti-NuMA sera, fluorescence from the polar region of the mitotic apparatus was observed. No other parts of the cell were fluorescent, suggesting that in mitosis NuMA is part of the spindle. Superficially, its distribution seemed comparable to certain mitotic apparatus (MA) proteins defined in similar analyses using appropriate antibodies.[5,6] There were clear dis-

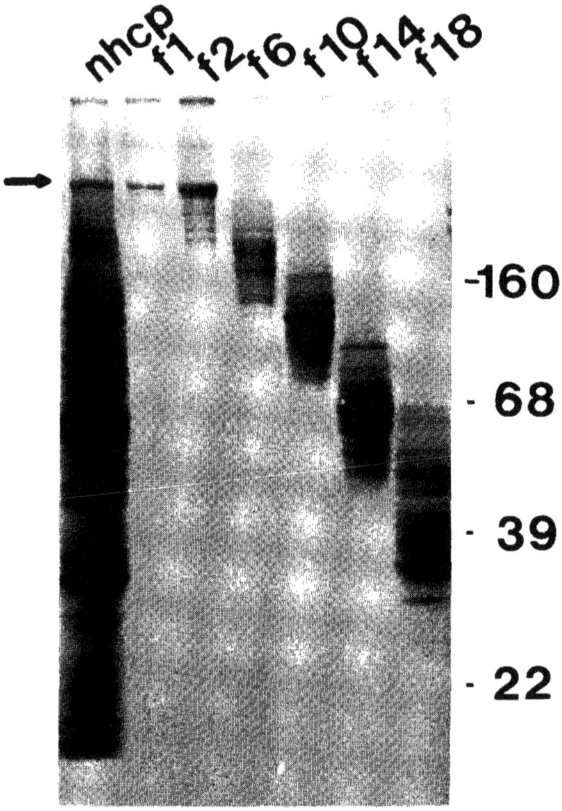

FIGURE 2. Purification of NuMA protein by gel filtration chromatography. Approximately 150 mg of purified total HeLa NHC proteins solubilized in solutions containing 0.2% SDS were applied to an 80-cm column of BioGel A5m agarose beads equilibrated with the same SDS-containing solution. Proteins were eluted in about 50 fractions and samples of some were analyzed by electrophoresis as shown above. Fractions f1 and f2 represent the largest proteins in the excluded volume; these fractions contain the NuMA protein. Higher-numbered fractions f6 through f18 contain lower molecular weight proteins separated from NuMA. The arrow on the left indicates the expected position of NuMA by comparison with molecular-weight markers run on the same gel (positions indicated on the right). (From Lydersen, B. K. and Pettijohn, D. E., *Cell*, 22, 489, 1982. Copyright MIT. With permission.)

tinctions, however, with the known MA proteins, which seemed to be associated with tubulin in microtubules wherever they were in the cell. Unlike the MA proteins there was no NuMA associated with cytoplasmic microtubules, microtubules near the central region of the spindle, or with microtubules in the intracellular bridge.[3] This was most convincingly shown in studies which visualized the distributions of tubulin and NuMA simultaneously in the same cells using anti-tubulin and anti-NuMA antibodies having different associated chromophores (see Figure 4c and 4d for one example).

Subsequent studies using anti-NuMA monoclonal antibodies showed that the above distribution in mitotic cells reflected only one of the two possible associations of NuMA protein in such cells. Using the monoclonal antibodies applied to suitably fixed mitotic cells, the immunofluorescence studies indicated that NuMA in metaphase cells was bound exclusively to chromosomes.[4] This variable localization of NuMA in mitotic cells is resolved below and explained in terms of a set of dual binding sites present on the protein.

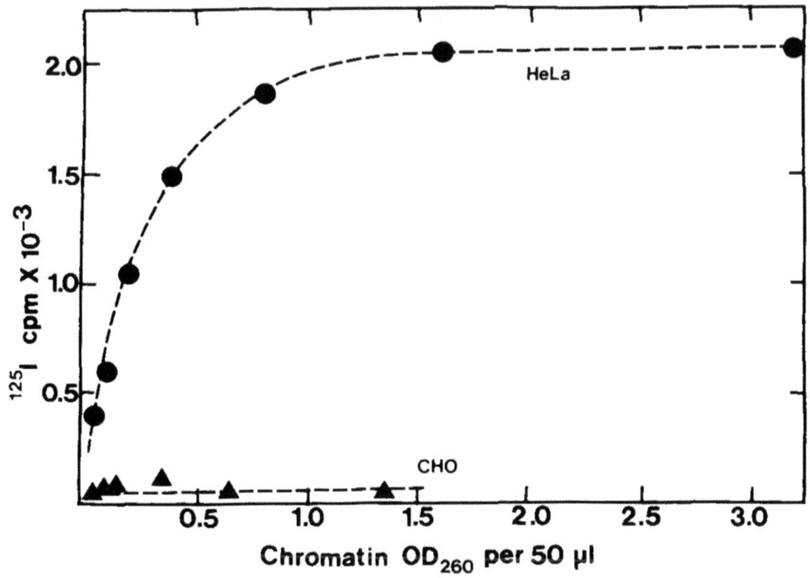

FIGURE 3. Radioimmunoassay of NuMA protein present in chromatin from human cells but absent in chromatin from other mammals. Isolated total chromatin was applied in amounts shown above and allowed to absorb to plastic surfaces of microtiter plate wells. The non-absorbed chromatin was washed away, and the wells were soaked extensively with a solution containing bovine serum albumin (BSA) then reacted with constant amounts of sera from a mouse immunized with purified NuMA protein. After washing away unabsorbed proteins, a constant amount of ^{125}I-labeled protein A was applied to each well. After incubation, unbound protein A was washed away and each well counted in a gamma counter.

Table 1
NuMA PROTEIN IN CELLS OF DIFFERENT SPECIES

		NuMA assayed[a]	
Species	**Cell type**	**Electrophoretically**	**Immunologically**
Human	HeLa	+	+
	Fibroblast	+	+
	KB	+	+
	HT-1080	nd[b]	+
	Lymphocyte	nd	+
	Raji	+	+
African green monkey	Vero	+	+
	CV-1	+	+
Chinese hamster	CHO-K1	−	−
	Fibroblast	−	−
	Liver	−	nd
	V-79	−	−
Syrian hamster	BHK-21	nd	−
Rabbit	Fibroblast	nd	−
Mouse	Liver	−	nd
	MTK	nd	−
Drosophila	SL2	nd	−

[a] The plus sign indicates the band at molecular weight 300,000 shown in Figure 1 is present; the minus sign indicates it is not. Immunological tests for NuMA have used polyclonal[3] as well as monoclonal antibodies in immunofluorescence,[3] immunoprecipitation,[3] radioimmune, and western blot assays. Not all of the antibody based assays were done on NHC proteins from each source.

[b] Indicates not done.

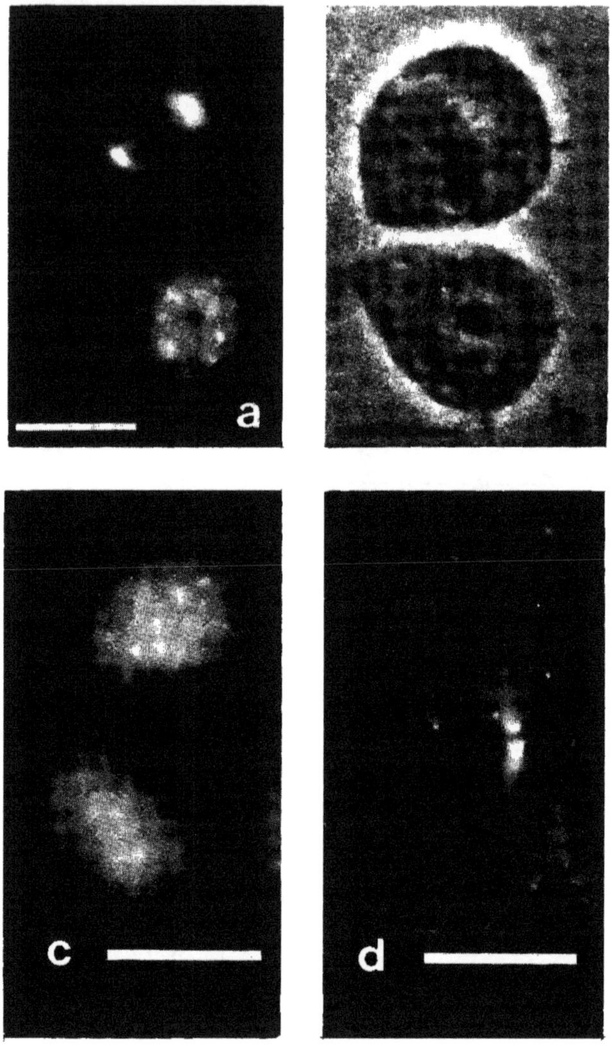

FIGURE 4. Indirect immunofluorescence photomicrographs of HeLa cells using anti-NuMA antiserum. HeLa cells grown attached to glass slides were fixed with formaldehyde/acetone and incubated with anti-NuMA sera, washed, and then incubated with fluorescein conjugated goat antibodies against mouse IgG. The cells were then photographed with a Zeiss RA fluorescence microscope. The same two cells, a metaphase cell on top and an interphase cell on the bottom, are shown here viewed fluorescence microscopy and (b) phase microscopy. The cells were incubated with rabbit anti-tubulin IgG at the same time as with mouse anti-NuMA and then with a mixture of rhodamine and fluorescein conjugated goat antibodies to mouse and rabbit IgG, respectively. The same late telophase cell with intracellular bridge connecting the daughter cell was photographed using different filters to detect only (c) rhodamine associated with anti-NuMA and (d) fluorescein associated with anti-tubulin. Bars in all panels represent 10 μm. (Reprinted by permission. Copyright M.I.T.)

D. Monoclonal Antibody 2E4 Detaches NuMA Protein from Purified Metaphase Chromosomes

Electrophoretic analysis of the NHC proteins of isolated metaphase chromosomes showed that NuMA protein was a prominent component.[4] When these chromosomes were treated with purified 2E4 type monoclonal antibody, specific for NuMA, there was a selective release from the chromosome of NuMA protein along with a few other high molecular weight NHC proteins. The release was most convincingly demonstrated in sedimentation studies of the purified chromosomes on sucrose gradients.[4] Human chromosomes were centrifuged through a band in the sucrose gradient containing the 2E4 antibody and were then pelleted in the bottom of the gradient. Greater than 99% of the ^{35}S-labeled NHC proteins co-sedimented with the chromosomes into the pellet, but NuMA protein was dissociated into the 2E4 band. Nothing was dissociated by this monoclonal antibody from Chinese hamster chromosomes which lack the 2E4 antigen. Moreover, other antibodies specific for different NHC proteins caused no significant release in this assay of the proteins of human chromosomes. The results showed that the 2E4-induced release of NuMA was a process dependent on the specificity of this antibody.

E. Binding Sites on NuMA Protein for Chromosomes and Mitotic Spindle Poles

When the 2E4 monoclonal antibody was applied to fixed, permeabilized, mitotic human cells in immunofluorescence studies, the distribution of NuMA protein indicated by the observed fluorescence was dependent on how long the cells were incubated with the 2E4 antibody and how the cells were fixed. Cells fixed with agents such as acetone or alcohol, which do not cross-link proteins, showed fluorescence exclusively from the mitotic spindle poles. Just as in the case of the studies described above using anti-NuMA sera, the chromosomes were not stained. When, however, cells were fixed prior to antibody addition with comparatively strong cross-linking agents such as glutaraldehyde or paraformaldehyde/lysine/periodate, only the chromosomes were fluorescent and poles were rarely fluorescent even after many hours of incubation with the 2E4-type antibody. With fixatives such as paraformaldehyde/acetone, which are believed to be less effective cross-linking agents, the immunofluorescent distribution was dependent on the time of incubation with 2E4. Initially only the chromosomes were fluorescent; gradually these lost their fluorescence as the polar regions became brighter. Finally, after hours of incubation the chromosomes were dark and the poles maximally fluorescent (see Figure 5 for representative photomicrographs). At this time no other sites except the spindle pole region were fluorescent (Figure 5c). It appeared from these studies that after its displacement from metaphase chromosomes, the NuMA protein was specifically attaching to the spindle pole region in a complex with the 2E4 antibody.

Several different control experiments and additional other experiments have supported this interpretation. First it is clear that the 2E4 antibody is required to obtain the observed transfer of fluorescence. Incubation of unfixed cells or paraformaldehyde/acetone fixed cells in the absence of antibody 2E4 had no effect on the transfer. Such cells after additional fixation and application of antibody 2E4 showed fluorescent chromosomes rather than fluorescent spindle poles. Moreover, incubation of paraformaldehyde/acetone-treated cells with antibodies specific for other NHC proteins had no effect; additional prolonged incubation with 2E4 was required to produce fluorescent spindle poles. Also, the results were independent of the fluorescent second antibody used in the indirect immunofluorescence. Neither chromosomes nor poles became fluorescent when fixed cells were incubated with secondary antibody alone. Moreover, when fluorescein-2E4 antibody was used without any secondary antibody, the same fluorescent transfer to poles was observed. The results described above showed that cells fixed with stronger protein cross-linkers did not transfer fluorescence to spindle pole regions, even after prolonged incubation with antibody 2E4. Thus, it seems that a release from mitotic chromosomes of the NuMA-associated antigens was required to obtain the observed transfer to mitotic spindle poles. The results show that the NuMA protein in metaphase cells is exclusively attached to chromosomes, but that if displaced from the chromosome, it can bind specifically at sites located in the polar region of the mitotic spindle.

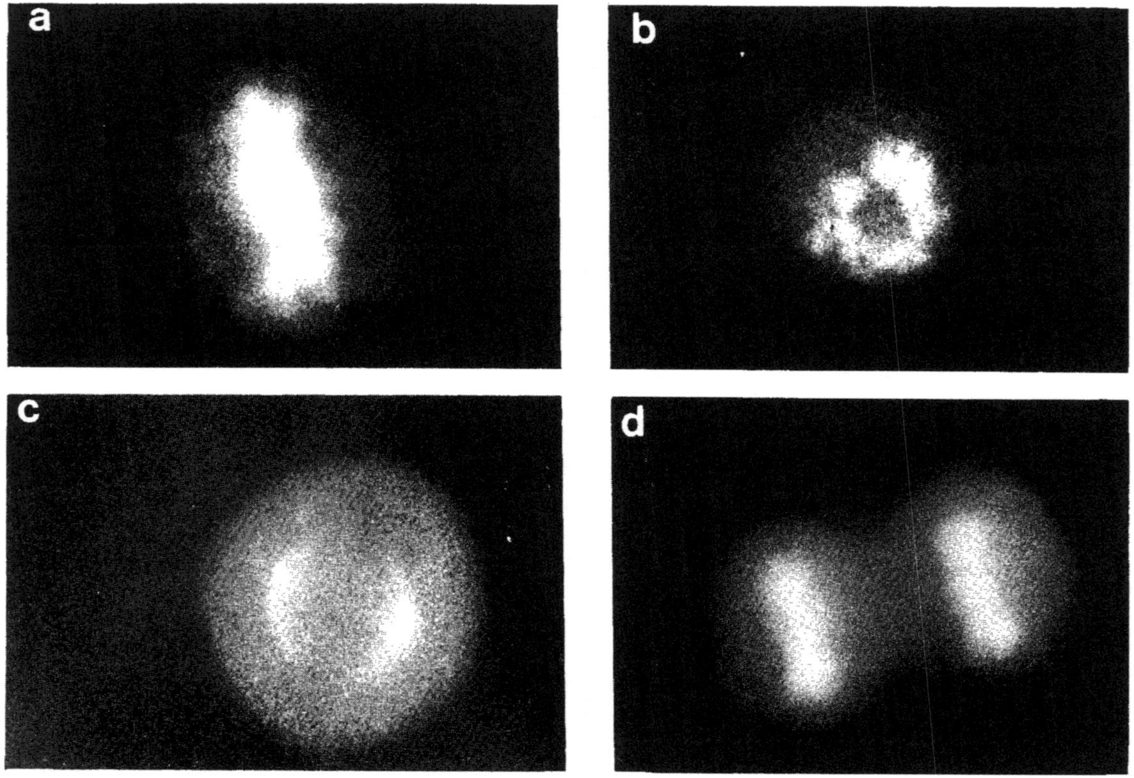

FIGURE 5. Indirect immunofluorescence photomicrographs of HeLa cells treated with monoclonal antibody 2E4. Cells growing attached to microscope slides were fixed with paraformaldehyde/acetone, washed, incubated with monoclonal antibody 2E4 for 1 hr, then washed and incubated in wash solutions for variable times before all slides were incubated 1 hr with fluorescein conjugated goat anti-mouse IgG. Mitotic cells were viewed and photographed with a Zeiss standard RA fluorescence microscope. (a) A typical metaphase cell incubated 1 hr or less after application of 2E4 (mitotic chromosomes only are fluorescent); (b) another mitotic cell incubated similarly as that of (a) but with an end-on view of the mitotic plate; (c) a metaphase cell incubated 24 hr after treatment with antibody 2E4 (the poles of the mitotic spindle are fluorescent and the chromosomes are dark); (d) a telophase cell incubated 1 hr after 2E4 application (the segregated chromosomes are fluorescent but size and shape is clearly distinct from fluorescent poles seen in mitotic cells).

II. MODEL FOR THE INVOLVEMENT OF NuMA PROTEIN IN NUCLEAR REASSEMBLY

A. The Model

The above results are strongly suggestive that the NuMA protein has two distinct sites with affinities for two different structures in the mitotic cell. The first site, the C site, is the one by which the protein is normally attached to metaphase chromosomes. The second site, the SP site, has an affinity for receptors called SP_r which are located in the spindle pole region of the mitotic apparatus. Apparently NuMA-specific antibodies of the 2E4 type can bind to their antigenic site and interfere with the C-site-mediated attachment of NuMA to chromosomes. Thus, the NuMA protein with bound 2E4 antibody is released from chromosomes. The complex is then free to attach anywhere in the cell, which leads to binding at the SP_r receptors via the SP site on NuMA protein.

A hypothesis for how a NHC protein with such dual attachment sites could function in post-mitotic nuclear reassembly is outlined in Figure 6. It is proposed that the NuMA protein normally functions during early telophase to link the segregated chromosomes into nuclear organizing centers located in the polar regions of the mitotic spindle. This protein would then hold in close proximity SP_r sites at the organizing center with corresponding C receptor sites on telophase chromosomes, and thus set the relative positions of the reassociating chromosomes. From the model (illustrated in Figure 6c) the effects of the 2E4 antibody would be to release the linkage of C sites and therefore uncouple the system.

Several uncertainties should be mentioned concerning this model. It is possible that the proposed SP and C sites are not structurally part of the NuMA protein itself, but rather are located on other proteins normally bound in a complex with NuMA. In this case the antigenic sites for antibody 2E4, which are on the NuMA proteins, would have to be coupled to the C site so that binding of the antibody could affect the C site. While there is no evidence supporting this possibility, none of our findings can rule it out. Also, there is no result establishing the fact that the C and SP sites are distinct. Thus, it cannot be excluded that the same site on NuMA (or a NuMA-associated NHC protein) is used in binding to both chromosomes and poles. While this seems unlikely, since the 2E4-type antibody can abolish the C-site-mediated binding without blocking the SP site, further research will be required to investigate the possibility.

B. Evidence for the Model

The model is compatible with all the immunofluorescence and biochemical studies of chromosomes probed with NuMA-specific antibodies as cited above. The 2E4-induced dissociation from chromosomes of NuMA antigens and NuMA protein observed by the respective techniques are accounted for, as is the observed attachment of the antigen-Ab complex at the spindle poles. Other findings to be described now are also compatible with the model.

The transfer of antigen-Ab complex from chromosomes to poles could be readily detected in metaphase and anaphase cells from the changes in immunofluorescence distribution. As seen in Figure 4 and 5, the chromosomes and polar regions viewed from a point perpendicular to the pole-to-pole axis had characteristic sizes and shapes. The chromosome distribution at the spindle pole had a relatively large bar-like shape, while the polar distribution had a concave appearance with open sides facing each other. The change from one of these distributions to the other could be readily seen after incubation of metaphase cells with antibody 2E4, as it could also be when anaphase cells were similarly incubated. In late anaphase when chromosomes were nearly at the poles, the distributions were particularly striking when chromosomes were independently stained with propidium to simultaneously visualize both poles and chromosomes.[4] In this case a clear juxtaposition could be seen of

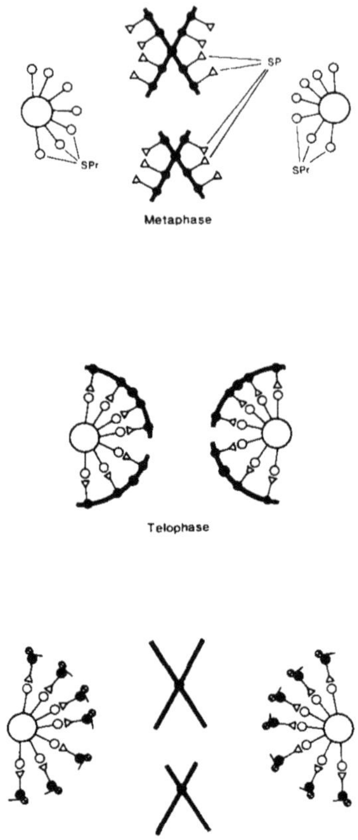

FIGURE 6. Schematic diagrams representing the NuMA protein during mitosis. The NuMA protein has two distinct binding sites — one a C site which normally is associated with chromosomes, and one an SP site which has affinity for receptors (SP_r) in the spindle pole region. The top diagram shows chromosomes at metaphase with associated NuMA proteins having C sites bound to the chromosomes but with SP sites open. The SP_r receptors at the spindle pole regions are also free. At telophase shown in the center diagram, the chromosomes have been drawn to the spindle poles and are attached to the nuclear organizing center via interactions between the SP sites and the SP_r receptors. The lower diagram illustrates the effects of 2E4-type monoclonal antibodies on chromosomes and spindle poles in metaphase cells. The antibody destabilizes the C site mediated attachment of NuMA to chromosomes, resulting in release of the Ag-Ab complex from chromosomes. Subsequently, the complex can attach to the poles via the SP site on NuMA.

the anaphase chromosomes with the fluorescent polar distribution due to reattached NuMA protein. Thus, the relative distribution of chromosomes and the apparent SP_r sites inferred from this kind of picture is as proposed in Figure 6.

While in late anaphase cells a clear displacement of the NuMA-associated polar structures and the closely associated anaphase chromosomes could be seen, when telophase cells (having well-defined intracellular bridges visualized by phase microscopy) were studied, no change in the fluorescent distribution of chromosomes was observed, even after prolonged incubation with antibody 2E4. The chromosomes maintained the bar-like shape (Figure 5) from the fluorescein-associated anti-NuMA antibodies, never acquiring the concave-disk appearance of the polar receptors. This was most clearly demonstrated in micrographs of fields showing both metaphase and telophase cells side by side. In the former fluorescence transfer to poles was complete, while in the latter only bar-shaped chromosomes were fluorescent. This result could be explained by the model (Figure 6), since at telophase the SP_r receptors and the NuMA C site loci on chromosomes should be in close proximity. Thus, a dissociation of the C site-mediated attachment by antibody 2E4 should result in negligible translocation of NuMA, because the protein would be independently held in position by its SP site.

C. Uncertainties and Further Considerations

While the evidence in support of the model of Figure 6 is not sufficient to establish it, the model does provide a basis for organizing future research designed to test its validity. It will be of some interest for example, to examine the effects of antibody 2E4 microinjected into living human cells. If such cells are blocked at telophase, it would be supportive of this model.

It should be emphasized that nothing indicates that NuMA protein is the only protein involved in attaching human chromosomes into the reforming nucleus; indeed it seems likely that other NHC proteins would also have roles in the process. Certainly in nonprimates other structurally distinct proteins would carry out related functions.

It is apparent that NuMA protein is not even indispensible for nuclear reassembly of human chromosomes, at least in certain specialized cases. In some human/hamster hybrid cell lines having defined complements of human chromosomes but missing a critical human chromosome, no NuMA is made (Figure 1); yet the human chromosomes after many generations of growth can be essentially as stable in the karyotype as hamster chromosomes.[2] However, it is well known that in the first series of mitosis following fusion of a human and hamster cell, a preferential loss of human chromosomes usually occurs.[7,8] This may reflect poor recognition of the SP sites on NuMA (or other similar human-specific proteins) with the hamster chromosome receptors in the nuclear reorganizing center. Stability of a human chromosome in a human/hamster hybrid may require replacement of NuMA protein (and other human-specific NHC proteins) with an equivalent hamster protein or, conversely, may require modification of receptors in the nuclear reorganizing center. Species-specific chromosomal markers such as NuMA may also provide an advantageous barrier on interspecies fertilization and on accidental integration of foreign chromosomes into the nuclei of both somatic and sex cells.

III. OTHER NuMA-LIKE NHC PROTEINS

Recent studies using both monoclonal and polyclonal antibodies specific for NHC proteins other than NuMA have shown that there are additional NHC proteins distributed in cells like NuMA.[9-12] That is, these proteins which seem to be exclusively in the nucleus of interphase cells become part of the mitotic apparatus in interphase cells. Some of these proteins are distributed in the polar regions of the spindle-like NuMA. At least one, the 0-2D5 antigen, has a distribution continuously along spindle microtubules like MAP proteins, but is bound only minimally to cytoplasmic microtubules in the manner of MAP proteins.[9] The 0-2D5 protein is clearly distinct from NuMA protein in that its molecular weight (about 200,000) is less and its distribution in the cell inferred from immunofluorescence studies is different. There is no polar distribution of this antigen in the mitotic spindle and it is concentrated in the midbody of telophase cells in complete distinction to NuMA.

The 0-2D5 antigen associated with this protein does not have antigenically related homologues in nonhuman eukaryotic cells. In this sense it is like NuMA. However, other antigenic sites on the protein appear to exist on a similar protein in other eukaryotes.[10] There is evidence that monoclonal antibodies specific for this protein, when microinjected into living cells, delay chromosome movement during mitosis. This protein may therefore have a role in karyokinesis rather than nuclear reassembly, although this is not yet certain.

A diverse group of autoantibodies having specificity for NHC proteins of the mitotic spindle have been identified in patients and normal subjects.[11] The distributions of some of these antigens at the spindle poles defined in immunofluorescence studies is indistinguishable from that of NuMA after its 2E4-induced translocation. Preliminary studies of these antigens using western blot analysis have indicated that many of these proteins are distinct from NuMA.[12] Thus, it seems that immunological tools are available for investigating other components of the putative nuclear organizing center.

ACKNOWLEDGMENT

This research was supported by research grants from the National Science Foundation (PCM-7921406) and the National Institutes of Health (GM-18243-11).

REFERENCES

1. **Lydersen, B. K., Kao, F.-T., and Pettijohn, D. E.,** Expression of genes coding for non-histone chromosomal proteins in human-Chinese hamster cell hybrids, *J. Biol. Chem.*, 255, 3002, 1980.
2. **Kao, F.-T., Hartz, J. A., Law, M. L., and Davidson, J. N.,** Isolation and chromosomal localization of unique DNA sequences from a human genomic library, *Proc. Natl. Acad. Sci. U.S.A.*, 79, 865, 1982.
3. **Lydersen, B. K. and Pettijohn, D. E.,** Human-specific nuclear protein that associates with the polar region of the mitotic apparatus: distribution in a human/hamster hybrid cell, *Cell*, 22, 489, 1980.
4. **Van Ness, J. B. and Pettijohn, D. E.,** Specific attachment sites for NuMA protein on metaphase chromosomes and mitotic spindle poles: possible role in nuclear reassembly, *J. Mol. Biol.*, 171, 175, 1983.
5. **Connolly, J., Kalnius, V., Cleveland, D., and Kirschner, M.,** Immunofluorescent staining of cytoplasmic and spindle microtubules in mouse fibroblasts with antibody to tau protein, *Proc. Natl. Acad. Sci. U.S.A.*, 74, 2437, 1977.
6. **Lockwood, A. H.,** Tubulin assembly protein: immunochemical and immunofluorescent studies of its function and distribution in microtubules and cultured cells, *Cell*, 13, 613, 1978.
7. **Weiss, M. C. and Green, H.,** Human-mouse hybrid cell lines containing partial compliments of human chromosomes and functioning human genes, *Proc. Natl. Acad. Sci. U.S.A.*, 58, 1104, 1967.
8. **Ringertz, N. R. and Savage, R. E.,** in *Cell Hybrids*, Academic Press, New York, 1976, 162.
9. **Izant, J. G., Weatherbee, J. A., and McIntosh, J. R.,** A microtubule-associated protein in the mitotic spindle and the interphase nucleus, *Nature (London)*, 295, 248, 1982.
10. **Izant, J. G., Weatherbee, J. A., and McIntosh, J. R.,** A microtubule-associated protein antigen unique to mitotic spindle microtubules in PtK. cells, *J. Cell Biol.*, 96, 424, 1983.
11. **McCarty, G. A., Fritzler, M. J., Erickson, H., and Pettijohn, D. E.,** A unique autoantibody to the mitotic spindle apparatus, *J. Cell Biol.*, 95, 353a, 1982.
12. **Price, C. M., McCarty, G. A., and Pettijohn, D. E.,** *Arthritis Rheum.*, in press.

Chapter 8

A RAPID PURIFICATION OF DNA TOPOISOMERASE I, A CHROMATIN-BOUND NONHISTONE PROTEIN, AND ITS INHIBITION BY HEPARIN

Toshiwo Andoh, Kazuyuki Kshii, Kinzo Nagasawa, and Nobuko Seno

TABLE OF CONTENTS

I. ABSTRACT

We have previously shown that DNA topoisomerase I is inhibited by heparin. Taking advantage of this enzyme-heparin interaction we developed a rapid method of purification of this enzyme by two-step column chromatography of heparin- and phenyl-Sepharose®. The enzyme was purified to near homogeneity with an overall yield of 60%.

In commercially available heparin preparations more than 50% of the molecules lack the ability to bind to antithrombin III (AT-III) and are thus devoid of anticoagulant activity. We found, however, that heparin fraction with low affinity for AT-III has a comparable inhibitory activity on DNA topoisomerase I as the fraction with high affinity for AT-III. This finding strongly suggests that heparin molecules contain a second biologically active sequence of oligosaccharide which binds to topoisomerase I and which is distinct from the well characterized AT-III binding oligosaccharide.

II. INTRODUCTION

In the past decade there has been substantial progress in understanding the way in which DNA is arranged in the eukaryotic nucleus. The most basic organization of DNA is the nucleosome, which had been studied extensively.[1] Nucleosomes appear to be arranged into ordered arrays of solenoids,[2,3] which in turn are further packaged into supercoiled loops of DNA called nucleoid.[4-7] We have previously shown that upon dissociation of proteins from the nucleoid by gently lysing nuclei in the presence of strong detergents such as sodium dodecyl sulfate (SDS), only a few nonhistone proteins (NHPs) remained firmly attached to DNA and further, that such highly deproteinized DNA still possessed the supercoiled state.[4,5] This could mean that the supercoiled state of DNA depends on the fixation of DNA loop ends by a limited number of NHPs. Such proteins with the presumed function of anchoring DNA have been amply shown in nucleoid structures such as nuclear matrix,[8-10] isolated metaphase chromosomes,[10,11] and DNA-protein complexes isolated from interphase nuclei,[4,5,12-14] and are being currently purified and characterized.[12-15]

Turning to the supercoiling and relaxation of closed circular DNA, two types of DNA topoisomerases (type I and type II) involved in such events have been characterized. A type-I topoisomerase known as ω protein was first discovered in *Escherichia coli* by Wang and Liu,[16] and similar enzymes were subsequently found in various organisms, mostly in chromatin-bound state.[17] Type-I enzymes catalyze various DNA topoisomerase reactions in vitro such as relaxation of superhelical twists, knotting and unknotting of single-stranded circular DNA, the linking of two single-stranded DNA rings of complementary sequences, and the catenation and decatenation of double-stranded circular DNA, provided that at least one partner is nicked.[18,19] All these type-I reactions are consistent with a possible mechanism in which the enzyme introduces a transient single-stranded break through which another strand passes.[17,20] In contrast to these type-I reactions, type-II topoisomerases catalyze similar DNA topoisomerization reactions through a transient double-stranded break and reunion.[21] The biological functions of type-I DNA topoisomerases have been a subject of speculation.[17,20] Although they were originally assumed to form a swivel which would facilitate replication or transcription, recent discoveries have suggested additional roles. The finding that the bacteriophage λ *int* protein is a type-I DNA topoisomerase[22] and recent isolation and characterization of an *E. coli* mutant lacking topoisomerase I[23] suggested that one of the possible roles of this enzyme may be in the recombinational processes. In order to establish the role played by this type-I enzyme in eukaryotes we have been attempting to find a specific inhibitor of this enzyme. In the course of the search we found heparin as a potent inhibitor of a mouse DNA topoisomerase I.[24] In this chapter we first describe a rapid method of purification of the enzyme to near homogeneity utilizing heparin-Sepharose® as the affinity substrate. In subsequent sections we show some characterization of this heparin inhibition of this enzyme.

III. MATERIALS AND METHODS

Cultured mouse mammary carcinoma cell line FM3A was used as the source of DNA topoisomerase I. Chromatin extract containing all the enzyme activity of the cells was prepared by the method of Germond et al.[25] and used as a starting material for the purification of the enzyme. A unit of enzyme is defined as the activity that converts 1 μg of col E1 DNA I to fully relaxed form Ir after incubation at 37°C for 15 min. Further details of enzyme assay were described in a previous article.[24] SDS-PAGE (polyacrylamide gel electrophoresis) of proteins was performed by the prescription of Laemmli[26] and Blattler et al.[27] Recovery of the enzyme from SDS-polyacrylamide gel was performed according to the method of Hager and Burgess.[28] Heparin-Sepharose® was prepared as described.[29] Phenyl-Sepharose® was purchased from Pharmacia Fine Chemicals, Uppsala, Sweden. Hog intestinal heparin was obtained from Cohelfred Laboratories, Chicago, and purified free of dermatan sulfate by the method of Rodén et al.[30] Affinity-chromatography of heparin on antithrombin III-Sepharose® was performed essentially as described by Laurent et al.[31] Gel-filtration chromatography of heparin samples was performed on Ultrogel® AcA44 agarose-acrylamide gel obtained from LKB-Produkter AB, Broma, Sweden. Heparin samples were electrophoresed on cellulose acetate membrane by the methods of Seno et al.[32] and Wessler.[33] Anticoagulant activity of heparin was assayed by the whole-blood assay method of the USP,[34] and the activity was expressed as units per milligram. Chemical analyses of heparin samples were performed as described.[35]

IV. RESULTS

A. Purification of Topoisomerase I

In a previous report we described a potent inhibition of topoisomerase I by heparin.[24] Taking advantage of this interaction of heparin with the enzyme we attempted to purify the topoisomerase by affinity chromatography on heparin-Sepharose®. Figure 1A shows an elution profile of the enzyme from the column. Results shown in insets a and b establish that the topoisomerase I is completely free of any nuclease contamination. By this single step of purification fifty- to hundredfold purification was attained. As shown in Figure 1B, a prominent band with apparent molecular weight of 96,000 (p96) seemed to be intimately associated with enzymatic activity. In order to establish the identity of the protein p96 with the enzyme we followed two-step identification procedures consisting of extraction of a candidate protein band from SDS-polyacrylamide gel followed by denaturation by high concentration of guanidine and renaturation by subsequent dilution. Figures 2 and 3 show the results of these experiments. Results clearly show that the protein extracted from the major band p96 (band B, Figure 3) was the enzyme. Weak enzymatic activity demonstrated by the protein eluted from the minor band A could be due to the contamination of the band-B protein, or the band-A protein itself being a modified form of the p96 enzyme. The heparin-Sepharose® enzyme fraction pooled as indicated in Figure 1A was further purified by a hydrophobic interaction chromatography on phenyl-Sepharose®. Results shown in Figure 4 clearly demonstrated that further purification of the enzyme was achieved, but that a minor contaminating band was still associated with the major enzyme band. However, it appears that about 70% purification was achieved. Overall yield and specific activities of the enzyme through the purification steps are summarized in Table 1. Yield was 59% with approximately 8000-fold purification as compared with whole cell lysate.

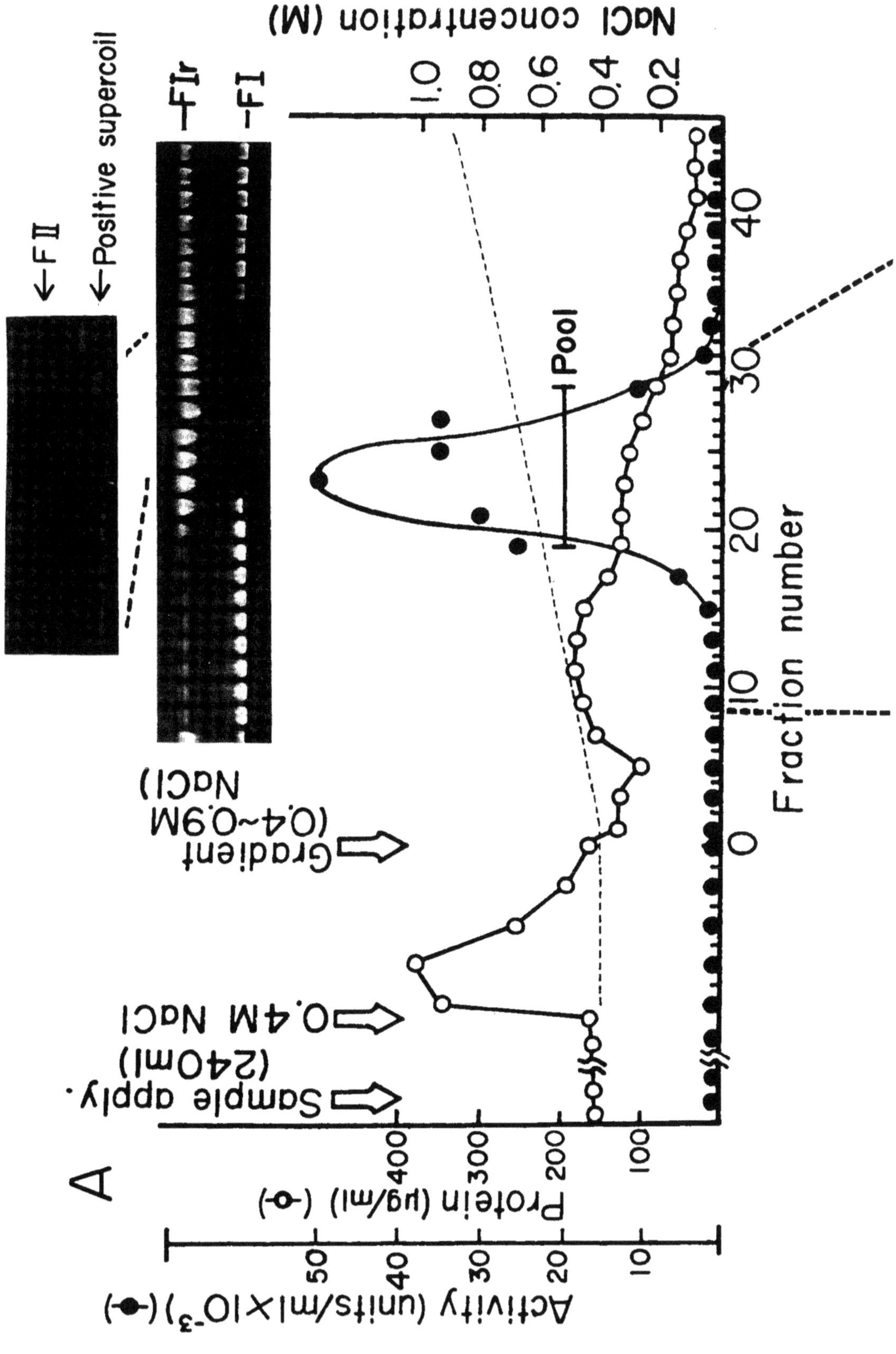
A
Activity (units/ml×10⁻³) (-●-)
10
20
30
40
50
Protein (µg/ml) (-o-)
100
200
300
400
Sample apply. (240ml)
0.4M NaCl
Gradient (0.4~0.9M NaCl)
Pool
Fraction number
0
10
20
30
40
NaCl concentration (M)
0.2
0.4
0.6
0.8
1.0
-FI
-FIr
←FII
←Positive supercoil

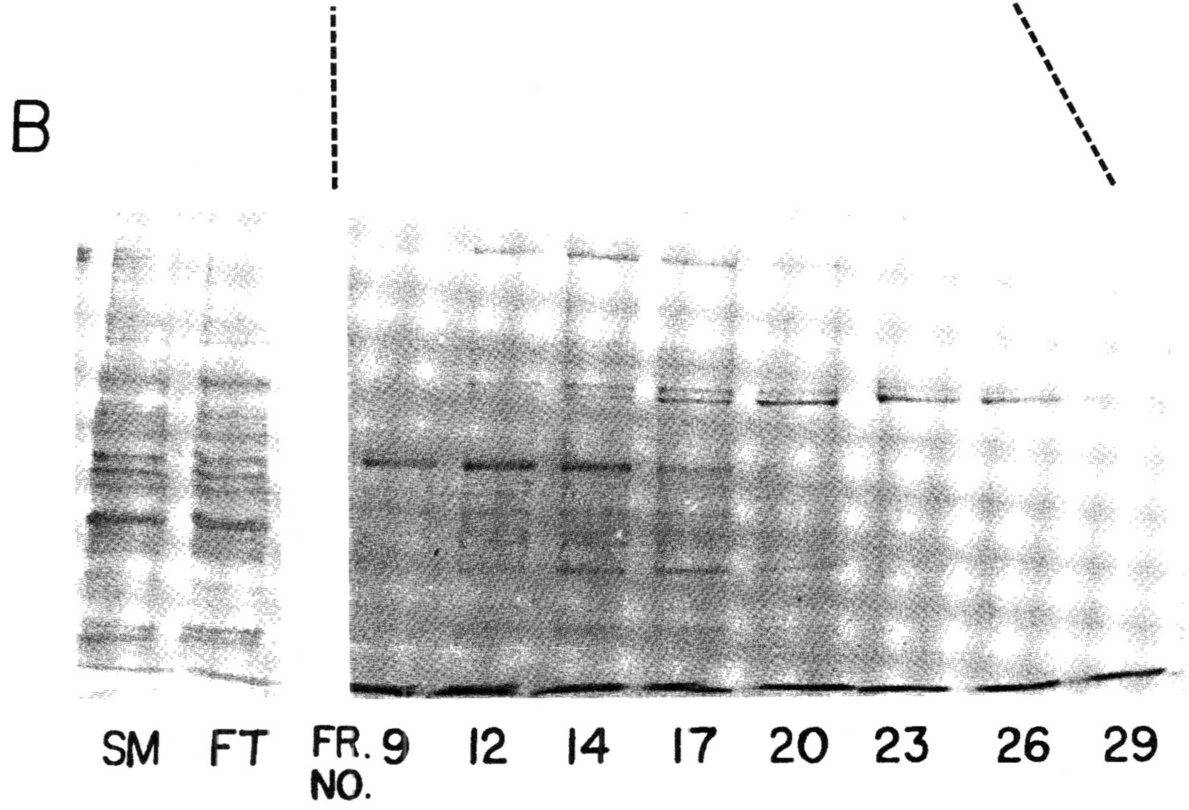

FIGURE 1. Heparin-Sepharose® affinity chromatography of topoisomerase I. (A) Chromatin extract was applied on heparin-Sepharose® column. After washing with 0.4 *M* NaCl in elution buffer the column was developed with a linear gradient from 0.4 to 0.9 *M* NaCl. Protein concentration and enzyme activity were assayed. Inset a shows the topoisomerase I assays by agarose gel electrophoresis. Form I is converted to form Ir with active fractions. Inset b shows the agarose gel electrophoresis of the same products in agarose gel containing chloroquine phosphate. Form Ir was converted to positively supercoiled DNA obviating the possibility of nuclease contamination. (B) Protein compositions of enzymatically active fractions were determined by SDS-PAGE.

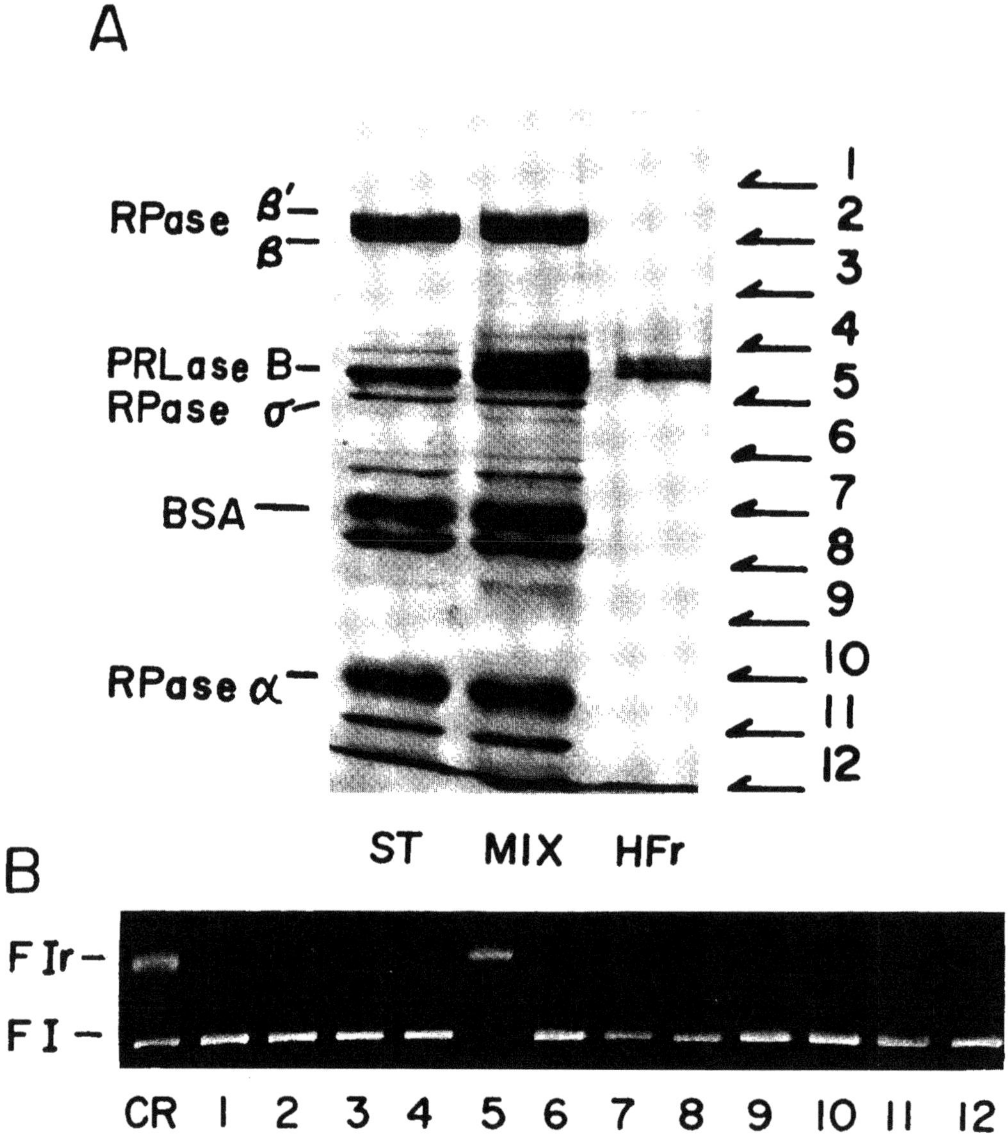

FIGURE 2. Identification of topoisomerase I on SDS-polyacrylamide gels. (A) Active fraction (SMPL) from heparin-Sepharose® was electrophoresed alone or as mixture (MIX) with standard proteins (ST) on SDS-polyacrylamide gel. Sample lane was cut into 12 segments as indicated and each segment was extracted and assayed for topoisomerase I activity as described in Section III. (B) Agarose gel pattern of DNA after incubation with each extract. Segment 5 shows the enzymatic activity.

B. Inhibition of Topoisomerase I by Heparin

Figure 5 shows a typical profile of topoisomerase I inhibition by heparin, with the 50% inhibition dose being about 1.75 μg/mℓ. This inhibition seems to be specific for heparin, as various other acid mucopolysaccharides including heparan sulfate, dermatan sulfate, chondroitin sulfates A and C, and hyaluronic acid were all without effect.[24] Heparin is a highly sulfated mucopolysaccharide which plays a well-characterized role in inhibition of the proteolytic cascade of blood coagulation through an interaction with AT-III.[36] In commercially available heparin preparations more than 50% of the molecules lack this ability and are essentially devoid of anticoagulant activity.[37] The AT-III binding sequence of the

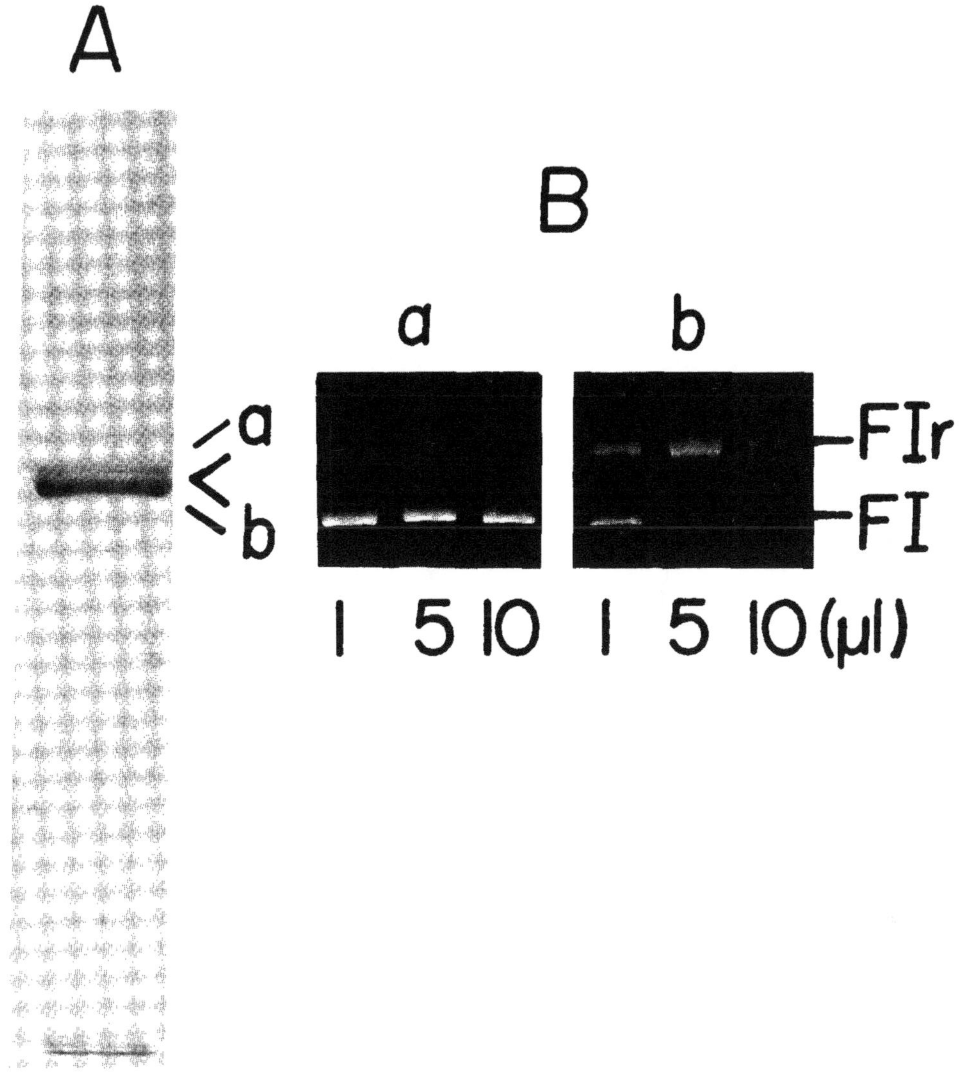

FIGURE 3. Identification of topoisomerase I on SDS-polyacrylamide gels. (A) Active fraction from heparin-Sepharose® was electrophoresed on SDS-polyacrylamide gel. The major (a) and minor bands (b) were separately extracted and assayed for topoisomerase activity.

active heparin molecules was investigated by isolating oligosaccharides with high affinity for AT-III after partial depolymerization of heparin. A tentative structure of the smallest consensus oligosaccharide with the following structure of hexasaccharide was proposed: L-iduronic acid-*N*-acetyl-D-glucosamine-D-glucuronic acid-*N*-sulfo-3-*O*-sulfo-D-glucosamine-2-*O*-sulfo-L-iduronic acid-*N*-sulfo-D-glucosamine.[38-40] The heparin species with low or no affinity for AT-III lacked this sequence.

As shown in Figure 6, we fractionated a heparin preparation according to its affinity for AT-III into three fractions: no affinity (NA)-, low affinity (LA)-, and high affinity (HA)-, the latter being subdivided into subfractions S-I, S-II, and S-III. We then asked whether these heparin fractions with varying degrees of affinity for AT-III inhibit topoisomerase I equally well or differentially in accordance with its affinity for AT-III. Before we went into this problem, however, some of the physicochemical characteristics of these fractions were investigated. Each of these heparin fractions was examined for its molecular size by passing

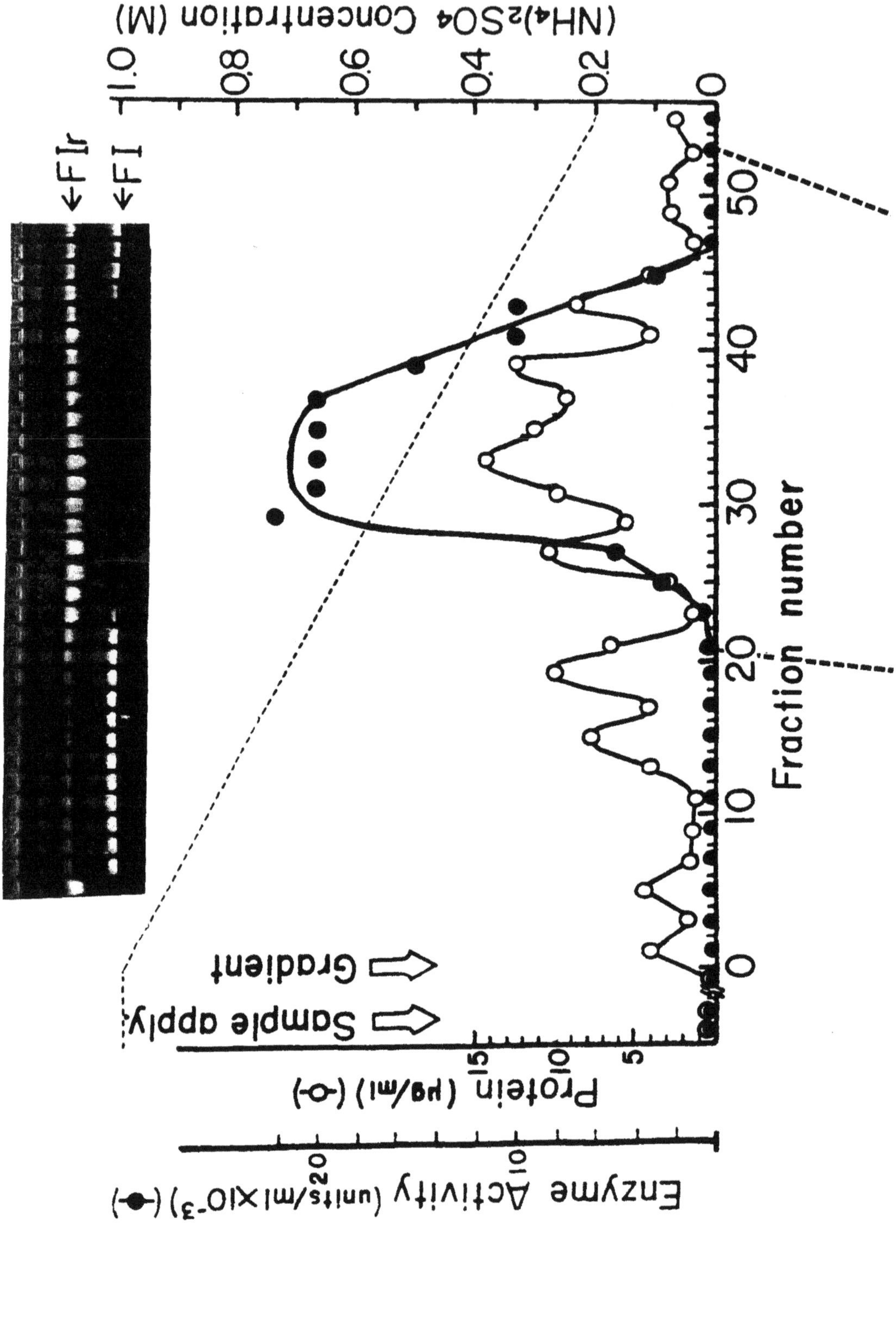
(NH4)2SO4 Concentration (M)
←FIr
←FI
Sample apply
Gradient
Protein (μg/ml) (-○-)
Enzyme Activity (units/ml×10-3) (-●-)
Fraction number

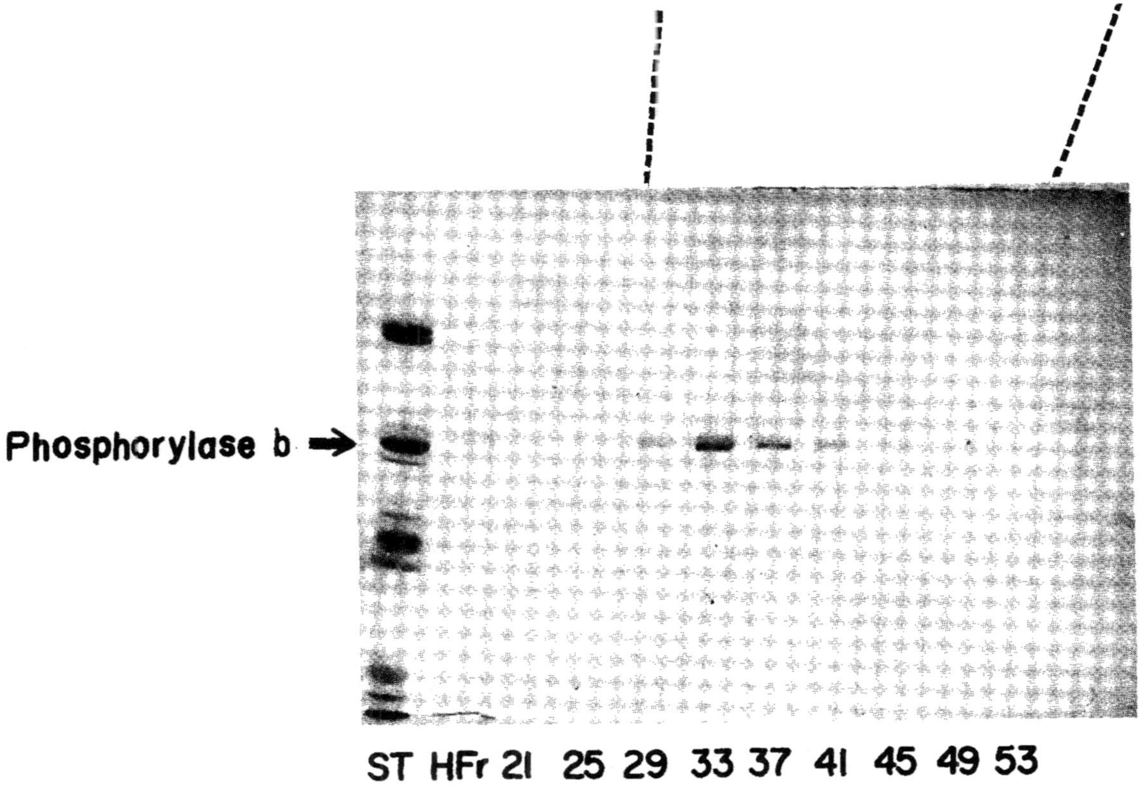

FIGURE 4. Phenyl-Sepharose® chromatography of topoisomerase I. (A) Enzymatically active fractions were pooled as indicated in Figure 3A, made 1 M $(NH_4)_2SO_4$ with addition of equal volume of 2 M $(NH_4)_2SO_4$ in elution buffer, and applied on phenyl-Sepharose® column. After washing with starting buffer the proteins were eluted with a linear gradient from 1.0 to 0.2 M $(NH_4)_2SO_4$. Fractions were assayed for protein and enzyme activity. Inset shows agarose gel electrophoresis of DNA after incubation with various fractions. (B) The determination of protein compositions of enzymatically active fractions by SDS-PAGE is shown.

Table 1
PURIFICATION OF DNA TOPOISOMERASE I FROM FM3A CELLS

Purification step	Total proteins (mg)	Total activity (units × 10^{-3})	Spec. act. (units/mg × 10^{-2})	Yield (%)
Whole cell lysate	1,500	222	1.48	(100)
Chromatin extract	36	240	66.6	108
Heparin-Sepharose® fraction	0.59	175	2,966	79
Phenyl-Sepharose® fraction	0.11	130	11,820	59

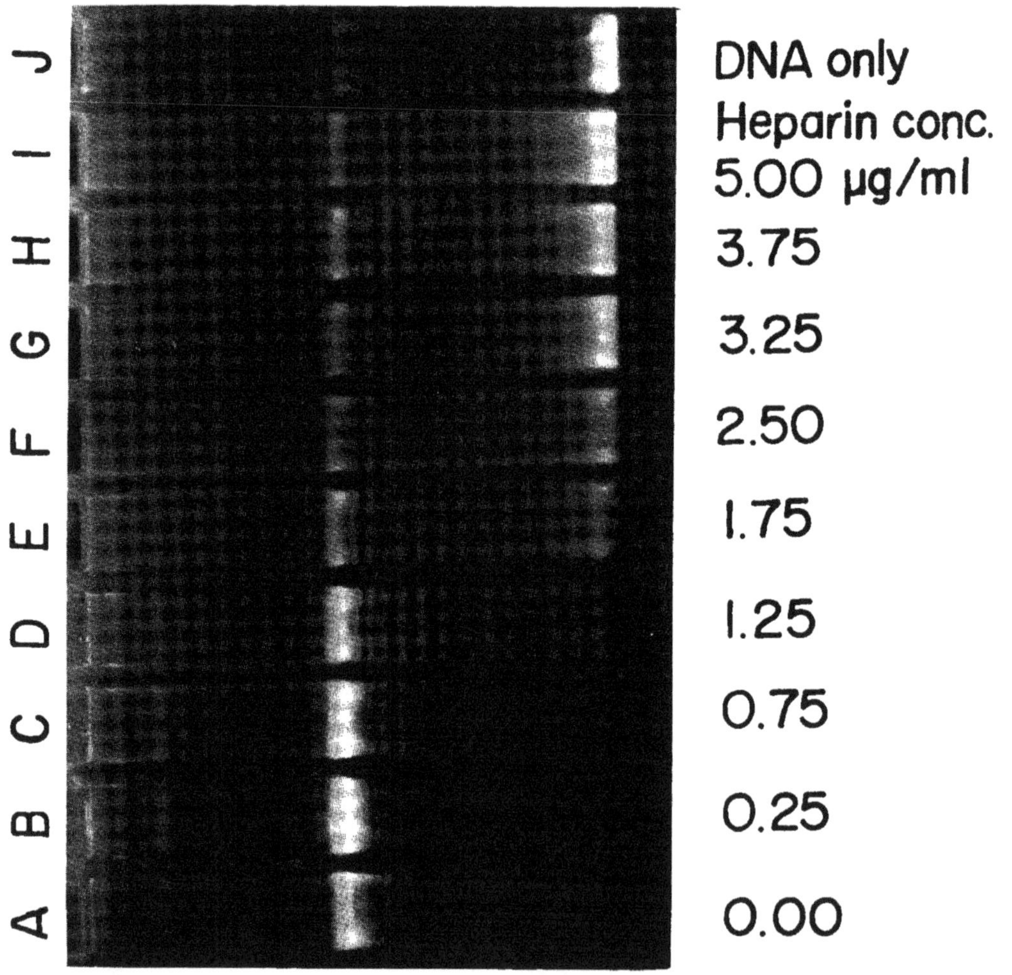

FIGURE 5. Inhibition of topoisomerase I by heparin. A crude topoisomerase I preparation (chromatin extract) was incubated with various amounts of heparin as indicated and the products were analyzed by agarose gel electrophoresis.

through Ultrogel® AcA44 (Figure 7). NA fraction contained the smallest molecular weight species; others were almost the same, except for the S-III fraction which contained the largest of all. Figure 8 shows an electrophoretic profile of these fractions, indicating that all these molecules were essentially equally charged. Table 2 shows some chemical analyses of these

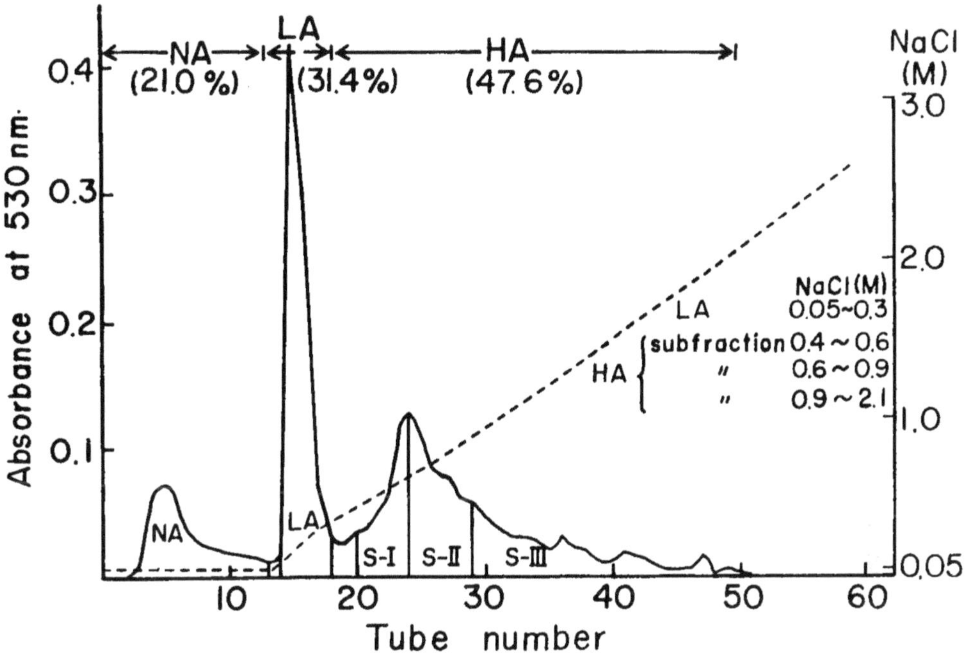

FIGURE 6. Fractionation of heparin by antithrombin III-Sepharose® chromatography. Heparin was applied on a column of AT-III-Sepharose®. After washing heparin was eluted with a linear gradient from 0.05 to 3.0 *M* NaCl.

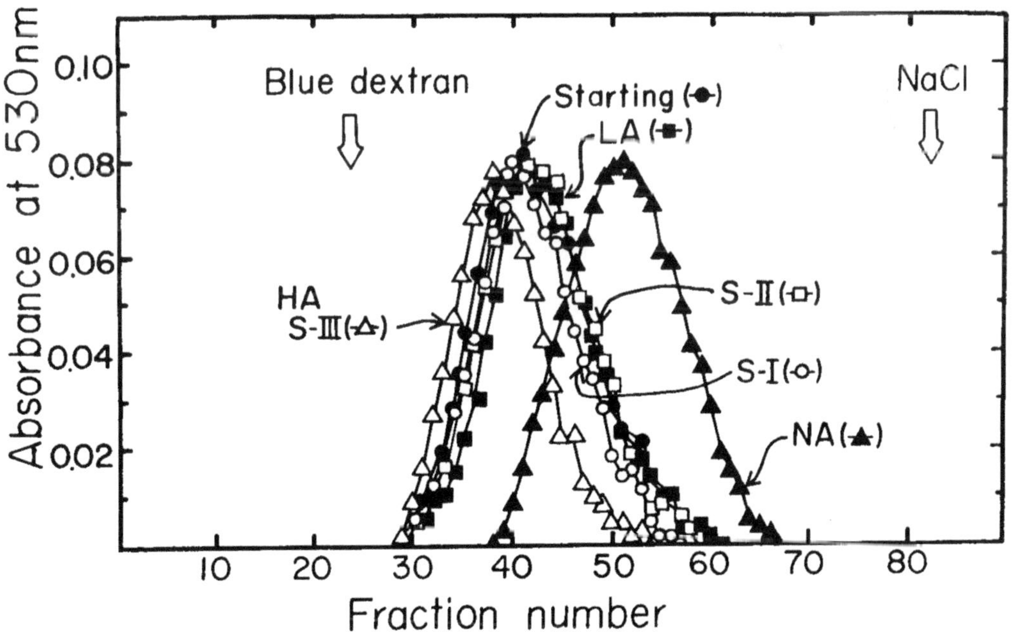

FIGURE 7. Gel-filtration chromatography of heparin fractions on Ultrogel® AcA44. Heparin fractions from AT-III-Sepharose® chromatography were filtered through Ultrogel® AcA44 column. Uronic acid content was determined on each fraction.

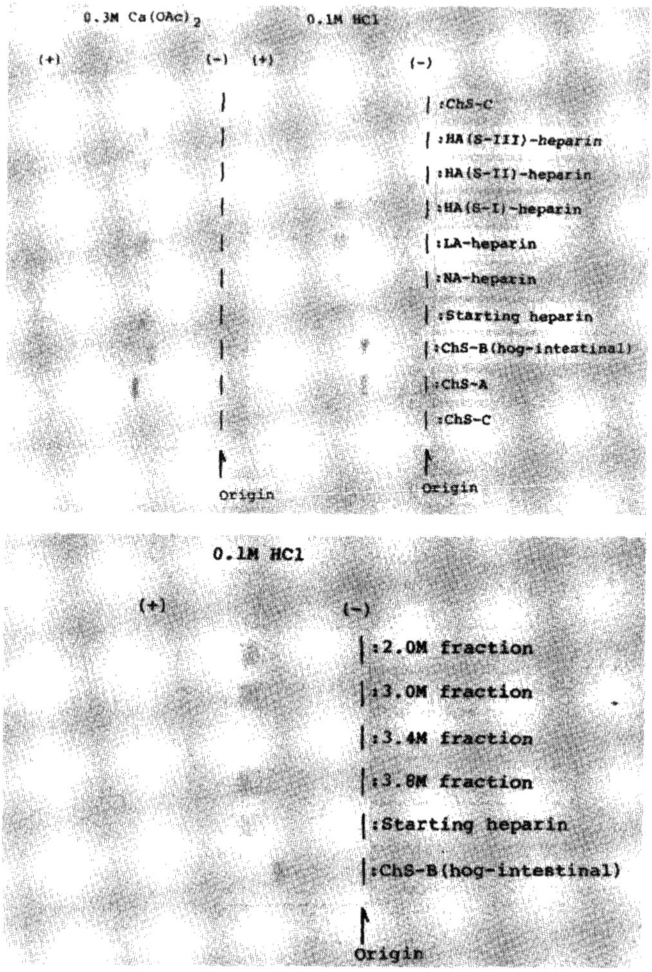

FIGURE 8. Electrophoresis of heparin fractions on cellulose acetate membrane. Heparin fractions from AT-III-Sepharose® chromatography were electrophoresed on cellulose acetate membranes in three different buffer systems and visualized with toluidine blue.

heparin fractions. No essential differences in total S, N-S, N-Ac, and free NH_2 groups were observed among these fractions. With all these properties in mind we examined the inhibitory activities of the heparin fractions on DNA topoisomerase I in comparison with their anticoagulant activities. As clearly seen in Table 3, NA and LA fractions are completely devoid of anticoagulant activity, and this activity was confined to HA subfractions. However, with respect to topoisomerase I inhibition all fractions except NA were active. It is to be noted that while LA fraction is inactive as anticoagulant, its inhibition of topoisomerase I activity was almost as strong as HA fractions. This peculiar nature of heparin species in LA fraction was confirmed by the following experiments. As pointed out above, HA subfractions were somewhat larger in molecular weight (Mr) than LA fraction. So, lower Mr fraction within the S-II fraction (subsequently referred to as modified or mod. S-II) was pooled and compared with LA fraction with respect to the Mr and the inhibitory activities on topoisomerase I. As shown in Figure 9 and in the inset table, modified S-II, now slightly less in Mr than LA, exhibited the same inhibitory activity as LA fraction. Taking into account the finding that the LA fraction is completely devoid of anticoagulant activity, these data strongly suggest

Table 2
CHEMICAL ANALYSES OF HEPARIN FRACTIONS

Heparin fraction	Total S[a] (% mol/UA[e])	N-S[b] (% mol/UA)	N-Ac[c] (% mol/UA)	TNP-NH_2^d
Starting materials (SM)	12.20 2.27	4.40 0.82	0.85 0.12	0.3300 (100)
Nonadsorbed fraction (NA)	10.94 1.91	4.26 0.74	0.61 0.08	0.4450 (135)
Low-affinity fraction (LA)	11.54 2.08	4.26 0.76	0.56 0.07	0.3790 (115)
High-affinity fraction (HA)				
Subfraction S-I	12.48 2.36	4.32 0.81	0.69 0.10	0.3230 (98)
Subfraction S-II	12.35 2.32	4.48 0.84	0.97 0.14	0.2624 (80)
Subfraction S-III	12.60 2.40	4.28 0.82	1.00 0.14	0.3203 (97)

[a] Total S includes O-sulfate and N-sulfate.
[b] N-S is N-sulfate only.
[c] N-Ac is N-acetyl group.
[d] TNP-NH_2 is free NH_2 groups reactive with trinitrofluorobenzene.
[e] Uronic acid.

Table 3
INHIBITORY ACTIVITIES OF HEPARIN FRACTIONS FROM AT-III-SEPHAROSE® AFFINITY CHROMATOGRAPHY

Heparin fraction	IC_{50}[a] (μg/mℓ)	USP[b] (unit/mg)	Kav[c]
Starting materials (SM)	1.75	167	0.34
Nonadsorbed fraction (NA)	100.00	2	0.54
Low-affinity fraction (LA)	3.75	7	0.34
High-affinity fraction (HA)			
Subfraction S-I	2.25	214	0.32
Subfraction S-II	2.50	244	0.34
Subfraction S-III	1.25	299	0.28

[a] IC_{50}: concentration giving 50% inhibition as judged from densitometer tracing.
[b] USP: Blood coagulation inhibition units according to USP.
[c] Kav: calculated from elution position from Ultrogel® filtration.

that the topoisomerase-binding sequence of oligosaccharide is quite distinct from the well-characterized AT-III binding oligosaccharide and, further, that this topoisomerase-binding sequence is found equally frequently among LA and HA subfractions.

V. DISCUSSION

Topoisomerase I is ubiquitous in nature and its enzymological properties have been extensively studied. However, its biological role in DNA metabolism remains a subject of speculation. From the enzymological properties it is assumed to form a swivel through which replication, transcription, and recombination is facilitated. In order to approach this problem we set out to search for a specific inhibitor of this enzyme which would unequivocally establish the role(s) played by this enzyme. In the course of the search we found heparin

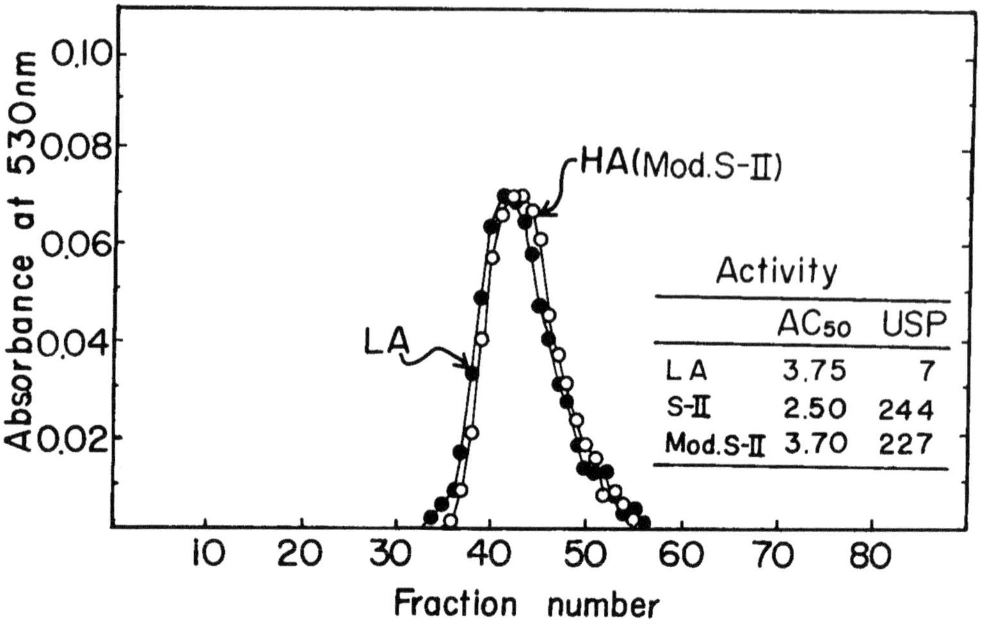

FIGURE 9. Gel-filtration chromatography of heparin fractions on Ultrogel® AcA44. Modified S-II and LA fractions were separately filtered through Ultrogel® AcA44 column. Uronic acid content was determined. Inset table shows anticoagulant activity and inhibitory activity on topoisomerase I of modified S-II and LA fractions.

as a potent inhibitor of the enzyme. Heparin has long been known as an anticoagulating agent in the blood-coagulation system. In addition it has been shown to inhibit the activity of several other mammalian enzymes including DNA-dependent RNA polymerase,[41,42] DNA polymerases and reverse transcriptase,[43] casein kinase II,[44] and initiation and elongation factors of protein synthesis.[45,46]

Taking advantage of the interaction of heparin with topoisomerase I we developed a rapid method purification of the enzyme. By two-step column chromatography on heparin-Sepharose® followed by phenyl-Sepharose® the enzyme was purified to near homogeneity. In the next section we established that the heparin inhibition of topoisomerase I was due entirely to different oligosaccharide structure from the well-established hexasaccharide sequence active in antithrombin binding in the proteolytic cascade of blood coagulation. The structure of this topoisomerase interacting oligosaccharide sequence is not known at present, but this sequence seems to be distributed evenly among heparin fractions with varying anticoagulant activities.

Heparin is a naturally occurring highly sulfated mucopolysaccharide, and has recently been shown to have a wide variety of biological activities. Intravenous injection of heparin releases two lipases into the circulating blood which act on lipoproteins in the plasma.[47] Heparin facilitates the induction of luteinizing hormone receptors by follicle stimulating hormone (FSH) in granulosa cells.[48] Phagocytosis by a murine macrophage line is dependent on combined action of fibronectin and heparin.[49] Heparin is also suspected to play an important role in regulated angiogenesis by stimulating migration of capillary endothelial cells on one hand[50] and by inhibiting the growth of vascular smooth muscle cells on the other.[51] Thus, heparin could be regarded a natural biomodulator in various aspects of biological processes. Heparin inhibition of a variety of enzymes is to be integrated in this concept.

REFERENCES

1. **Felsenfeld, G.,** Chromatin, *Nature (London)*, 271, 115, 1978.
2. **Finch, J. T. and Klug, A.,** Solenoidal model for superstructure in chromatin, *Proc. Natl. Acad. Sci. U.S.A.*, 73, 1897, 1976.
3. **Worcel, A. and Benyajati, C.,** Higher order coiling of DNA in chromatin, *Cell*, 12, 83, 1977.
4. **Ide, T., Nakane, M., Anzai, K., and Andoh, T.,** Supercoiled DNA folded by nonhistone proteins in cultured mammalian cells, *Nature (London)*, 258, 445, 1975.
5. **Nakane, M., Ide, T., Anzai, K., Ohara, S., and Andoh, T.,** Supercoiled DNA folded by nonhistone proteins in cultured mouse carcinoma cells, *J. Biochem.*, 84, 145, 1978.
6. **Cook, P. R. Brazell, I. A., and Jost, E.,** Characterization of nuclear structures containing superhelical DNA, *J. Cell Sci.*, 22, 303, 1976.
7. **Benyajati, C. and Worcel, A.,** Isolation, characterization and structure of the folded interphase genome of Drosophila melanogaster, *Cell*, 9, 393, 1976.
8. **Hodge, L. D., Mancini, P., Davis, F. M., and Heywood, P.,** Nuclear matrix of HeLa S_3 cells. Polypeptide composition during adenovirus infection and in phases of the cell cycle, *J. Cell Biol.*, 72, 194, 1977.
9. **Berezney, R. and Coffey, D. S.,** Nuclear matrix. Isolation and characterization of a framework structure from rat liver nuclei, *J. Cell Biol.*, 73, 616, 1977.
10. **Detke, S. and Keller, J. M.,** Comparison of the proteins present in HeLa cell interphase nucleoskeltons and metaphase chromosome scaffolds, *J. Biol. Chem.*, 257, 3905, 1982.
11. **Adolph, K. W., Cheng, S. M., and Laemmli, U. K.,** Role of nonhistone proteins in metaphase chromosome structure, *Cell*, 12, 805, 1977.
12. **Krauth, W. and Werner, D.,** Analysis of the most tightly bound proteins in eukaryotic DNA, *Biochem. Biophys. Acta*, 564, 390, 1979.
13. **Capesius, I., Krauth, W., and Werner, D.,** Proteinase K-resistant and alkali-stably bound proteins in higher plant DNA, *FEBS Lett.*, 110, 184, 1980.
14. **Spieß, E., Neuer, B., and Werner, D.,** Isolation and visualization of alkali stable protein/DNA complexes, *Biochem. Biophys. Res. Commun.*, 104, 548, 1982.
15. **Ohara, S., Anzai, K., and Andoh, T.,** Purification of folding proteins of DNA from cultured mouse mammary carcinoma cells, *J. Biochem.*, 92, 315, 1982.
16. **Wang, J. C.,** Interaction between DNA and an *Escherichia coli* protein ω, *J. Mol. Biol.*, 55, 523, 1971.
17. **Wang, J. C. and Liu, L. F.,** DNA topoisomerases: enzymes that catalyze the concerted breaking and rejoining of DNA backbone bonds, in *Molecular Genetics, Part III, Chromosome Structure*, Taylor, H., Ed., Academic Press, New York, 1979, 65.
18. **Tse, Y. C. and Wang, J. C.,** *E. coli* and *M. luteus* DNA topoisomerase I can catalyze catenation or decatenation of double-stranded DNA rings, *Cell*, 22, 269, 1980.
19. **Brown, P. O. and Cozzarelli, N. R.,** Catenation and knotting of duplex DNA by type I topoisomerases, a mechanistic parallel with type 2 topoisomerases, *Proc. Natl. Acad. Sci. U.S.A.*, 78, 843, 1981.
20. **Champoux, J. J.,** Proteins that affect DNA conformation, in *Annu. Rev. Biochem.*, 47, 449, 1979.
21. **Cozzarelli, N. R.,** DNA topoisomerases, *Cell*, 22, 327, 1980.
22. **Kikuchi, Y. and Nash, H. A.,** Nicking-closing activity associated with bacteriophage λ *int* gene product, *Proc. Natl. Acad. Sci. U.S.A.*, 76, 3760, 1979.
23. **Sternglanz, R., DiNardo, S., Voelkel, K. A., Nishimura, Y., Hirota, Y., Becherer, K., Zumstein, L., and Wang, J. C.,** Mutations in the gene coding for *Escherichia coli* DNA topoisomerase I affect transcription and translation, *Proc. Natl. Acad. Sci. U.S.A.*, 78, 2747, 1981.
24. **Ishii, K., Katase, A., Andoh, T., and Seno, N.,** Inhibition of topoisomerase I by heparin, *Biochem. Biophys. Res. Commun.*, 104, 541, 1982.
25. **Germond, J. E., Hirt, B., Oudet, P., Gross-Bellard, M., and Chambon, P.,** Folding of the DNA double helix in chromatin-like structures from simian virus 40, *Proc. Natl. Acad. Sci. U.S.A.*, 72, 1843, 1975.
26. **Laemmli, U. K.,** Cleavage of structural proteins during the assembly of the head of bacteriophage T4, *Nature (London)*, 227, 680, 1970.
27. **Blattler, D. P., Garner, F., van Slyke, K., and Bradley, A.,** Quantitative electrophoresis in polyacrylamide gels of 2-40%, *J. Chromatogr.*, 64, 147, 1972.
28. **Hager, D. A. and Burgess, R. R.,** Elution of proteins from sodium dodecyl sulfate-polyacrylamide gels, removal of sodium dodecyl sulfate, and renaturation of enzymatic activity: results with Sigma subunit of *Escherichia coli* RNA polymerase, wheat germ DNA topoisomerase, and other enzymes, *Anal. Biochem.*, 109, 76, 1980.
29. **Funahashi, M., Matsumoto, I., and Seno, N.,** Preparation of three types of heparin-Sepharose and their binding activities to thrombin and antithrombin III, *Anal. Biochem.*, 126, 414, 1982.

30. **Rodén, L., Baker, J., Cifonelli, J. A., and Mathews, M. B.,** Isolation and characterization of connective tissue polysaccharides, *Meth. Enzymol.*, 28, 73, 1972.
31. **Laurent, T. C., Tengblad, A., Thunberg, L., Höök, M., and Lindahl, U.,** The molecular-weight-dependence of the anticoagulant activity of heparin, *Biochem. J.*, 175, 691, 1978.
32. **Seno, N., Anno, K., Kondo, K., Nagase, S., and Saito, S.,** Improved method for electrophoretic separation and rapid quantitation of isomeric chondroitin sulfates on cellulose acetate strips, *Anal. Biochem.*, 37, 197, 1970.
33. **Wessler, E.,** Electrophoresis of acidic glycosaminoglycan in hydrochloric acid: a micromethod for sulfate determination, *Anal. Biochem.*, 41, 67, 1971.
34. *The United States Pharmacopoeia*, 18th Rev., U.S. Pharmacopoeial Convention, Inc., New York, 1970, 629.
35. **Ogamo, A., Uchiyama, H., and Nagasawa, K.,** Separation of heparin into fractions with different anticoagulant activity by hydrophobic interaction chromatography, *Biochim. Biophys. Acta*, 626, 477, 1980.
36. **Rosenberg, R. D.,** Chemistry of the hemostatic mechanism and its relationship to the action of heparin, *Fed. Proc. Fed. Am. Soc. Exp. Biol.*, 36, 10, 1977.
37. **Lam, L. H., Silbert, J. E., and Rosenberg, R. D.** The separation of active and inactive forms of heparin, *Biochem. Biophys. Res. Commun.*, 69, 570, 1976.
38. **Choay, J., Lormeau, J.-C., Petitou, M., Sinay, P., and Fareed, J.,** Structural studies on a biologically active hexasaccharide obtained from heparin, *Ann. N. Y. Acad. Sci.*, 370, 644, 1981.
39. **Casu, B., Oreste, P., Torri, G., and Zoppetti, G.,** The structure of heparin oligosaccharide fragments with high anti-(factor Xa) activity containing the minimal antithrombin III-binding sequence, *Biochem. J.*, 197, 599, 1981.
40. **Thunberg, L., Bäckström, G., and Lindahl, U.,** Further characterization of the antithrombin-binding sequence in heparin, *Carbohydrate Res.*, 100, 393, 1982.
41. **Walter, G., Zillig, W., Palm, P., and Fuchs, E.,** Initiation of DNA-dependent RNA synthesis and the effect of heparin on RNA polymerase, *Eur. J. Biochem.*, 3, 194, 1967.
42. **Pfeffer, S. R., Stahl, S. J., and Chamberlin, M. J.,** Binding of *Escherichia coli* RNA polymerase to T7 DNA, *J. Biol. Chem.*, 252, 5403, 1977.
43. **DiCioccio, R. A. and Srivastava, B. I. S.,** Inhibition of deoxynucleotide-polymerizing enzyme activities of human cells and of simian sarcoma virus by heparin, *Cancer Res.*, 38, 2401, 1978.
44. **Hathaway, G. M., Lubben, T. H., and Traugh, J. A.,** Inhibition of casein kinase II by heparin, *J. Biol. Chem.*, 255, 8038, 1980.
45. **Waldman, A. A. and Goldstein, J.,** Inhibition by heparin of globin messenger ribonucleic acid translation in a mammalian cell-free system, *Biochemistry*, 12, 2706, 1973.
46. **Slobin, L. I.,** The inhibition of elongation factor 1 activity by heparin, *Biochem. Biophys. Res. Commun.*, 73, 539, 1976.
47. **Olivecrona, T., Bengtsson, G., Marklund, S.-E., Lindahl, U., and Höök, M.,** Heparin-lipoprotein lipase interactions, *Fed. Proc. Fed. Am. Soc. Exp. Biol.*, 36, 60, 1977.
48. **Nimrod, A. and Lindner, H. R.,** Heparin facilitates the induction of LH receptores by FSH in granulosa cells cultured in serum-enriched medium, *FEBS Lett.*, 119, 155, 1980.
49. **van de Water, L., III, Schroeder, S., Crenshaw, E. B., III, and Hynes, R. O.,** Phagocytosis of gelatin-latex particles by a murine macrophage line is dependent on fibronectin and heparin, *J. Cell Biol.*, 90, 32, 1981.
50. **Azizkhan, R. G., Azizkhan, J. C., Zetter, B. R., and Folkman, J.,** Mast cell heparin stimulates migration of capillery endothelial cells *in vitro*, *J. Exp. Med.*, 152, 931, 1980.
51. **Castellot, J. J., Jr., Addonizio, M. L., Rosenberg, R., and Karnovsky, M. J.,** Cultured endothelial cells produce heparinlike inhibitor of smooth muscle cell growth, *J. Cell Biol.*, 90, 372, 1981.

Chapter 9

CURRENT RESEARCH ON EUKARYOTIC DNA METHYLTRANSFERASES

Jean-Numa Lapeyre, Mathuros Ruchirawat, and Frederick F. Becker

TABLE OF CONTENTS

I. INTRODUCTION

The genomes of eukaryotes contain, in addition to the four nucleobases, a single minor modified base 5-methylcytosine (m^5C) that occurs in differing amounts in the DNAs of organisms and their tissues.[54,95,96,106] This modified base arises from a transmethylation reaction catalyzed by DNA methyltransferase (DNA methylase, DMase) utilizing *S*-adenosylmethionine (SAM) as the methyl donor. The reaction depends on a steady supply of SAM synthesized by SAM synthetase isozymes, and elimination of a by-product of the reaction, *S*-adenosylhomocysteine (SAH), a competitive inhibitor of the reaction ($K_i \simeq 1 \mu M$) by SAH-hydrolase isozymes, all of which are found in close proximity with methylases in the rodent liver.[23]

In mammalian cells, enzymatic methylation and replication of DNA are closely coupled.[1,53] Methylation of *de novo* synthesized progeny strands takes place within minutes of replication but may continue for several hours.[1,2,53] In murine Ehrlich's ascites cells, methylation of new replicated DNA in 9S Okazaki fragments is not detected until ligation of these molecules to 30-60S intermediates.[2,35] Similarly, damaged tracts of DNA involving a single strand which have undergone repair synthesis are then "repair methylated", in most instances, to levels similar to those before damage.[36,45] On these templates, DMase operates in a semi-conservative methylating mode which serves to maintain a heritable pattern of genomic methylation. This has been verified from the experiments on *Xenopus* rDNA by Bird[10] and Southern and Bird,[11] showing the absence of hemimethylated as opposed to fully symmetrically methylated sites in DNA before replication, as well as following through many generations expressly methylated DNAs introduced into cells by gene transfer techniques.[71,93,101,105,106] Thus, in vivo, the major function of DMase is to recognize and methylate hemimethylated sites on duplex DNA consisting of unmethylated daughter and methylated parental strands.

A second form of DMase activity, "initiating methylase", has been theorized to operate in nondividing cells. This activity would be necessary to restore the pattern of methylation after being lost or eliminated by damage affecting both strands, or in the event that a totally new site is to be methylated in response to unknown signals.

In all likelihood, the methyl group in m^5C, which protrudes in the major groove of B-DNA and is known to be the determinant for such protein-DNA interactions as specific lac repressor binding and the correct functioning of the prokaryotic modification-restriction system, provides a mechanism for eukaryotic cells to modulate specific protein-DNA interactions for biological function(s) that are as yet not understood.

Only a small proportion of cytosines, typically from 1 to 7% m^5C to total cytosines are modified in sequences containing mainly[17,77] but not exclusively[30,44] the dinucleotide 5′CpG. It has been estimated that from 50 to 90% of CpGs are methylated depending on the source of eukaryotic tissues and cells.[106] The fact that the dinucleotide is the most underrepresented and is nonrandomly distributed in higher eukaryotes has led to the suggestion that mutation by deamination of m^5C to T (so-called minor thymine because it derives its label from the methyl group in methionine while pyrimidines, unlike purines, are not labeled by one carbon pool transfer) must otherwise be selected against since they are potentially lethal when methylated and cannot be repaired by glycolyases such as for cytosine deamination to uracil.[13,85]

Scarano et al. proposed that m^5C deamination to minor T which, after one round of replication would replace a GC base pair by an AT in one daughter cell, to be a mechanism for controlling differentiation.[85] Holliday and Pugh further refined this mechanism based on protein recognition of base-pair changes or methylated or unmethylated sites in repetitive sequences to explain the timing of cell differentiation and switching of developmental programs.[48] Without specifying any mechanisms, Srinivasan and Borek,[92] and later Vanyushin et al. from measurements of tissue-specific differences of m^{5}C content in DNA[98,99] suggested that DNA methylation may be a regulatory mechanism for controlling cell differentiation. Indeed, several lines of evidence (discussed later) from recent experiments strongly implicate DNA methylation in the regulation of gene activity. Riggs[77] proposed that dosage compensation in females by uniparental X chromosome inactivation was passed on to progeny via a mechanism of DNA methylation, based on the analogy with bacterial modification-restriction systems that are also controlled by methylation.

Recently, an analog of cytosine, 5-aza cytosine (AzaC), that is incapable of being methylated has been employed to reactivate inactivated X chromosomes segregated into heterokaryons.[69] Further experiments involving gene transfer of either inactivated X chromosomes or AzaC reactivated X chromosomes have demonstrated a functional role of DNA methylation in the heterochromatization and inactivation of certain but not all genes on X chromosomes.[101] Sager and Kitchin[81] proposed the operation of a modification-restriction-like system for the selective silencing of genomic information in eukaryotes. Except for one report of an eukaryotic restriction-like enzyme recognizing AGMK alphoid satellite sequences,[20] and the enigmatic selective degradation of amplified rDNA sequences in Amphibia (which are unmethylated compared to somatic cell sequences), no known restriction enzyme activities have been isolated from eukaryotes. Nevertheless, maternal inheritance and the genetic exclusion of male chloroplast genomes in the sexual green alga *Chlamydomonas* appears to be regulated by a type of modification-restriction system.[19,49,83] Sneider et al., from studies showing that sperm DNA is about 50% less methylated and appears to be hemimethylated as a result of meiotic reduction divisions, proposed that methylation may be a basis for genetic recombination in germ cells.[55,90] The experimental evidence, however, for germ-cell hemimethylation has been disputed, and the reduction in methylation may well be attributed to unequal distribution of methylation levels in genomic domains.[96] Holliday[47] proposed a new theory of carcinogenesis based on the concept of infidelity of repair methylation which could change the methylation pattern associated with a stable differentiated phenotype and thereby initiate novel phenotypes; these, by malignant evolution, may lead to neoplasms. Support for this theory has been generated from studies showing carcinogen effects on genomic methylation levels,[14,16,59,61] direct DMase inactivation,[21,25] carcinogen altered templates which impair DMase function,[70,82] and changes in specific methylation of sites which are correlated with gene function of cancer and normal cell markers.[57]

II. DNA METHYLATION AND GENE FUNCTION

Recent advances using gene probes and specific restriction enzymes whose cleavage at a site is conditional on m^{5}C modification at that site (so-called CpG restriction enzymes)[11,102] have given rise to burgeoning numbers of reports in the last several years that support the hypothesis that there is an inverse correlation between methylation in a eukaryotic gene and its activity (recently summarized and reviewed in References 38, 39, 73, and 74). This large body of evidence suggests that DNA methylation plays a functional role as one out of several possible switching mechanisms for a cell to regulate the transcriptional activity of a gene. In support of this, genomic compartmentalization of m^{5}C into enriched and impoverished domains in the sea urchin has been correlated with gene activity in the latter compartment.[12] Secondly, m^{5}C-containing sequences are primarily found in inverted and tandem repeats (palindromic and satellite sequences) that characteristically do not support transcription.[4,14,37,40] This may suggest an additional role of methylation in chromosome structure. Indeed, methylation of alternating CpG sequences in poly(dC-dG) facilitates a transition from B to Z forms of DNA, an intramolecular rearrangement of the helix from a right- to left-handed structure.[9] Lastly, it may be appreciated that demethylation of certain DNA sequences has not only been found to occur during rabbit embryo development,[66] but also during estrogen stimulation of the chick vitellogenin gene;[107] and thus may be a more widespread phenomenon. A possible mechanism for loss of methyl groups, not postulating a site-selective DMase inhibition upon replication, has been provided by the finding of a specific m^5C demethylating activity in Friend's erythroleukemia cells.[43]

Although a large number of specific genes,[29,46,56,57,65,68,86,110] and integrated viral genes,[28,31,62,93,94,100] have been shown to be active when undermethylated at particular sites (mainly Hpa II), the evidence for a molecular mechanism is lacking. For *Xenopus* rDNA multigene family[64] and the α2(1) collagen gene,[67] this correlation, which has depended exclusively on CpG type restriction endonuclease data, has not been borne out. It should be stressed that even using a complete battery of known CpG type restriction endonucleases (cf, the table in Reference 39) will not ensure that all methylation sites are detected as evidenced from the discovery of methylated sequences in rat and monkey alphoid satellite DNAs, which can only be ascertained by direct sequencing of both strands by the Gilbert-Maxam method.[60,84] Thus, the general practice of molecular cloning of nonrepeated coding sequences into a prokaryotic background precludes obtaining this valuable information.[42]

A second line of evidence that strongly links methylation with regulation of gene expression, though indirect, comes from the use of the cytosine analogues 5-azacytidine or 5-azadeoxycytidine (AzaC and AzadC), in which the presence of a triazine ring has been postulated to interfere with transmethylation at the 5 position.[51,69] In several systems, genes shown to be present in the genome and intact (with respect to intron/exon organization and junctions) have been reactivated ("rescued") by AzaC with a concomitant restoration of methylated sites (detected by Hpa II and Hha I cleavage) to an unmethylated status.[27,29,101] These findings strongly implicate DNA methylation as responsible for the former state of gene inactivity. Nonetheless, the effects of AzaC may also be ascribed to a generalized impairment of finely tuned protein-DNA interactions with AzadC substituted DNA in accord with the experiments assessing the effect of AzaC on cell growth and DNA methylation.[6]

III. GENERAL PROPERTIES OF DNA METHYLTRANSFERASE

DMase can be isolated from chromatin that is prepared from nuclear lysates by pelleting through dense sucrose gradients at low ionic strength,[26] but is generally isolated from nuclei. The enzyme has been solubilized in extracts of nuclei as low as 0.3 *M* NaCl to prevent complications of histone-nonhistone chromatin (NHC) protein aggregation. DMase appears to be predominantly associated with Mg^{++} insoluble, condensed heterochromatin;[26] although the enzyme binds to and methylates DNA-cytosines, its ease in solubilization suggests that it is a rather loosely bound NHC protein rather than a tightly bound NHC protein that would require high salt or high salt and 5 *M* urea for extraction.[58]

In regenerating liver, the amount of DMase trebles during DNA replication. The basal level in nondividing cells or nonregenerating liver is quite high (from 0.5 to 1.5 pmol CH_3/hr/mg nuclear protein or about 12 to 15 pmol CH_3/hr/mg NHC protein), suggesting that DMase is stored unengaged in a loose complex with chromatin.[61,103] DMase has been partially purified in a number of laboratories and from various sources including rat liver,[32-34,102] regenerating rat liver,[87,103] rat brain,[70,82] HeLa cells,[78] *Xenopus* kidney cell,[5] Friend erythroleukemia cell,[26,27] Novikoff rat ascites,[89] Krebs II murine ascites,[97] and mouse spleen and liver.[4] Sneider et al.[89] reported that more than one DMase species could be extracted from Novikoff cells. All other studies, however, report the existence of a single major species. This may be due to the insensitivity of the [^{3}H]-SAM assay, low initial specific activity, interferences by protein and/or RNA methylases, and perhaps a poor choice of methyl acceptor templates obscuring minor activities. This latter shortcoming was the major basis for the slow resolution of *Escherichia coli* DNA polymerases and mammalian RNA polymerases. It should not be discounted that mammalian DMase can form aggregates in low salt,[78,87] in addition to spurious aggregates in crude extracts. In our hands, this phenomenon has led to a poor separation by DEAE ion-exchange chromatography when this has been chosen as the first purification step.

A. DNA Methyltransferase Assay

Various assays for DNA methylase activity to date rely on the transmethylation of a [^{3}H]-methyl or a [^{14}C]-methyl group in SAM to DNA and the quantitation thereof by radioactivity measurements. The assay must overcome the problem of nonspecific nucleases and RNA and protein methylase contaminants in crude extracts. The standard assay used in this laboratory is a modification of the procedure outlined by Drahovsky and Morris.[32] Standard reactions contain in 200 $\mu\ell$, 3 μCi [^{3}H]-methyl SAM (~2 μM, 15 Ci/mmol ICN, in dilute acetic acid, immediately neutralized with Na_2HPO_4 before use) 1 to 10 μg DNA depending on its acceptor activity; KCl at less than 0.1 *M* for ds DNA or less than 0.15 *M* for ss DNA substrates, 1000 cpm of high specific activity ^{32}P labeled DNA to serve as internal control for DNA loss and degradation, enzyme protein, and assay buffer consisting of 20 m*M* Tris-HCl pH 7.4, 1 m*M* EDTA, 0.5 m*M* DTT. The incubation is carried out for 30 min to 2 hr, depending on the acceptor ability of the substrate and the problem of substrate degradation. Divalent cations which inhibit the reaction and may activate nuclease must be chelated. However, with highly purified DMase, EDTA may be omitted from the reaction mixture. The reaction is terminated by addition of 12 $\mu\ell$ of 5 *M* NaOH, followed by 10 $\mu\ell$ of 1 mg/mℓ sonicated calf thymus DNA. Blanks are zero time controls with base added before the enzyme. RNA methylation contaminants are then removed by heating for 15 min at 60°C and cooled on ice. If no protein methylation is suspect, the reactions are directly precipitated with an equal volume of 20% TCA, 40 *m*M sodium pyrophosphate after the base hydrolysis step. Otherwise, 100 $\mu\ell$ of 0.5 *N* Na_2HPO_4 is added for neutralization, and protein is removed by two phenol-Sevag (1:1 v/v) extractions, followed by precipitation with 20% TCA as before. After 30 min on ice, the DNA is trapped by filtration on GF/C (Whatman) glass filter discs which are washed three times with 5% TCA, twice with 95% alcohol, and air dried before counting by liquid scintillation on ^{3}H and ^{32}P channels. Methylation is expressed as picomoles incorporated per milligram of protein per hour which is readily calculated from the specific activity of [^{3}H]-SAM, the efficiency of ^{3}H counting, microgram of protein in the assay, and recovery of ^{32}P compared to zero time control. The concentration of rat liver methylase is determined by assaying the activity in 0.02 *M* Tris HCl, pH 7.4, 0.5 m*M* DTT, 1 m*M* EDTA under enzyme limiting conditions on double-stranded poly(dC-dG). Poly(dC-dG), a homogenous polymer of CpG sites, has been found to be the methyl-acceptor template of choice since base sequence effects found in natural DNAs are obviated and thus permit by direct estimation from kinetic parameters expressed as concentration of CpG sites (or modified CpG sites), on the effects of template modification. Other methyl acceptor templates may be used but higher concentrations of these DNAs will have to be used since the K_m is approximately 0.5 to 3.0 μM of CpG dinucleotide depending on the salt concentration and whether the template is in single- or double-stranded form.

IV. DMASE ISOLATION AND PURIFICATION

Nuclei from liver, primary hepatocellular carcinoma, or 252 ascites tumor cells are purified in the following manner. With all operations performed at 0 to 4°C, the tissue is minced in 0.01 *M* Tris-HCl, pH 7.4, buffered saline containing 1 m*M* EDTA and 0.5 m*M* PMSF. The tissue or cells are washed twice in this buffer by low-speed centrifugation. In nuclear wash (NW) buffer (0.3 *M* sucrose, 0.01 *M* Tris-HCl, pH 7.4, 1.5 m*M* $MgCl_2$, 1 m*M* DTT, 0.5 m*M* PMSF, 0.2% Nonidet P-40 [Shell]) 10 to 15% homogenates are made with 10 to 15 strokes of a Teflon® glass homogenizer. Crude nuclei are pelleted at 2000 rpm for 10 min.

The pellet is gently homogenized in NW buffer, and 4 volumes of 2.5 *M* sucrose, 1.5 m*M* $MgCl_2$, 1 m*M* DTT, 0.5 m*M* PMSF are mixed in and stirred thoroughly. Purified nuclei are obtained by sedimentation at 100,000 × *g* for 1 hr. The pellet is gently homogenized in 1.0 *M* sucrose, 1 m*M* $MgCl_2$, 1 m*M* DTT, 0.5 m*M* PMSF and sedimented at 6000 rpm for 10 min. The purified nuclei are stored frozen at −80°C until further use.

DMase is extracted from the purified nuclei by a modification of the procedure developed by Simon et al.[87] The nuclear pellet is homogenized in 3.7 volumes of 0.8 *M* KCl, 50 m*M* Tris-HCl, pH 7, 0.5% Triton® X-100, 1 m*M* DTT, 0.5 m*M* PMSF. The KCl concentration should be adjusted to 0.63 *M* as determined by refractive index measurements. The extract is periodically stirred for 30 min, at which time the KCl concentration is lowered to 0.3 *M* by slowly mixing in 10 m*M* Tris-HCl, pH 7.4, 1 m*M* DTT, 0.5 m*M* PMSF. After another 30 min of stirring, the insoluble chromatin is sedimented at 13,000 rpm for 15 min. About 95% of methylase activity is extracted by this step. This corresponds to fraction 0 (F0). DMase is quite stable frozen at −20°C and contains large amounts of protein and RNA methylating activities.

The crude methylase extract is diluted with an equal volume of 40 m*M* Tris-HCl, pH 7.4, 1 m*M* $MgCl_2$, 0.1 m*M* DTT, 0.5 m*M* PMSF, 0.1% Triton® X-100, 20% glycerol and treated with 5 μg/mℓ DNase I (Sigma DN-C1) for 10 min at 38°C. EDTA is added to 5 m*M* and the extract is batch absorbed onto Whatman P-11 phosphocellulose at 5 mg protein per milliliter P-11. Protein concentration is determined by the Coomassie blue dye binding method.[61] The P-11 cellulose is equilibrated in 0.15 *M* KCl buffer A (buffer A = 20 m*M* Tris-HCl, pH 7.4, 1 m*M* EDTA, 0.5 m*M* DTT, 0.5 m*M* PMSF, 0.1% Triton® X-100, 10% glycerol. After batch absorption, the P-11 is packed by gravity in a large-diameter column and the nonabsorbed protein is eluted with 0.15 *M* KCl in buffer A. The bulk of absorbed protein is then eluted stepwise by 0.3 *M* KCl in buffer A, followed by the elution of DMase with 0.6 *M* KCl in buffer A (F1). The DMase activity is determined after appropriate dilution of enzyme solution to <0.15 *M* KCl. The activity peak is diluted to 0.2 *M* KCl with buffer A and loaded onto a much smaller hydroxylapatite (HAP) column (BioRad, HA) at 20 mg protein per milliliter of bed. The HAP column is equilibrated in buffer B (0.05 *M* KPO_4, pH 7.5, 1 m*M* DTT, 1 m*M* EDTA, 0.1 m*M* PMSF, 10% glycerol). After washing off the flow through fraction with four column-volumes of buffer B, a ten-column-volume gradient is developed from 0.05 *M* KPO_4 to 0.4 *M* in buffer B. A single major methylase activity peak elutes between 0.14 and 0.18 *M* KPO_4 and is purified approximately 80-fold over the initial crude extract (F-II).

The concentrated HAP eluted enzyme is dialyzed once against buffer A (no salt) for 4 hr, then diluted with buffer A until the conductivity is less than that for buffer A + 0.025 *M* NaCl. The fraction-II enzyme is loaded on a DEAE-Sephacel® (Pharmacia) ion-exchange column equilibrated in buffer A + 0.025 *M* NaCl at 10 mg/mℓ bed volume. A linear 20-bed volume gradient from starting buffer to 0.3 *M* NaCl is developed. A single peak of activity eluting at approximately 0.05 to 0.08 *M* NaCl is immediately concentrated by absorption onto a small HAP column as in F-II purification which is eluted with 0.25 *M* KPO_4 buffer B. Purification steps which are dispersive in nature (e.g., result in dilution of the enzyme or prolonged dialysis at 4°C) lead to a rapid 80% or greater loss of activity with further purification. Although gel filtration chromatography is extremely dispersive, the fact that large amounts of contaminating proteins have already been removed coupled with the rapidity of newer matrices make it possible to recover a large proportion of DMase activity at this stage with good purification, which is not possible at earlier stages of purification. Fraction-IV DMase is obtained by gel filtration on AcA 34 equilibrated in buffer A + 0.1 *M* NaCl. As shown in Figure 1, the major peak of activity (peak I) emerges near the void volume and ahead of the major protein peak which enclosed a second peak of activity (peak II). We have noted a consistent but variable amount of peak II as a function of aging of the

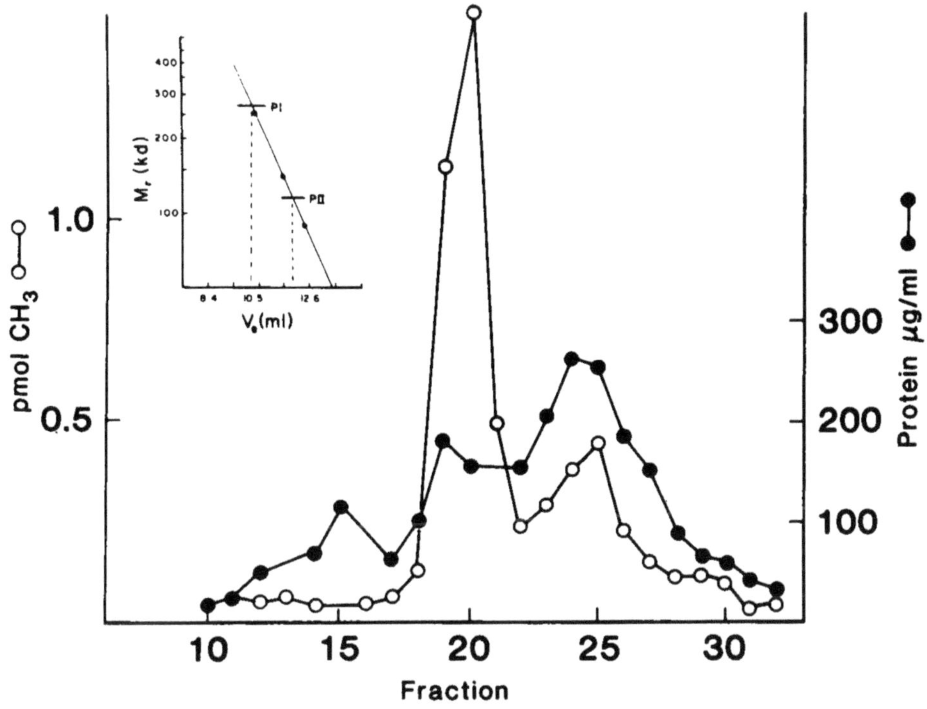

FIGURE 1. Gel filtration of fraction-III DMase on Ultrogel® AcA34. (-●-●) Protein profile; (-O-O-) enzyme activity. Insert: calibration plot of Mr vs. elution volume using catalase, yeast alcohol dehydrogenase, and *E. coli* alkaline phosphatase.

preparation and possible occurrence of proteolytic degradation, in particular with the preparations from 252 ascites tumor cells where peak II may become the major peak of activity (but with lower yield and specific activity). A calibration of the molecular weight of proteins eluted from AcA 34 is shown in the insert to Figure 1. Using catalase molecular weight 250 Kd, yeast alcohol dehydrogenase molecular weight 140 Kd, and *E. coli* alkaline phosphatase molecular weight 89 Kd, the estimated molecular weight for peak I falls in the range of 280 to 260 Kd and peak II 130 to 120 Kd for four different determinations. From the ratio of molecular weights, it is likely that dissociation or breakdown of an active dimer (PI) takes place yielding variable amounts of an active monomer. One of the salient properties of PI DMase is its threefold higher rate of methylation on duplex as opposed to single-stranded DNAs; this is a property also observed for murine Krebs II ascites DMase purified by an improved protocol by Adams et al. having an estimated molecular weight of 184 Kd,[4] compared to an earlier scheme which yielded from murine Ehrlich's ascites a DMase having a molecular weight of 80 Kd.[97] DMase purified from HeLa cells[78] and also from regenerating rat liver[87] were estimated by sedimentation in 20 to 40% glycerol gradients and 5 to 20% sucrose gradients containing 0.5 *M* KCl to have molecular weights of approximately of 120 Kd. Preparations yielding an enzyme of this size characteristically have greater template activity on single-stranded as opposed to duplex DNAs. Fraction IV is immediately concentrated by absorption on HAP which is washed with 0.1 *M* KPO_4 buffer B, then eluted in 0.25 *M* KPO_4 buffer B, and further concentrated by dialysis against buffer A + 50% glycerol. Both fractions III and IV DMase when stored at −20°C at a protein concentration of >500 µg/mℓ have been found to be stable for over 12 months with less than 20% loss of activity.

Table 1
PURIFICATION OF RAT LIVER DNA METHYLTRANSFERASE

Fraction	Total protein (mg)	Recovery (%)	Spec. act. (pmol CH_3/mg protein/hr)	Purification (fold)
FO crude nuclear extract	2770.0	100.0	0.5—1.5	1
FI, P-11 phosphocellulose	342.0	12.3	4—7	4—7
FII, HAP	64.0	2.3	81—120	81—120
FIII, DEAE-Sephacel	5.7	0.2	350—790	350—790
FIV, AcA 34, Peak I	0.3	0.01	1584	1584

Further progress in purification should be possible with less dispersive means such as obtained with a specific absorbent followed by rapid concentration on HAP. In Table 1, representative results for rat liver DMase isolation showing the specific activity, purification, and yield at each stage are shown.

The modified procedure currently employed has proven to be a reliable method for purification of DMase from several tissues including normal and regenerating liver, tumors, and 252 ascites cells. DMase purification through the P-11 phosphocellulose step removed over 85% of the nuclear protein yielding fraction I DMase which was purified four- to sevenfold over the initial crude extract. Recovery of total enzyme activity was not markedly reduced through subsequent steps which involved chromatography on HAP and DEAE-Sephacel® columns. Loss of most of the total activity occurred in the final step, i.e., after gel filtration on Ultrogel® AcA 34 (fraction IV). Fraction IV enzyme represents only 0.01% of the total protein in the nuclear extract and has a specific activity of 1584 units per milligram of protein with a recovery of 25% of total initial enzyme activity. Thus, DMase represents, at most, 0.04% of total nuclear protein from rat liver. Although a 1584-fold purification was achieved at this step, it was not sufficient for unambiguously determining the subunit structure on denaturing polyacrylamide gels.

The improved protocol by Adams et al., which involves salt extraction followed by ion exchange chromatography on DEAE-cellulose, then phosphocellulose, ammonium sulfate precipitation, and gel filtration on Ultrogel® AcA 34 resulted in a 35-fold purification of Krebs II ascites DMase which has a specific activity of 980 units per milligram of protein.[7] This enzyme preferentially methylated double over single-stranded DNA and had a molecular weight of 184 Kd. Purification of another high specific activity DMase from regenerating rat liver in which the crude enzyme extract from regenerating rat liver was purified by batch then gradient elution on DEAE-Sephadex® A50 as the first step has been reported by Simon et al.[87] This was followed by a phosphocellulose chromatography and a final purification step by glycerol gradient centrifugation. This resulted in a 660-fold purification with a specific activity of 4680 units/mg protein. Recovery of the activity was 29% of initial total activity at the final stage. DMase prepared by this method preferentially methylated single- over double-stranded DNA and had a reported molecular weight in the range of 120 Kd.

A. Physical Properties of DMase

Flat bed isoelectric focusing on granulated gel was used to study the heterogeneity of fraction-III enzyme. The enzyme was mixed into an Ultrodex® (LKB) slurry which was swollen in a solution containing 1 m*M* DTT, 5 m*M* β-mercaptoethanol, 10% glycerol, and ampholines (LKB) at a final concentration of 1% (pH 9 to 11 and pH 3 to 10 mixed in a 1:1 ratio). Cytochrome *c* was included as a marker which also served to indicate when electrofocusing was completed. Focusing was performed at 4°C for 15 hr at 0.5 kV, then

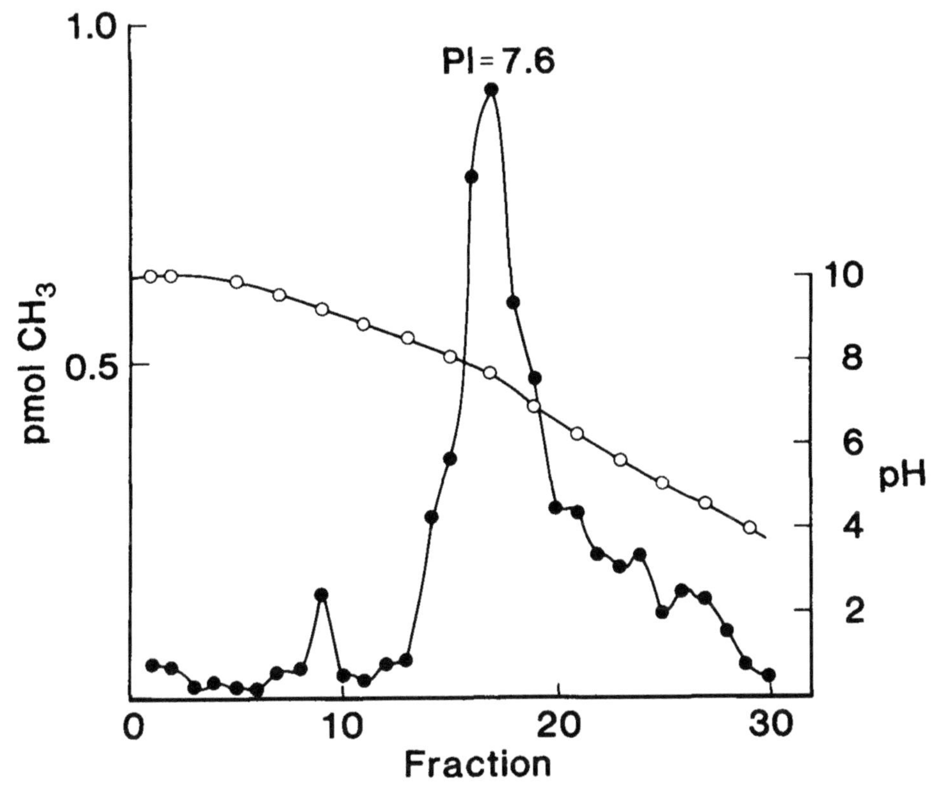

FIGURE 2. Flat bed isoelectric focusing of fraction-III DMase in Ultrodex. (-●-●-) Enzyme activity; (-○-○-) pH.

for 1 hr at 1 kV. Figure 2 illustrates the profile of fraction-III DMase activity as a function of pH. A single major peak of enzyme activity with an isoelectric point of 7.6 is observed, suggesting that peak I and -II DMase have very similar isoelectric points and may be part of the same functional enzyme.

To investigate the mechanism of reaction of DMase, the assay was carried out by varying the order of addition of the reaction components which comprise poly(dC-dG) as the methyl acceptor template, [^{3}H]-SAM, and fraction-III DMase. Two of the three reaction components were preincubated for 3 min at 37°C prior to the addition of the remaining components. Altered kinetics would indicate if DNA methylation is an ordered bi-substrate reaction in which either enzyme-template or enzyme-SAM complex must be formed prior to the trans-methylation step. Figure 3 demonstrates that a permutation of the sequence of addition of the reaction components and their preincubation does not affect the kinetics of methylation. Secondly, within the time frame of the assay, the effect of spontaneous methylation by SAM by decomposition is negligible (<1% of enzymatic activity).[8]

DNA methyltransferase has been reported to show a differential inhibitory effect of salt on the methylation of single-vs. double-stranded DNAs.[32] Moreover, a relatively salt-resistant complex has been observed when the enzyme is preincubated with the template before the addition of high salt and SAM.[33] It has been proposed that this complex is formed as a result of the enzyme "walking" or being engaged on the template prior to the actual transmethylation step. In Figure 4, preincubation of the enzyme with poly(dC-dG) at 37°C for 3 min prior to the addition of NaCl shows no salt-resistant complex formation with this DNA nor with *M. lysodecticus* DNA (data not shown). These results are in accord with the

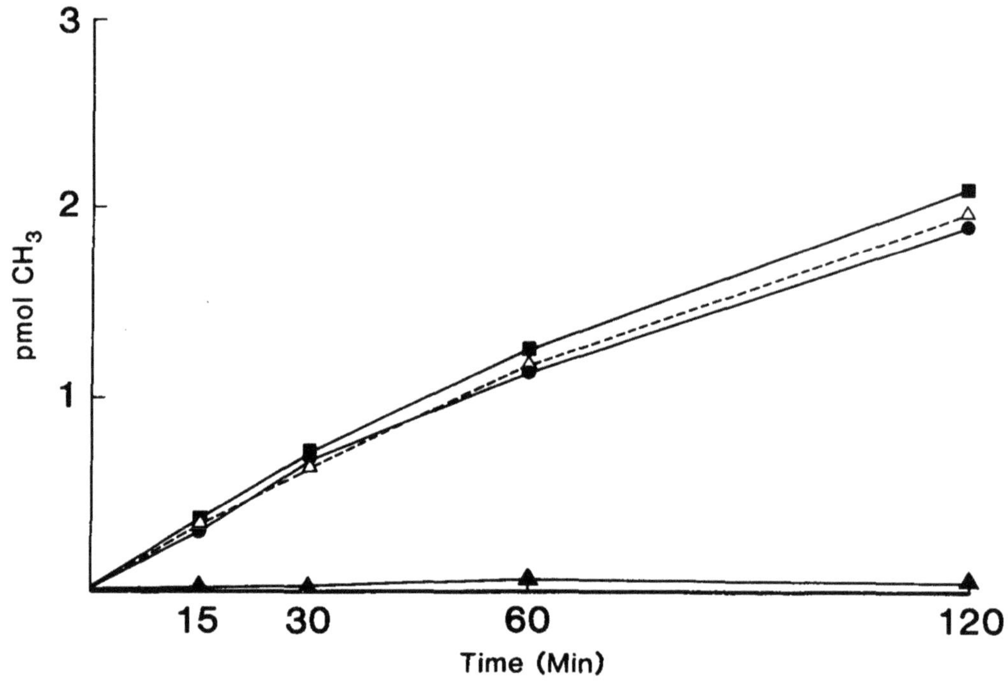

FIGURE 3. DMase reaction kinetics on poly(dC-dG) as a function of preincubation of reaction components. (-■-■-) Enzyme preincubated with SAM prior to the addition of poly(dC-dG); (-△-△-) enzyme preincubated with poly(dC-dG) prior to the addition of SAM; (-●-●-) poly(dC-dG) preincubated with SAM prior to the addition of enzyme; (-▲-▲-) SAM incubated with poly(dC-dG) without enzyme (basal level of spontaneous SAM methylation reaction).

findings of Adam et al. obtained with mouse DMase which also did not appear to walk on methyl-acceptor templates, nor methylate preferentially single-stranded DNA.[4]

B. Evidence for a Single Functional Enzyme

As shown in Figure 2, a single peak of activity is discerned on flat bed isoelectric focusing of fraction-III enzyme suggesting a single functional enzyme species. When temperature denaturation curves of fraction-III DMase were run, as shown in Figure 5, no break in the inhibition curve nor change in the temperature of DMase inactivation in the presence or absence of EDTA or DTT could be discerned. This leads one to conclude that DMase is not a metallo-enzyme and apparently behaves as a single functional species with respect to temperature inactivation. Further evidence for this is derived from the smooth inhibition curves with iodoacetamide and *N*-ethylmaleimide, sulfhydryl group alkylators as shown in Figure 6. The very low concentration required for inactivation of DMase by these agents (0.1 and 0.5 m*M*, respectively), suggests that a sulfhydryl group(s) is necessary for the catalytic activity of DMase.

Inorganic metal salts and divalent cations inhibit DMase. More than 90% of the enzyme activity is inhibited by the mutagenic and carcinogenic heavy metal ions: Ag^{++}, Hg^{++}, Cd^{++}, Zn^{++}, Ni^{++}, and Co^{++} at concentrations ranging between 0.1 to 0.2 m*M* (data not shown). With Cu^{++}, Mg^{++}, Mn^{++}, Ca^{++}, Pb^{++}, Sr^{++}, and $CrO_4^{=}$ salts, higher concentrations were required to produce the same level of inhibition as shown in Figure 7. However, millimolar concentrations of these divalent cations completely inhibit DMase with the possible exception of Sr^{++}.

The effects of NAD, NADP, and ATP have also been tested on this enzyme. NAD and NADP in a concentration range of 0.2 to 5 m*M* have no effect on the activity. ATP, which

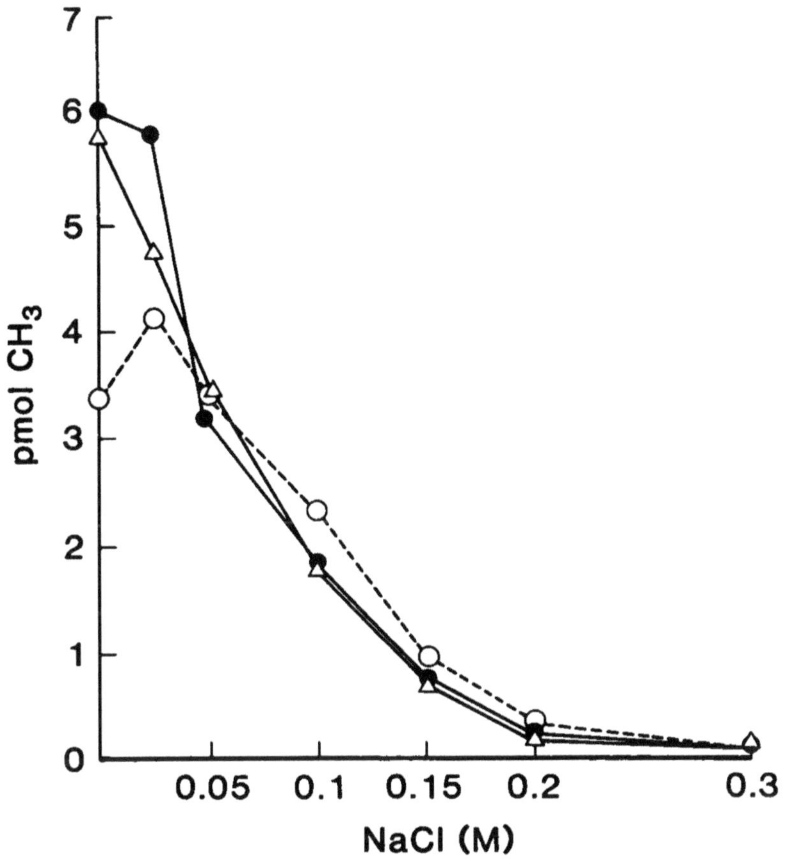

FIGURE 4. Salt dependence and effect of preincubation of DMase with double- and single-stranded templates. (-●-●-) Double-stranded poly (dC-dG) not preincubated; (-○-○-) double-stranded poly(dC-dG) preincubated with DMase 3 min at 37°C prior to the addition of NaCl; (-△-△-) heat-denatured poly(dC-dG) not preincubated.

has been shown to be a cofactor for type-I restriction modification methylase from *E. coli* K, slightly stimulated DMase in the concentration range from 0.5 to 1 m*M*.[18] Unlike type-I methylases which require millimolar Mg^{++} as a cofactor, rat liver DMase is inhibited by Mg^{++}.

Polyamines, including spermine, spermidine, and putrescine, have been reported to inhibit DNA methylation in vitro by a crude preparation of DMase.[24] When 0.05 m*M* spermine (10 nmol) or 0.25 m*M* spermidine (50 nmol) is included in the standard assay that contains 1.5 μg poly(dC-dG) (4.5 nmol), 50% inhibition of DMase activity is obtained. The inhibition by polyamines is thought to be mediated through their binding to the DNA helix although a direct inhibitory effect on DMase has not been ruled out experimentally.

C. DMase Activity on Hemimethylated Templates

To investigate whether fraction-IV DMase has the postulated properties of a maintenance methylase, poly(dC-dG) was nick translated essentially by the procedure of Rigby,[76] using dm^5CTP (P-L Labs) instead of dCTP to introduce hemimethylated sites on poly(dC-dG). By varying the ratio of dm^5CTP to dCTP in a preliminary experiment, it was observed that the rate of dm^5CTP incorporation was nearly the same as for dCTP, thus permitting the use of a tracer amount of $\alpha[^{32}P]$-dCTP (at 1% of dm^5CTP) to quantitate the degree of m^5C substitution in poly(dC-dG).

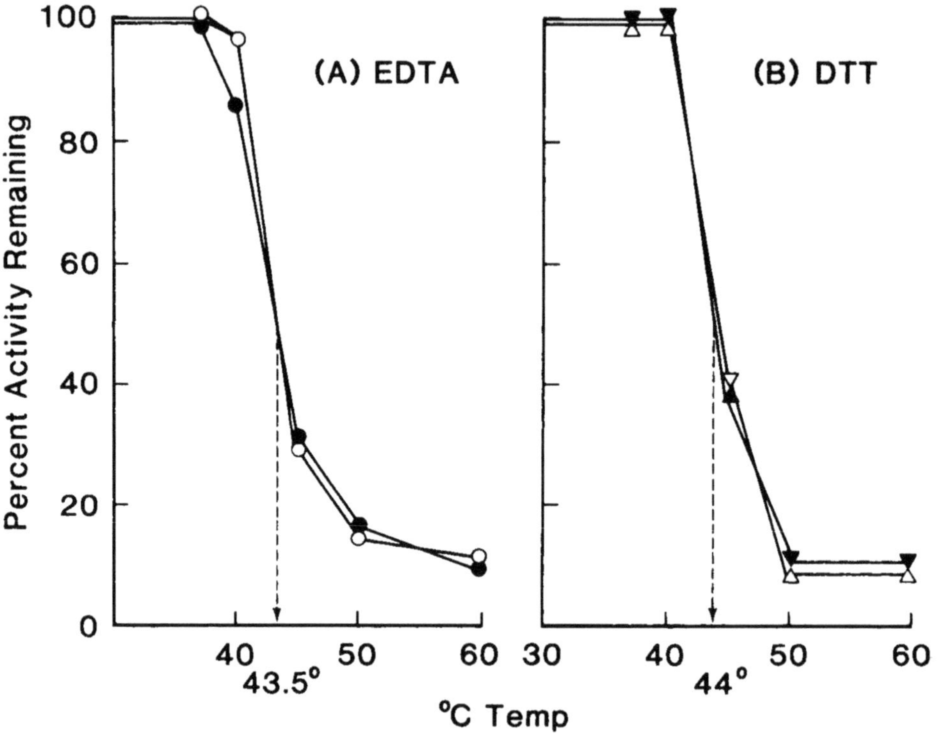

FIGURE 5. Effect of temperature on fraction-III DMase activity. (A) Thermal denaturation profile in presence of 1 m*M* EDTA (-○-○-); thermal denaturation profile with no EDTA present in assay (-●-●-); temperature for 50% inactivation of DMase in presence or absence of EDTA 43.5°C. (B) Thermal denaturation profile in presence of 1 m*M* DTT (-▽-▽-); thermal denaturation profile in absence of DTT in assay (-▲-▲-). Temperature for 50% inactivation of DMase in presence or absence of DTT 44°C.

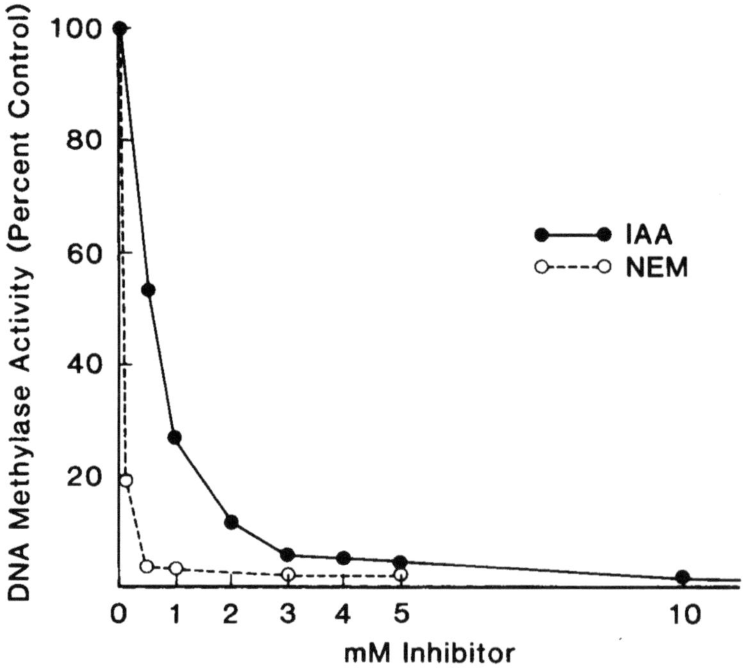

FIGURE 6. Inhibition curve for sulfhydryl group alklyation on fraction-III DMase activity. (-●-●-) Iodoacetamide; (-○-○-) *N*-ethylmaleimide.

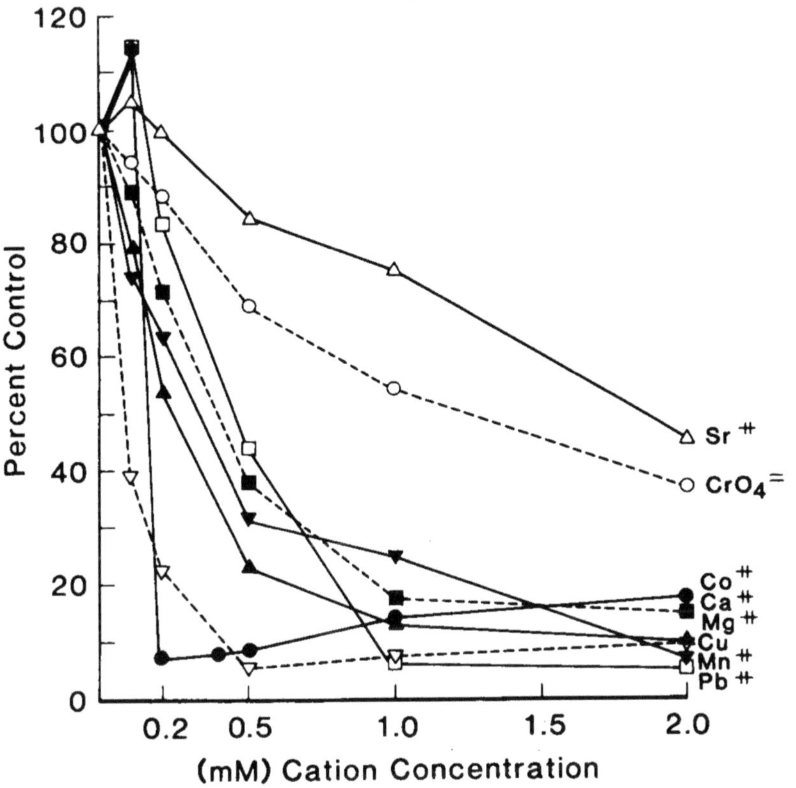

FIGURE 7. Effect of divalent cations on fraction-III DMase activity. (-▽-▽-) Various concentrations of Sr^{++}; (-○-○-) $CrO_4^{=}$; (-●-●-) Co^{++}; (-■-■-) Ca^{++}; (-▲-▲-) Mg^{++}; (-▽-▽-) Cu^{++}; (-▼-▼-) Mn^{++}; (-□-□-) Pb^{++} were added to the standard assay and the extent of transmethylation measured.

One to two micrograms of DNA to be nick translated is incubated in 50 m*M* Tris-HCl, pH 7.4, 5 m*M* $MgCl_2$, 1 m*M* DTT, 80 μ*M* unlabeled dNTP (generally dATP, dGTP, dTTP) as recommended by Balmain and Burnie,[7] and 80 μ*M*α[^{32}P]-dCTP (2000 to 3000 Ci/mmol, aqueous stabilized (Amersham) diluted tenfold from initial specific activity) and 10 μℓ of 1 pg/μℓ activated DNase I for most DNA substrates (25 μℓ for poly(dC-dG)). The reaction mix is brought to 14°C for 2 min, then 10 units of DNA polymerase I (6000 units/mg, New England Biolabs) is added. The time course of incorporation is checked by withdrawing and spotting ½-μℓ aliquots on 3-MM paper discs which are rinsed in 5% TCA. Incorporation proceeds linearly for 2 to 4 hr but at times may peak earlier due to DNA degradation by DNase I. This is remedied by lowering the amount of DNase I in the reaction. For nick translation of poly(dC-dG), dATP and dTTP are omitted from the reaction mix, and the reactants and enzyme were scaled up tenfold in a volume of 500 μℓ. In Figure 8, a time course of methylation of control nick translated poly(dC-dG) with dCTP is compared with 40% m^{5}C substituted poly(dC-dG) and poly(dC-dG) that has not been nick translated.

As shown in Figure 8, nick translation (hence the introduction of unsealed nicks or gaps in the template) has about a twofold stimulatory effect over non-nick-translated poly(dC-dG). However, 40% substitution with m^{5}C results in an additional threefold stimulation over its nick-translated control or sixfold over intact poly(dC-dG). In this instance, 60% of the sites remain unmethylated and, if one assumes an equal probability of formation of totally methylated to hemimethylated sites by nick translation, then 20% hemimethylated sites have been created to a total 80% unmethylated plus hemimethylated sites. Then, one in four sites

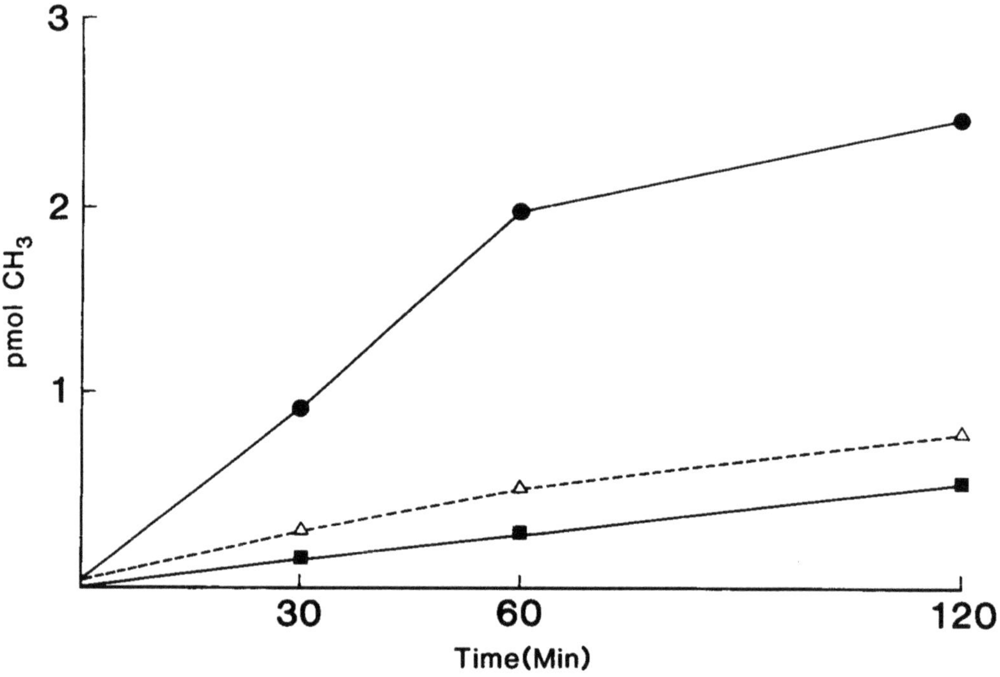

FIGURE 8. Effect of hemimethylation on kinetics of methylation of poly(dC-dG). (-●-●-) Nick-translated poly(dC-dG) with dm⁵CTP (40% substituted); (-△-△-) control nick-translated poly(dC-dG) with dCTP; (-□-□-) intact poly(dC-dG).

is hemimethylated and available for methylation. This extrapolates to roughly a 12- to 24-fold increase in rate of methylation of hemimethylated to unmethylated CpG sites or an order of magnitude greater selectivity for maintenance-type over initiation-type methylation by DMase. Homologous DNAs that are hypomethylated as a result of ethionine administration,[22] methionine starvation,[89] or isolated from carcinogen or tumor-induced hypomethylation,[103] also provide a more efficient template but not approaching this rate. It is also impossible in these systems to ascribe this effect to decreasing the concentration of formerly inactive symmetrically methylated sites to active hemimethylated or totally unmethylated sites. Jones and Taylor have observed that AzaC-treated cells had DNAs which were far better templates for methylation.[52] Newly replicated DNA that was initially unifiliarly (heavy-light, HL) BrdU-substituted while containing variable amounts of AzaC substitution in the progeny strand, was isolated to test more directly the hypothesis that the major DMase activity in cells (DMase essentially prepared by procedure of Simon et al.[87]) corresponds to a maintenance methylating activity operating more efficiently on hemimethylated sites. By increasing AzaC doses, they were able to effectively decrease the m^5C content in 10T-$^1/_2$ cell DNA, and the ratio of HL to LL CsCl density gradient purified templates could be used to measure the postulated DMase preference for hemimethylated sites on DNA. What was observed was a twofold increase for control HL to LL but a two- to threefold increase for AzaC-substituted DNA. The HL and LL ratio for DNAs isolated from ethionine- or cycloleucine-treated cells was on the order of fivefold, a very significant increase, but paradoxically, these DNAs showed no significant decrease in m^5C content measured by HPLC. The report by Creusot and Christman on the mechanism of DMase inhibition by AzaC may shed light on the above results.[27] The high potency by which AzaC and AzadC inhibit DNA methylation was found to correlate with a large decline in the intracellular level of extractable DMase. They, as well as Friedman,[41] working on AzaC-treated *E. coli* and its methylases, have provided evidence that because DMase is resistant to even very harsh methods for solubilization, it is likely irreversibly bound to AzaC-substituted DNA, and is prevented from further movement on or off the template to carry out further methylation reactions.

D. DNA Modification and DMase Activity

Drahovsky and Morris[34] investigated the template properties of a number of DNAs and demonstrated that DNA templates could be classified as methyl and nonmethyl accepting. A partial list of the latter category includes all nonalternating CpG containing DNA copolymers, deaminated, depurinated, N7-guanine, and N^6-adenosine alkylated DNAs. By running competition experiments with these inactive templates, they found that depurinated and alkylated DNAs did not lower the rate of methylation and hence bind DMase, whereas glycosylated and deaminated DNAs disproportionately bound the enzyme and were very efficient at reducing the rate of methylation of normal active templates. These initial findings suggested that exogenous alterations of a DNA template could modify its efficiency to be methylated by DMase in a striking manner. Although the rate of methylation is GC base pair concentration-dependent in natural DNAs, Drahovsky and Morris demonstrated in competition experiments that DMase binding affinity to DNA of various GC content was independent of the GC base pair concentration suggesting that the enzyme does not necessarily bind to a site which it methylates. Salas et al.,[82] using in vitro N-AcAAF modified calf-thymus DNA, showed a progressive inhibition of template methyl-acceptor activity with arylamidation. Upon further investigation, Pfohl-Leszkowicz et al.[70] indicated that AAF-modified DNA has a much higher binding efficiency for DMase than native DNA. The affinity for AAF-modified DNA was so high that irreversible inhibition was produced suggesting that not only is movement of DMase along the helix blocked by bulky AAF adducts, but also it prevents DMase dissociation for methylation of unsubstituted DNA templates added in competition experiments.

Experiments using another model carcinogen, 7β, 8α, dihydroxy-9α, 10α, epoxy-7,8,9,10-tetrahydrobenzo[*a*]pyrene (anti-B[*a*]P diol epoxide, anti-BPDE), the ultimate carcinogenic form of B[*a*]P which covalently reacts with cellular macromolecules including DNA,[95] were performed to assess the validity of the theory that carcinogen adducts may, in general, prevent or adversely affect specific DNA methylation at the template level.

Poly(dC-dG) was reacted under nitrogen in a buffer consisting of 0.02 *M* NaCl, 2 m*M* Na_3 citrate, 1% DMSO, pH 7.0, at 37°C according to Straub et al.,[95] with varying ratios of 7-^{14}C-anti-BPDE (IIT Midwest Research Institute, 58 mCi/mmol). Unreacted BPDE was removed by several ethyl acetate-benzene (1:1) and ethyl acetate extractions, followed by heating to 60°C to remove any intercalated BPDE, ether extracted, and finally dialyzed against 1 × SCC followed by 0.1 × TE. Figure 9 shows a time course of fraction-IV DMase methylation of 0.5 to 12.1% dG modified poly(dC-dG) (BPDE reacts principally at N^2 of dG in DNA and negligibly with dC).[95] As a function of the increasing degree of BPDE-dG modification, a decreasing methylation-acceptor efficiency is observed. This suggests that DMase movement on the helix to methylate a next site is blocked or that the enzyme is incapable of dissociation from such modified DNAs. In Figure 10, a competition experiment was set up to test the template selectivities for binding to 5.8 or 12.1% dG modified poly(dC-dG). After a 30-min incubation of these templates with DMase and SAM, an equal amount of unmodified poly(dC-dG) was added and the rate of methylation (which is linear with time) was compared with that of the modified template and poly(dC-dG) alone. As shown in panels A and B, the rate of methylation of control poly(dC-dG) is not reestablished even after 30 min. This suggests that DMase is nearly irreversibly bound to B[*a*]P-modified CpG site, very much like phenomena encountered with AcAAF and AzaC-substituted DNAs. In the converse experiment (not shown), DMase freely dissociates from poly(dC-dG) and interacts with the modified templates. This results in a progressive decline

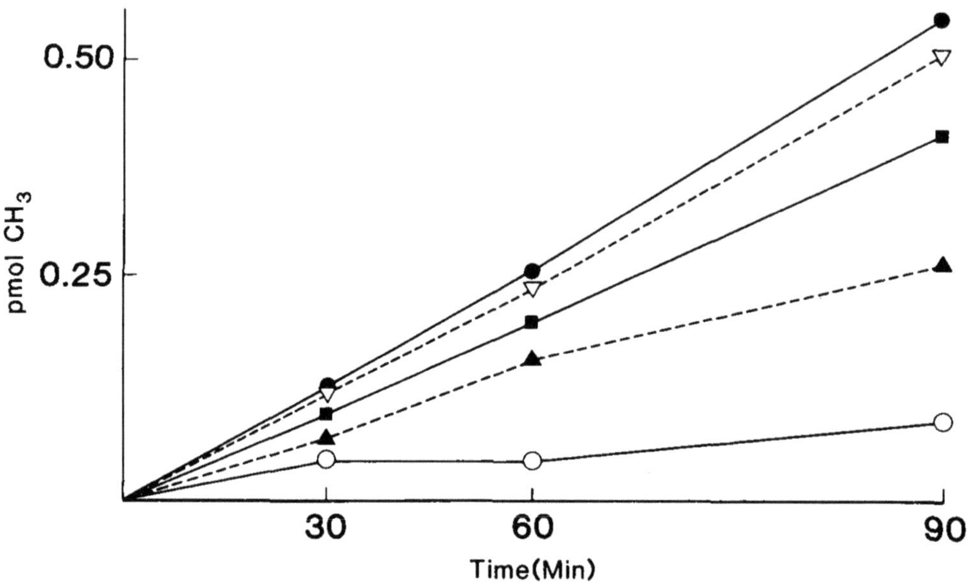

FIGURE 9. Time course of the enzymatic methylation of anti-BPDE modified poly(dC-dG) template. (-●-●-) Unmodified poly(dC-dG); (-▽-▽-) 0.5% dG modified template; (-■-■-) 2.1% dG modified template; (-▲-▲-) 5.8% dG modified template; (-○-○-) 12.1% dG modified template.

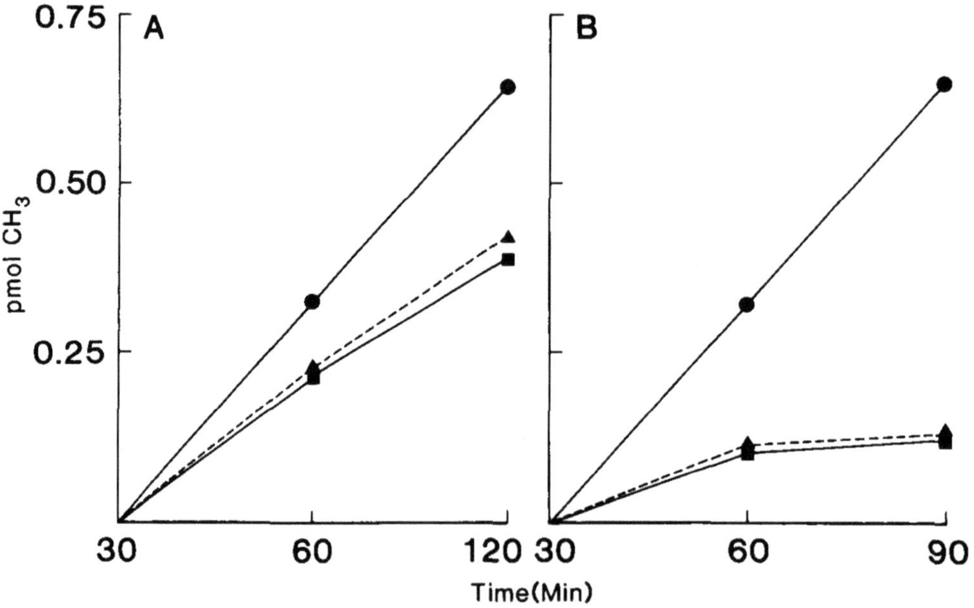

FIGURE 10. Time course of competitive DNA methylation of BPDE modified poly(dC-dG) templates. (A) (-●-●-), Unmodified poly(dC-dG); (-▲-▲-), BPDE 5.8% modified dG template allowed to react with DMase for 30 min prior to the addition of 1 μg unmodified poly(dC-dG); (-■-■-), BPDE (5.8% modified dG) template no addition. (B) (-●-●-), unmodified poly(dC-dG); (-▲-▲-), BPDE (12.1% modified dG). Template allowed to react with DMase for 30 min prior to the addition of 1 μg unmodified poly)dC-dG); (-■-■-), BPDE (12.1% modified dG) template no addition.

in the overall rate of methylation with time as the DMase is sopped up by the B[*a*]P-modified templates. These findings provide a possible mechanism for the reports that DNA isolated from a number of carcinogen-induced neoplasms shows significant undermethylation or aberrations in the genomic methylation patterns.[14,16,57,59,61]

E. Chromatin Organization and DMase Activity

In the nucleus of eukaryotes, DNA is packaged into nucleosomes which in turn are organized into higher-order structures of varying degrees of complexity and compaction. In addition, various individual NHC proteins are postulated to be distributed either widely or selectively in chromatin domains on the basis of their function. The distribution of the bulk of DMase is not random, but appears to be primarily associated with DNase II-sensitive, Mg^{++}-insoluble and late phase of micrococcal nuclease digestion-sensitive regions in chromatin, presumably heterochromatin.[26] Like the majority of NHC protein, DMase is rapidly released from chromatin by micrococcal nuclease action and does not appear to be associated with small nucleosome arrays. DMase, however, like the bulk of NHC protein, forms spurious complexes with aggregates of insoluble nucleosome arrays during digestion. Interestingly, in this configuration, Creusot and Christman[26] showed that DMase can be resolubilized with 8 m*M* Mg^{++}. Similarly, Adams et al. observed the rapid release of about 80% of DMase in initial stages of nuclease treatment, but after 30 min, half of the released enzyme sediments with insoluble nucleosome structures.[3] According to these authors, using double-labeling techniques (^{3}H-methyl methionine and ^{14}C-deoxycytosine), the proportion of m^5C liberated in CHO cell chromatin as a function of digestion time does not vary, nor is there any large difference in m^5C content in nucleosomes up to tetramer separated by agarose gel electrophoresis. Additionally, during the earliest phase of digestion with micrococcal nuclease or with DNase I, the liberated oligonucleosomes and acid-soluble DNA, respectively, were found to be enriched in m^5C content. Razin and Cedar[72] analyzed by mass spectrometry the m^5C content of various micrococcal nuclear digests at 50% acid-soluble DNA (limit digest conditions which essentially trim away linker regions, completely leaving core particles). In sharp contrast to the aforementioned results, the nuclease-resistant fraction was from 1.3- to 1.7-fold enriched in m^5C, whereas nuclease-sensitive DNA (open DNA, e.g., poly-d-lysine complexable) contain about one half the m^5C.[72] Solage and Cedar,[91] using a double labeling technique similar to that of Adams et al.[3] obtained during early digestion time points, very little m^5C release but enrichment in core nucleosomes. This strongly suggests that the linker DNA is preferentially deficient in m^5C compared to core DNA. Applying the same technique to DNase I digests yielded results that indicated no selective DNase I release of m^5C-enriched DNA. One possible explanation for these diametrically inconsistent results lies with the labeling time for marking m^5C. Chemical carcinogens are known to initially and preferentially attack linker regions, but with time they become associated with nuclease-resistant fraction and covered by core structures.[49,50,111] Similarly, by labeling repair tracts, such tracts are initially in a nuclease-sensitive configuration and with time become covered, suggesting that the bulk of nucleosomes per se are not rigidly bound to the DNA helix but can be displaced.[88]

DNase I is known to digest active gene region at a higher rate than bulk or inactive gene regions in nuclei or chromatin.[104] DNase I preferentially attacks certain well-defined sites in chromatin (so-called DNase I-hypersensitive sites) that are generally upstream but also downstream from a sensitive domain.[108] Kuo et al.[56] analyzed the status of the DNase I sensitivity of Hha I methylatable sites in conalbumin and ovalbumin genes from active oviduct vs. inactive erythrocyte nuclei of the chick. Their results indicated that residual methylated sites (Hha I cleavage resistant sites) arose from cells whose chromatin contain DNase I insensitive and inactive genes. On the other hand, erythrocyte chromatin DNase I-insensitive regions contained unmethylated Hha I sites. In addition, Kuo et al. observed that hypersensitive sites at the 3′ end of the conalbumin gene were present in both tissue chromatins, although a greater number of Hpa II and Hha I sites are methylated in inactive erythrocyte chromatin regions than in oviduct. These findings suggest that there is no simple relation between methylation (at least in the Hha I type of site) and DNase I hypersensitivity. Since nucleosomes are assembled shortly after replication in addition to NHC protein dep-

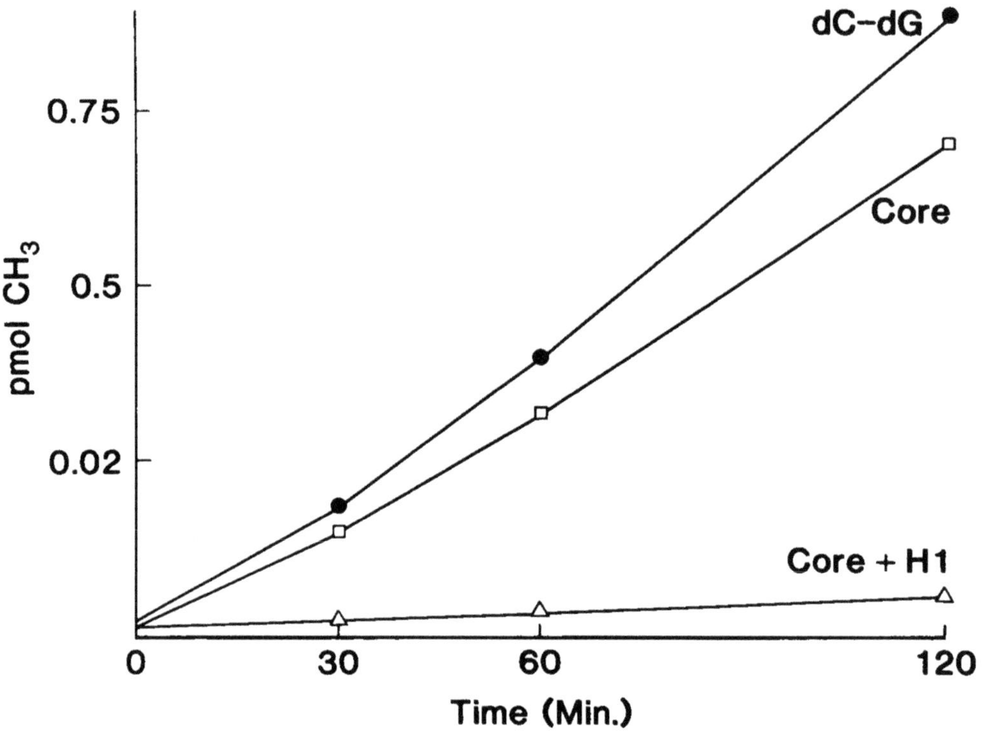

FIGURE 11. Time course of enzymatic methylation of poly(dC-dG) complexed with histones by fraction-III DMase. (-●-●-), poly(dC-dG); (-□-□-), poly(dC-dG) reconstituted with core octomer histones at 1.1 g histone per gram DNA; (-△-△-), poly(dC-dG) reconstituted with total histones at 1.1 g histone per gram DNA. Input 1 μg template.

osition, it is unclear how chromosomal proteins may perpetrate or initiate new patterns of methylation in eukaryotic genomes.

In Figures 11 and 12, poly(dC-dG) was reconstituted with purified core octomer histones or total histones (H1 + octomer histones) at a stoichiometry of 1.1 and 1.5 g histone per gram DNA, respectively. Histones were prepared from HAP chromatography pass through of 2 *M* NaCl, 5 *M* urea, 25 m*M* $NaPO_4$, pH 7.0, 1 m*M* PMSF rat liver chromatin extract,[63] which was repurified over BioRex 70 (BioRad).[58] H1 was purified similarly from two consecutive extracts of 0.5 *M* NaCl, 25 m*M* $NaPO_4$, pH 7.0, 1 m*M* PMSF passed through HAP and absorbed to BioRex at 0.3 *M* NaCl, 0.1 *M* $NaPO_4$, pH 7.0, 1 m*M* PMSF. After washing the absorbed histones on BioRex with the above buffer, histones were eluted at 2 *M* NaCl, 0.01 *M* Tris-HCl, pH 7.4, 0.1 m*M* DTT. Reconstitution was performed by gradient dialysis at 4°C over a 16-hr period from 2 *M* NaCl to 50 m*M* NaCl, at which time "chromatins" were isolated by sedimentation at 15,000 × g for 10 min and resuspended in methylase assay buffer.[75] The time course of methylation of equal inputs of these templates (1 μg) is shown in Figure 11. The rate of methylation of core histone complexed poly(dC-dG) is about 80% the rate of uncomplexed templates. The effect of adding Hl is to drastically reduce the rate by about 90%, presumably by reducing the amount of accessible linker DNA between nucleosome cores and by complexing and stabilizing this DNA region. The relative rate of micrococcal nuclease digestion of these chromatins and poly(dC-dG) at 37°C at 1 μg nuclease per OD_{260} DNA was 0.15:0.4:1.0, respectively (data not shown).

In Figure 12, a competition experiment was set up to investigate if DMase has a higher affinity for nucleohistone or naked DNA templates. As before with BPDE-modified templates, after 30 min incubation the reactions were split and half received the additional 1

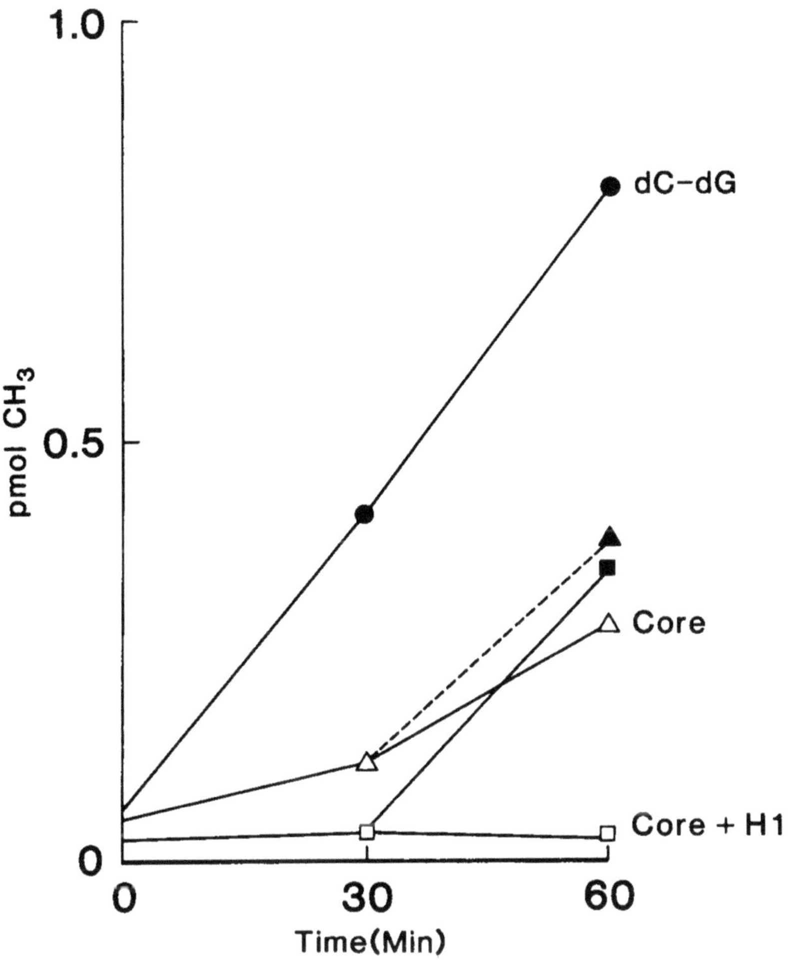

FIGURE 12 Time course of competitive DNA methylation of poly(dC-dG) complexed with histones. (-●-●-), Methylation of control poly(dC-dG); (-△-△-), methylation of 1 μg poly(dC-dG) complexed with core histones at 1.5 g/g DNA; (-△-▲-), methylation of poly(dC-dG) reconstituted with core histone with the addition of 1 μg poly(dC-dG) at 30 min; (-□-□-), methylation of 1 μg poly(dC-dG) reconstituted with total histones at 1.5 g/g DNA. (-□-■-), methylation of poly(dC-dG) reconstituted with total histone with the addition of 1 μg poly(dC-dG) at 30 min.

μg of poly(dC-dG); whereas in controls the rate of methylation of the "chromatins" continues at the previous rate, the effect of addition of exogenous uncomplexed template is to increase the rate to that observed for poly(dC-dG) alone and not the summation of both rates if the enzyme was in excess. These results suggest that DMase can rapidly dissociate from both the nearly completely complexed H1 + core histone template or from the core histone complexed template containing accessible regions. DMase can rebind very efficiently to DNA that has open and exposed sites. Thus, the transient opening up of chromatin to permit DNA replication and repair, as shown from the greater sensitivity of these regions to micrococcal nuclease than regions not undergoing replication, would provide the necessary accessibility requirement for DMase action in chromatin and may explain the paradoxical results on m^5C distribution between linker and core DNA regions.

Further progress in the elucidation of the enzymology and regulation of DMase will be crucial for understanding the function of DNA methylation in eukaryotic cells. The key to the regulation of this enzyme may lie in the supramolecular organization and composition of chromatin domains, especially with respect to NHC protein. This may determine the unequal distribution and in all likelihood the unique pattern of methylation with cell specialization. It may also be fruitful to consider the potential role of DNA methylation in the selection and/or timing of DNA replication in eukaryotes, an area that has not attracted experimental attention.

ACKNOWLEDGMENTS

This work was supported by grants CA 20657 and CA 31487 from the National Cancer Institute. We are grateful for the [^{14}C]-anti-BPDE supplied by the Midwest Research Institute, NCI Chemical Repository. We thank Dr. D. Finkel for his comments and critical reading of this manuscript. We are indebted to Jamileh Noshari and Paula Holton for their excellent technical assistance and to Catherine Johns for preparation of the manuscript.

REFERENCES

1. **Adams, R. L. P.,** The relationship between synthesis and methylation of DNA in mouse fibroblasts, *Biochim. Biophys. Acta*, 254, 205, 1971.
2. **Adams, R. L. P.,** Newly synthesized DNA is not methylated, *Biochim. Biophys. Acta*, 335, 365, 1974.
3. **Adams, R. L. P., McKay, E. L., Douglas, J. T., and Burdon, R. H.,** Methylation of nucleosomal and nuclease sensitive DNA, *Nucleic Acids Res.*, 4, 3097, 1977.
4. **Adams, R. L. P., McKay, E. L., Craig, L. M., and Burdon, R. H.,** Mouse DNA methylase: methylation of native DNA, *Biochim. Biophys. Acta*, 561, 345, 1979.
5. **Adams, R. L. P., Burdon, R. H., Gibbs, S., and McKay, E. L.,** DNA methylase during *Xenopus laevis* development, *Biochim. Biophys. Acta*, 655, 329, 1981.
6. **Adams, R. L. P., Fulton, J., and Kirk, D.,** The effect of 5-azadeoxycytidine on cell growth and DNA methylation, *Biochim. Biophys. Acta*, 697, 286, 1982.
7. **Balmain, A. and Birnie, G. D.,** Nick translation of mammalian DNA, *Biochim. Biophys. Acta*, 561, 155, 1979.
8. **Barrows, L. R. and Magee, P. N.,** Nonenzymatic methylation of DNA by S-adenosylmethionine *in vitro*, *Carcinogenesis*, 3, 349, 1982.
9. **Behe, M. and Felsenfeld, G.,** Effects of methylation on a synthetic polynucleotide: the B-Z transition in poly(dG-m^{5}dC), *Proc. Natl. Acad. Sci. U.S.A.*, 78, 1619, 1981.
10. **Bird, A. P.,** Use of restriction enzymes to study eukaryotic DNA methylation. II. The symmetry of methylated sites supports semi-conservative copying of the methylation pattern, *J. Mol. Biol.*, 118, 49, 1978.
11. **Bird, A. P. and Southern, E. M.,** Use of restriction enzymes to study eukaryotic DNA methylation. I. The methylation patterns in ribosomal DNA from *Xenopus laevis*, *J. Mol. Biol.*, 118, 27, 1978.
12. **Bird, A. P., Taggert, M. H., and Smith, B. A.,** Methylated and unmethylated DNA compartments in the sea urchin genome, *Cell*, 17, 889, 1979.
13. **Bird, A. P.,** DNA methylation and the frequency of CpG in animal DNA, *Nucleic Acids Res.*, 8, 1499, 1980.
14. **Boehm, T. L. J. and Drahovsky, D.,** Effect of carcinogen ethionine on enzymatic methylation of DNA sequences with various degrees of repetitiveness, *Eur. J. Sci.*, 15, 1167, 1979.
15. **Boehm, T. L. J. and Drahovsky, D.,** Enzymatic hypomethylation of inverted repeats in transcribed DNA regions of mouse P815 mastocytoma cells, *Int. J. Biochem.*, 12, 523, 1980.
16. **Boehm, T. L. J. and Drahovsky, D.,** Hypomethylation of DNA in Raji cells after treatment with N-methyl-N-nitrosourea, *Carcinogenesis*, 2, 39, 1981.
17. **Browne, M. J. and Burdon, R. H.,** The sequence specificity of vertebrate DNA methylation, *Nucleic Acids Res.*, 4, 1025, 1977.

18. **Burckhardt, J., Weismann, J., and Yuan, R.,** Characterization of the DNA methylase activity of the restriction enzyme from *Escherichia coli* K, *J. Biol. Chem.*, 256, 4024, 1981.
19. **Burton, W. G., Grabowy, C. T., and Sager, R.,** Role of methylation in the modification and restriction of chloroplast DNA in *Chlamydomonas*, *Proc. Natl. Acad. Sci. U.S.A.*, 76, 1390, 1979.
20. **Cato, A. C. B. and Burdon, R. H.,** Mammalian DNA methylation and a nuclear S-adenosyl-L-methionine-dependent nuclease activity, *FEBS Lett.*, 99, 33, 1979.
21. **Chan, J. Y. H., Ruchirawat, M., Lapeyre, J.-N., and Becker, F. F.,** The protective role of thiol reducing agents in the in vitro inhibition of rat liver DNA methylase by direct acting carcinogens, *Carcinogenesis*, 4, 1097, 1983.
22. **Christman, J. K., Price, P., Pedrinan, L., and Acs, G.,** Correlation between hypomethylation of DNA and expression of globin genes in Friend erythroleukemia cells, *Eur. J. Biochem.*, 81, 53, 1977.
23. **Cox, R., Prescott, C., and Irving, C. C.,** Effects of S-adenosylhomocysteine on DNA methylation in isolated rat liver nuclei, *Biochim. Biophys. Acta*, 474, 493, 1977.
24. **Cox, R.,** Polyamines inhibit DNA methylation *in vitro*, *Biochem. Biophys. Res. Commun.*, 86, 594, 1979.
25. **Cox, R.,** DNA methylase inhibition *in vitro* by N-methyl-N'-nitro-N-nitroso-guanidine, *Cancer Res.*, 40, 61, 1980.
26. **Creusot, F. and Christman, J. K.,** Localization of DNA methyltransferase in the chromatin of Friend erythroleukemia cells, *Nucleic Acids Res.*, 9, 5359, 1981.
27. **Creusot, F., Acs, G., and Christman, J. K.,** Inhibition of DNA methyltransferase and induction of Friend erythroleukemia cell differentiation by 5-azacytidine and 5-Aza-2'-deoxycytidine, *J. Biol. Chem.*, 257, 2041, 1982.
28. **Cohen, J. C.,** Methylation of milk-borne and genetically transmitted mouse mammary tumor virus proviral DNA, *Cell*, 19, 653, 1980.
29. **Compere, S. J. and Palmiter, R. D.,** DNA methylation controls the inducibility of the mouse metallothionein-1 gene in lymphoid cells, *Cell*, 25, 233, 1981.
30. **Doskocil, J. and Sorm, F.,** Distribution of 5-methylcytosine in pyrimidine sequences of DNA, *Biochim. Biophys. Acta*, 55, 953, 1962.
31. **Desrosiers, R. C., Mulder, C., and Fleckenstein, B.,** Methylation of Herpesvirus saimiri DNA in lymphoid tumor cell lines, *Proc. Natl. Acad. Sci. U.S.A.*, 75, 3839, 1979.
32. **Drahovsky, D. and Morris, R. N.,** Mechanism of action of rat liver DNA methylase. I. Interaction with double-stranded methyl-acceptor DNA, *J. Mol. Biol.*, 57, 475, 1971.
33. **Drahovsky, D. and Morris, R. N.,** Mechanism of action of rat liver DNA methylase. II. Interaction with single-stranded methyl-acceptor DNA, *J. Mol. Biol.*, 61, 343, 1971.
34. **Drahovsky, D. and Morris, R. N.,** The mechanism of action of rat liver DNA methylase. III. Nucleotide requirements for binding and methylation, *Biochim. Biophys. Acta*, 277, 245, 1972.
35. **Drahovsky, D. and Wacker, A.,** Enzymatic methylation of replication intermediates in Ehrlich ascites tumor, *Naturwissenschaften*, 62, 189, 1975.
36. **Drahovsky, D., Lacko, I., and Wacker, A.,** Enzymatic DNA methylation during repair synthesis in nonproliferating human peripheral lymphocytes, *Biochim. Biophys. Acta*, 447, 139, 1976.
37. **Dahovsky, D., Boehm, T. L. J., and Kreis, W.,** Distribution pattern and enzymatic hypermethylation of inverted repetitive DNA sequences in P815 mastocytoma cells, *Biochim. Biophys. Acta*, 563, 28, 1979.
38. **Drahovsky, D. and Boehm, T. L. J.,** Enzymatic DNA methylation in higher eukaryotes, *Int. J. Biochem.*, 12, 523, 1980.
39. **Ehrlich, M. and Wang, R. Y.-H.,** 5-Methylcytosine in eukaryotic DNA, *Science*, 212, 1350, 1981.
40. **Eremenko, T., Timofeeva, M. Y., and Volpe, P.,** Organization, replication and modification of the human genome: synthesis and methylation of palindromic repeated and unique HeLa nDNA sequences during the S-phase, *Mol. Biol. Rep.*, 6, 131, 1980.
41. **Friedman, S.,** The inhibition of DNA (Cytosine-5) methylases by 5-azacytidine. The effect of azacytosine-containing DNA, *Mol. Pharmacol.*, 19, 314, 1981.
42. **Gauther, F., Bunimann, H., and Grotian, L.,** Analysis of calf thymus satellite DNA: evidence for specific methylation of cytosine in CG sequences, *Eur. J. Biochem.*, 80, 175, 1977.
43. **Gjerset, R. A. and Martin, D. W.,** Presence of a DNA demethylating activity in the nucleus of murine erythroleukemic cells, *J. Biol. Chem.*, 257, 8581, 1982.
44. **Harbers, K., Harbers, B., and Spencer, J. H.,** Nucleotide clusters in deoxyribonucleic acids. XII. The distribution of 5-methylcytosine in pyrimidine oligonucleotides of mouse L-cell satellite DNA and main band DNA, *Biochem. Biophys. Res. Commun.*, 66, 738, 1975.
45. **Hilliard, J. K. and Sneider, T. W.,** Repair methylation of parental DNA in synchronized cultures of Novikoff hepatoma cells, *Nucleic Acids Res.*, 2, 809, 1975.
46. **Hjelle, B., Phillips, J. A., III, and Seeburg, P. H.,** Relative levels of methylation in human growth hormone and chorionic somatommotropin genes in expressing and non-expressing tissues, *Nucleic Acids Res.*, 10, 3459, 1982.

47. **Holliday, R.**, A new theory of carcinogenesis, *Br. J. Cancer*, 40, 513, 1979.
48. **Holliday, R. and Pugh, J. E.**, DNA modification mechanisms and gene activity during development, *Science*, 187, 226, 1975.
49. **Ishiwata, K. and Oikawa, A.**, Chromatin structure interferes with excision of abnormal bases from DNA, *Biochim. Biophys. Acta*, 698, 15, 1982.
50. **Jack, P. L. and Brookes, P.**, Mechanism for the loss of preferential benzo[a]pyrene binding to the linker DNA of chromatin, *Carcinogenesis*, 3, 341, 1982.
51. **Jones, P. A. and Taylor, S. M.**, Cellular differentiation, cytidine analogs, and DNA methylation, *Cell*, 20, 85, 1980.
52. **Jones, P. A. and Taylor, S. M.**, Hemimethylated duplex DNAs prepared from 5-azacytidine-treated cells, *Nucleic Acids Res.*, 9, 2933, 1981.
53. **Kappler, J. W.**, The kinetics of DNA methylation in cultures of a mouse adrenal cell line, *J. Cell Physiol.*, 75, 24, 1970.
54. **Kappler, J. W.**, The 5-methylcytosine content of DNA: tissue specificity, *J. Cell Physiol.*, 78, 33, 1971.
55. **Kaput, J. and Sneider, T. W.**, Methylation of somatic vs. germ cell DNAs analyzed by restriction endonuclease digestion, *Nucleic Acids Res.*, 7, 2303, 1979.
56. **Kuo, M. T., Mandel, J. L., and Chambon, P.**, DNA methylation: correlation with DNase I sensitivity of chicken ovalbumin and conalbumin chromatin, *Nucleic Acids Res.*, 7, 2105, 1979.
57. **Kuo, M. T., Iyer, B., Wu, J. R., Lapeyre, J.-N., and Becker, F. F.**, DNA methylation of alpha fetoprotein gene in rat liver and hepatocellular carcinomas, *Cancer Res.*, in press, 1984.
58. **Lapeyre, J.-N. and Bekhor, I.**, Chromosomal protein interactions in chromatin and with DNA, *J. Mol. Biol.*, 104, 24, 1976.
59. **Lapeyre, J.-N. and Becker, F. F.**, 5-Methylcytosine content of nuclear DNA during chemical hepatocarcinogenesis and in carcinomas which result, *Biochem. Biophys. Res. Commun.*, 87(3), 698, 1979.
60. **Lapeyre, J.-N., Beattie, W. G., Dugaiczyk, A., Vizard, D., and Becker, F. F.**, Eco RI-generated reiterated components of the rat genome I. Sequence of two (92 and 93 bp) related DNA fragments, *Gene*, 10, 339, 1980.
61. **Lapeyre, J.-N., Walker, M. S., and Becker, F. F.**, DNA methylation and methylase levels in normal and malignant mouse hepatic tissues, *Carcinogenesis*, 2, 873, 1981.
62. **Larocca, D., Homisak, D., and Clough, W.**, Synthesis of hypomethylated Epstein-Barr viral DNA is stimulated by dimethylsulfoxide treatment of lymphoblastoic cells, *Biochem. Biophys. Res. Commun.*, 100, 559, 1981.
63. **MacGillivray, A. J., Cameron, A., Krauze, A., Rickwood, R. J., and Paul, J.**, Non-histone chromosomal proteins of chromatin: their isolation and characterization, *Biochim. Biophys. Acta*, 277, 384, 1972.
64. **Macleod, D. and Bird, A.**, DNase I sensitivity and methylation of active versus inactive rRNA genes in *Xenopus* species hybrids, *Cell*, 29, 211, 1982.
65. **Mandel, J. L. and Chambon, P.**, DNA methylation: organ specific variations in the methylation pattern within and around ovalbumin and other chicken genes, *Nucleic Acids Res.*, 7, 2081, 1979.
66. **Manes, C. and Menzel, P.**, Demethylation of CpG sites in DNA of early rabbit trophoblast, *Nature (London)*, 293, 589, 1981.
67. **McKeon, C., Ohkubo, H., Pastan, I., and de Crombrugghe, B.**, Unusual methylation pattern of the α2(I) collagen gene, *Cell*, 29, 203, 1982.
68. **McGhee, J. D. and Ginder, G. D.**, Specific DNA methylation sites in the vicinity of the chicken β-globin gene, *Nature (London)*, 280, 419, 1979.
69. **Mohandas, T., Sparkes, R. S., and Shaprio, L. J.**, Reactivation of an inactive human X chromosome: evidence for X inactivation by DNA methylation, *Science*, 211, 393, 1981.
70. **Pfohl-Leszkowicz, A., Salas, C., Fuchs, R. P. P., and Dirheimer, G.**, Mechanism of inhibition of enzymatic deoxyribonucleic acid methylation by 2-(acetylamino)fluorene bound to deoxyribonucleic acid, *Biochemistry*, 20(11), 3021, 1981.
71. **Pollack, Y., Stein, R., Razin, A., and Cedar, H.**, Methylation of foreign DNA sequences in eukaryotic cells, *Proc. Natl. Acad. Sci. U.S.A.*, 77, 11, 6463, 1980.
72. **Razin, A. and Cedar, H.**, Distribution of 5-methylcytosine in chromatin, *Proc. Natl. Acad. Sci. U.S.A.*, 74(7), 2725, 1977.
73. **Razin, A. and Riggs, A. D.**, DNA methylation and gene function, *Science*, 210, 604, 1980.
74. **Razin, A. and Friedman, J.**, DNA methylation and its possible biological roles, *Prog. Nucleic Acids Res. Mol. Biol.*, 25, 33, 1981.
75. **Rhodes, D.**, Nucleosome cores reconstituted from poly(dA-dT) and octomer of histones, *Nucleic Acids Res.*, 6, 1805, 1979.
76. **Rigby, P. W. J., Dieckmann, M., Rhodes, C., and Berg, P.**, Labeling of DNA to high specific activity *in vitro* by nick translation with DNA polymerase *I*, *J. Mol. Biol.*, 113, 237, 1977.

77. **Riggs, A. D.,** X inactivation, differentiation, and DNA Methylation, *Cytogenet. Cell Genet.*, 14, 9, 1975.
78. **Roy, P. H. and Weissbach, A.,** DNA methylase from HeLa cell, *Nucleic Acids Res.*, 2, 1669, 1975.
79. **Royer, H.-D. and Sager, R.,** Methylation of chloroplast DNAs in the life cycle of *Chlamydomonas*, *Proc. Natl. Acad. Sci. U.S.A.*, 76, 5794, 1979.
80. **Rybdereg, C. and Lindahl, T.,** Nonenzymatic methylation of DNA by the intracellular methyl group donor S-adenosyl-L-methionine is a potentially mutagenic reaction, *EMBO J.*, 2, 211, 1982.
81. **Sager, R. and Kitchin, R.,** Selective silencing of eukaryotic DNA, *Science*, 189, 426, 1975.
82. **Salas, C. E., Pfohl-Leszkowicz, A., Lang, M. D., and Dirheimer, G.,** Effect of modification by N-acetoxy-N-2-acetylaminofluorene on the level of DNA methylation, *Nature (London)*, 278, 71, 1979.
83. **Sano, H., Grabowy, C., and Sager, R.,** Differential activity of DNA methyltransferase in the life cycle of *Chlamydomonas reinhardi*, *Proc. Natl. Acad. Sci. U.S.A.*, 78(5), 3118, 1981.
84. **Sano, H. and Sager, R.,** Tissue specificity and clustering of methylated cytosines in bovine satellite I DNA, *Proc. Natl. Acad. Sci. U.S.A.*, 79, 3584, 1982.
85. **Scarano, E., Iaccarino, M., Grippo, P., and Parisi, E.,** The heterogeneity of thymine methyl group origin in DNA pyrimidine isostichs of developing sea urchin embryos, *Proc. Natl. Acad. Sci. U.S.A.*, 57, 1394, 1967.
86. **Shen, C.-K. J. and Maniatis, T.,** Tissue-specific DNA methylation in cluster or rabbit β-like globin genes, *Proc. Natl. Acad. Sci. U.S.A.*, 77, 6634, 1980.
87. **Simon, D., Grunert, F., v. Acken, U., Doring, H. P., and Kroger, H.,** DNA-methylase from regenerating rat liver: purification and characterization, *Nucleic Acids Res.*, 5, 2153, 1978.
88. **Smerdon, M. J., Tlsty, T. D., and Lieberman, M. W.,** Disribution of ultraviolet-induced DNA repair synthesis in nuclease sensitive and resistance regions of human chromatin, *Biochemistry*, 17, 2377, 1978.
89. **Sneider, T. W., Teague, W. M., and Rogachevsky, L. M.,** S-adenosylmethionine: DNA-cytosine 5-methyltransferase from a Novikoff rat hepatoma cell line, *Nucleic Acids Res.*, 2, 1685, 1975.
90. **Sneider, T. W., Kaput, J., Neiman, D., and Westmoreland, B.,** Hemimethylation of DNA: a basis for genetic recombination? in *Transmethylation*, Usdin, E., Borchardt, R. T., and Creveling, C. R., Eds., Elsevier/North Holland, Amsterdam, 1978, 473.
91. **Solage, A. and Cedar, H.,** Organization of 5-methylcytosine in chromosomal DNA, *Biochemistry*, 17, 2934, 1978.
92. **Srinivasan, P. R. and Borek, E.,** Enzymatic alteration of macromolecule structure, *Prog. Nucleic Acid Res.*, 5, 157, 1966.
93. **Stewart, C. L., Stuhlman, H., Jahner, D., and Jaenisch, R.,** *De novo* methylation, expression, and infectivity of retroviral genomes introduced into embryonal carcinoma cells, *Proc. Natl. Acad. Sci. U.S.A.*, 79, 4098, 1982.
94. **Stuhlmann, H., Jahner, D., and Jaenisch, R.,** Infectivity and methylation of retroviral genomes is correlated with expression in the animal, *Cell*, 26, 221, 1981.
95. **Straub, K. M., Meehan, T., Burlingame, A. L., and Calvin, M.,** Identification of the major adducts formed by reaction of benzo[a]pyrene diolepoxide with DNA *in vitro*, *Proc. Natl. Acad. Sci. U.S.A.*, 12, 5285, 1977.
96. **Sturm, K. S. and Taylor, J. H.,** Distribution of 5-methylcytosine in the DNA of somatic and germline cells from bovine tissues, *Nucleic Acids Res.*, 9, 4537, 1981.
97. **Turnbull, J. F. and Adams, R. L. P.,** DNA methylase: purification from ascites cells and the effect of various DNA substrates on its activity, *Nucleic Acids Res.*, 3, 677, 1976.
98. **Vanyushin, B. F., Tkacheva, S. G., Belozersky, A. N.,** Rare bases in animal DNA, *Nature (London)*, 225, 948, 1970.
99. **Vanyushin, B. F., Mazin, A. L., Vasilyev, V. K., and Belozersky, A. N.,** The content of 5-methylcytosine in animal DNA: the species and tissue specificity, *Biochim. Biophys. Acta*, 299, 397, 1973.
100. **Vardimon, L., Neumann, K., Kuhlman, I., Sutter, D., and Doerfler, W.,** DNA methylation and viral gene expression in adenovirus transformed and infected cells, *Nucleic Acids Res.*, 8, 2461, 1980.
101. **Venolia, L., Gartler, S. M., Wassman, E. R., Yen, P., Mohandas, T., and Shapiro, L. J.,** Transformation with DNA from 5-azecytidine-reactivated X chromosomes, *Proc. Natl. Acad. Sci. U.S.A.*, 79, 2352, 1982.
102. **Waalwijk, C. and Flavell, R. A.,** DNA methylation at a CCGG sequence in the large intron of the rabbit β-globin gene: tissue-specific variations, *Nucleic Acids Res.*, 5, 4634, 1978.
103. **Walker, M. S. and Becker, F. F.,** DNA methylase activity of normal liver, regenerating liver, and a transplantable hepatocellular carcinoma, *Cancer Biochem. Biophys.*, 5, 169, 1981.
104. **Weintraub, H. and Groudine, M.,** Chromosomal subunits in active genes have an altered conformation, *Science*, 193, 848, 1976.
105. **Wigler, M., Levy, D., and Perucho, M.,** The somatic replication of DNA methylation, *Cell*, 24, 33, 1981.
106. **Wigler, M. H.,** The inheritance of methylation patterns in vertebrates, *Cell*, 24, 285, 1981.

107. **Wilks, A. F., Cozens, P. J., Mattaj, I. W., and Jost, J.-P.,** Estrogen induces a demethylation at the 5′ end region of the chicken vitellogenin gene, *Proc. Natl. Acad. Sci. U.S.A.*, 79, 4252, 1982.
108. **Wu, C., Wong, Y. C., and Elgin, S. C. R.,** Chromatin structure of specific genes. II. Disruption of chromatin structure during gene activity, *Cell*, 16, 807, 1979.
109. **Wyatt, G. R.,** Recognition and estimation of 5-methylcytosine in nucleic acids, *Biochem. J.*, 48, 581, 1951.
110. **Yagi, M. and Koshland, M. E.,** Expression of J chain during B cell differentiation is inversely correlated with DNA methylation, *Proc. Natl. Acad. Sci. U.S.A.*, 78, 4907, 1981.
111. **Zolan, M. E., Smith, C. A., Calvin, N. M., and Hanawalt, P. C.,** Rearrangement of mammalian chromatin structure following excision repair, *Nature (London)*, 299, 462, 1982.

Chapter 10

ALKALINE AND NEUTRAL PROTEINASES BOUND TO CHROMATIN: PROPERTIES AND FUNCTIONS

Kaoru Miyazaki, Nobuhiko Yamashita, Yasunori Kihira, Hideaki Hagiwara, Yosinobu Nagao, Jinpei Yamashita, and Takekazu Horio

TABLE OF CONTENTS

I. INTRODUCTION

It has long been thought that histones and nonhistone proteins in the nuclei are functional in the gene expression of eukaryotic cells. Although this thought has been supported by numerous studies carried out in this decade, no conclusive evidence has been obtained for physiological functions of nuclear nonenzymatic proteins. Besides the nonenzymatic proteins, the nuclei contain enzymes such as DNA and RNA polymerases which catalyze the replication and the transcription of genes, respectively. The nuclei contain other enzymes such as proteinase, protein kinase, acetylase, methylase, poly(ADP) synthetase, etc.[1] These enzymes catalyze chemical modifications of nuclear proteins. Many workers have supposed that the protein kinases catalyze the phosphorylation of chromosomal proteins to regulate gene expression.[2] The enzymes also phosphorylate RNA polymerases and their regulator proteins.[3,4] Over 20 years ago, Philips and Johns[5] first reported that a proteinase was present in chromatin preparations from calf thymus. However, the enzyme, as well as most of the other nuclear enzymes, had been neither highly purified nor well characterized in the beginning of the present study. In addition, it was controversial whether the enzyme was originally present in the nuclei or a contaminant from other organelles.

We have recently solubilized, highly purified, and characterized two different proteinases from the fraction of chromatin of the cancerous tissue of rats, rhodamine sarcoma.[6-8] They are called alkaline and neutral proteinases after their optimum pH of 10 and 8, respectively. Earlier, we highly purified a protein with a molecular weight of about 60 k and an isoelectric point (pI) of about 5.0 from the tumor (60-k protein).[9] When 60-k protein is i.p. or s.c. injected into normal rats, the biosynthesis of catalase in the liver is depressed, whereas the protein does not inhibit the activity of the enzyme in vitro. This chapter deals mainly with the properties of alkaline and neutral proteinases and their function for the intranuclear processing of 60-k protein.

The animals, reagents, and methods used were described previously.[6-10] Various tumors including rhodamine sarcoma were transplanted on the backs of rats.[9]

II. EXISTENCE OF TWO DIFFERENT PROTEINASES IN NUCLEI

Philips and Johns[5] are the first to observe that histones are hydrolyzed by a proteinase in nuclei. This enzyme was extracted at an acidic pH together with histones from the fraction of calf thymus chromatin, and its activity was inhibited by diisopropyl fluorophosphate (DFP). Later, Sarkar and Dounce[11] reported that the nuclear fraction contained four proteinases different in optimum pH, 3.7, 5.6, 7.0, and 9.0. Furlan and Jericijo[12,13] partially purified a serine proteinase from the nuclear fraction. The purified enzyme specifically hydrolyzed histones at the neutral pH range. Since then, the histone-specific proteinase was studied by many workers.[14-26]

We measured the proteolytic activities of various subcellular fractions isolated from rhodamine sarcoma (Figure 1). The reagents added as the exogenous substrates were casein (Hammarsten grade, Merck) and histone (type II-S, Sigma). The buffers used were 20 m*M* GTA buffer (an equimolar mixture of 3,3-dimethylglutaric acid, Tris, and 2-amino-2-methyl-1,3-propanediol, pH 3 to 10) and 0.1 *M* borate buffer (pH 7.5 to 12). With casein, the pH-activity curves had a broad peak at pH 10 with the fractions of nuclei and chromatin. This peak was not observed with the other fractions. Occasionally with less pure fractions of nuclei, a smaller peak at about pH 4.5 also was observed. The main peaks were at pH 5.5 with the mitochondrial fraction, at pH 5.5 and 8 with the microsomal fraction, and at pH 6 and 8.5 with the lysosomal fraction. With histone, the fractions of nuclei and chromatin showed a peak at pH 8 (Figure 2). These results indicated that the fractions of nuclei and chromatin contained at least two kinds of proteinases, and that their activities were separately

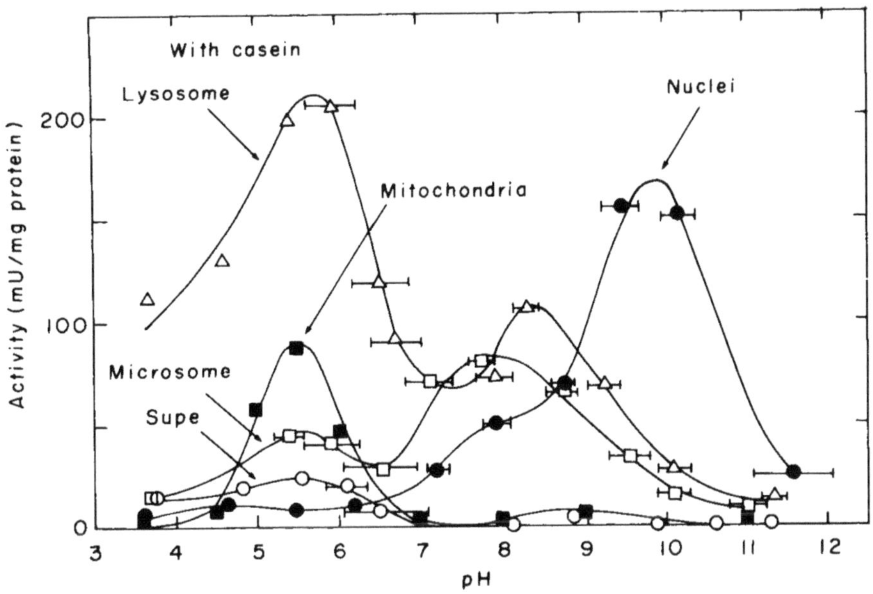

FIGURE 1. Effect of pH on casein-hydrolyzing activities in various subcellular fractions from rhodamine sarcoma.[6]

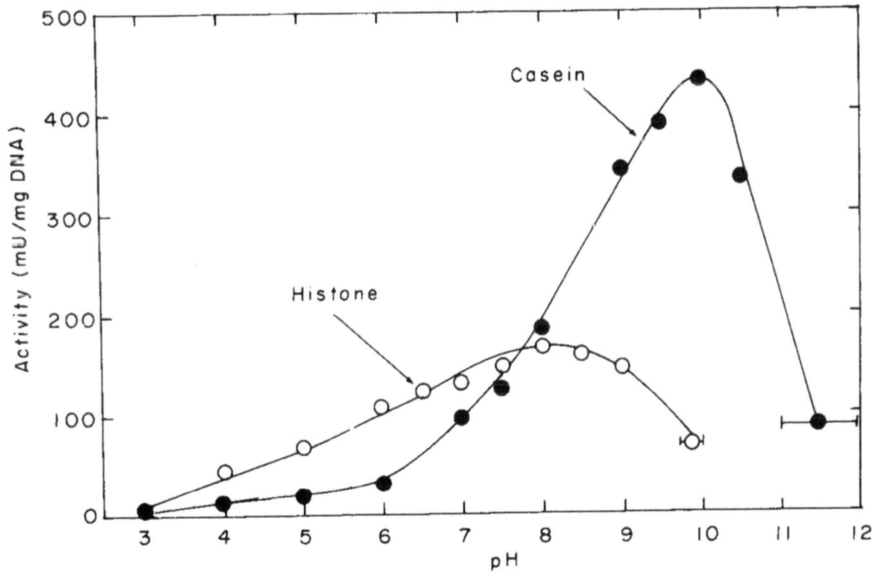

FIGURE 2. Effect of pH on histone and casein-hydrolyzing activities in chromatin fraction from rhodamine sarcoma.[6]

measurable by assay at pH 10 with casein (alkaline proteinase activity) and at pH 8 with histone (neutral proteinase activity). Both proteinase activities were more abundant in nuclear fractions from malignant tissues than in those from normal ones (Table 1). In the five normal tissues tested, the faster the rate of cell proliferation, the higher the alkaline proteinase activity; thymus > spleen > kidney ≧ liver > brain. This was the case also in the malignant tissues: Yoshida sarcoma > rhodamine sarcoma ≧ AH130 hepatoma. Although the neutral proteinase activity also was higher in the tumors than in the normal tissues, it was not so

Table 1
DISTRIBUTION OF ALKALINE AND NEUTRAL PROTEINASES IN NUCLEI FROM VARIOUS TISSUES OF RATS[6]

	Specific activity (mU/mg DNA)	
Tissue	**Alkaline proteinase[a]**	**Neutral proteinase[b]**
Tumors		
Yoshida sarcoma	1052	322
Rhodamine sarcoma	444	211
AH130 hepatoma	422	185
Normal rat tissues		
Thymus	367	78
Spleen	300	130
Kidney	196	56
Liver	181	100
Brain	52	33

[a] Assayed at pH 10 with casein.
[b] Assayed at pH 8 with histone.

proportional to the rate of cell proliferation as the alkaline proteinase activity. Both alkaline and neutral proteinase activities were irreversibly inhibited by DFP, indicating that they were of the class of serine proteinase.

Carter et al.[21] incubated chromatin fractions from normal tissues of rats with [^{3}H]DFP at a neutral pH and analyzed the proteins bound with the radioactivity by sodium dodecyl sulfate-polyacrylamide gel electrophoresis (SDS-PAGE). The radioactive proteins has molecular weights of about 70, 60, and 25 k. They reported that 25-k protein was converted into 20-k protein when the fractions were incubated at an acidic pH. In our experiments with rhodamine sarcoma, 27-k protein was significantly labeled with radioactivity when nuclear fractions were incubated with [^{3}H]DFP at pH 8.0. Other proteins with molecular weights of about 48, 43, and 18 k also were labeled, but to a significantly lower extent. When nuclear fractions were incubated at pH 10, 18-k protein was newly labeled to a significant extent, while 27-K protein was labeled to the same extent as in incubation at pH 8.0. The radioactivities due to 18- and 27-k proteins were practically the same. Other proteins with molecular weights of about 48, 43, and 16.5 k also were labeled, but to a significantly lower extent. The alkaline and neutral proteinase activities were extractable from fractions of nuclei and chromatin at an acidic pH in the presence of a high ionic strength (see below). When the acid extract was incubated with [^{3}H]DFP at pH 8.0, both 27- and 18-k proteins were labeled in nearly the same manner as when the acid extract was incubated at pH 10. The same experiments with pure samples of (neutral) 27- and (alkaline) 18-k proteinases showed that 27- and 18-k proteins were histone-specific and casein-specific (nonhistone-specific), respectively. It seems unlikely that 27-k protein was the precursor of 18-k protein, because the DFP-binding of 18-k protein in incubation at pH 10 was not accompanied by any decrease in the DFP-binding of 27-k protein at pH 8. Tsurugi and Ogata[26] have recently reported essentially the same results as ours, but with minor differences in molecular weight. In addition, they supposed that alkaline proteinase was formed from 52-k protein capable of binding with DFP.

Some workers insisted that neutral proteinase did not originate in nuclei, and it arose from other subcellular organelles such as mitochondria, lysosomes, or microsomes which had contaminated nuclear fractions.[27-30] In fact, as described above, less pure fractions of nuclei occasionally show an acidic proteinase activity with optimum pH of 4 to 6. This activity varies from one preparation of nuclear fraction to another, and it is removable by washing the fractions with isotonic saline or 1% Triton® X-100.[8,12] It seems likely that the acidic proteinase activity observed in some of the nuclear fractions is due to cathepsins of the lysosomes contaminating them.

As described above, the analysis for the intracellular distribution of the proteolytic activity in rhodamine sarcoma indicated that alkaline proteinase was strictly localized in the nuclei. On the other hand, a neutral proteinase activity with an optimum pH of 7 to 9 was observed with nuclear, microsomal, and lysosomal fractions, but not with mitochondrial fractions; the specific activities with casein as the substrate were in the order lysosomes > microsomes > nuclei (Figure 1). On the other hand, DFP-binding 18-k protein was found only with nuclear fractions, whereas DFP-binding 27-k protein was with nuclear and microsomal fractions to similar extents, but not with lysosomal fractions. The extraction of alkaline and neutral proteinases was attempted by suspending fractions of chromatin or nuclei in various solvents. Neither alkaline nor neutral proteinase was extracted with 0.15 *M* NaCl or 1% Triton® X-100 at the neutral pH range, although neutral, but not alkaline proteinase was in a small part extracted with 0.7 *M* NaCl. Both proteinases were extracted most effectively when the fractions were suspended in a mixture of 0.2 *M* H_3PO_4 and 0.5 *M* NaCl (pH 1.5). These results suggest that these two proteinases were tightly bound to the chromatin. Therefore, it is reasonable to conclude that alkaline proteinase originated in the nuclei, whereas neutral proteinase originated in both the nuclei and microsomes.

III. PURIFICATION AND CHARACTERIZATION OF ALKALINE AND NEUTRAL PROTEINASES

Several groups have partially purified histone-specific, neutral proteinase from fractions of nuclei and chromatin of animal tissues. However, there are serious discrepancies in their descriptions. The enzyme obtained from rat liver by Chong et al.[16] has a molecular weight of 200 k, is composed of two identical subunits, and is activated by Mn^{++}. A similar enzyme was obtained by Ramponi et al.[23] On the other hand, the enzymes obtained from calf thymus by Furlan and Jericijo[13] and Kuręcki and Toczko[17] had molecular weights of 24 and 15.4 k, respectively. Carter et al.[21] observed an enzyme with a molecular weight of 25 k in rat liver, supposing that the 200-k proteinase originated in extranuclear organelles.

We purified both alkaline and neutral proteinases from chromatin fractions of rhodamine sarcoma to a homogeneous purity by the method described below.[7] These two enzymes were extracted from the fractions with 0.2 *M* H_3PO_4/0.5 *M* NaCl (pH 1.5). This acid extract was subjected to two successive affinity chromatographies: one on a soybean trypsin inhibitor-Sepharose® column and the other on a casein-Sepharose® column. The resultant eluate contained both enzymes, and it was then subjected to another affinity chromatography on a histone-Sepharose® column. Neutral, but not alkaline, proteinase was adsorbed to this column, and fractions of the eluate showing the alkaline proteinase activity were collected. At this step, alkaline proteinase was purified 1400-fold from the chromatin fractions with a yield of 4%. The neutral proteinase adsorbed on the histone-Sepharose® column was hardly eluted with all solutions tested, thus, it was purified differently. The initial acid extract was applied on a soybean trypsin inhibitor-Sepharose® column previously equilibrated with 30 m*M* sodium citrate-HCl/0.2 *M* NaCl (pH 6.5). When the volume of the extract applied was appropriate, both alkaline and neutral proteinases were adsorbed as described above. However, when it was increased, a greater amount of neutral proteinase was adsorbed, while the alkaline proteinase once adsorbed was eluted. Therefore, it was applied to the column in such an excess volume that the eluate would show the neutral proteinase activity to an appreciable extent. The charged column was washed with sodium acetate-HCl/0.2 *M* NaCl (pH 1.5). The resultant column adsorbed only neutral proteinase, whereas all the other

Table 2
SUBSTRATE-SPECIFICITIES OF PURIFIED ALKALINE AND NEUTRAL PROTEINASES FROM RHODAMINE SARCOMA[7]

Substrate	Specific activity (mU/mg DNA)	
	Alkaline proteinase[a]	Neutral proteinase[b]
Natural		
Casein	91	12
Histone (mixtures)	16	37
Arginine-rich histone (H3)	7	25
Lysine-rich histone (H1)	4	17
Synthetic		
Acetyl-L-phenylalanine ethyl ester	1.6	19
Acetyl-L-tyrosine ethyl ester	1.5	31
Benzoyl-L-arginine ethyl ester	0.5	4.0

[a] Assayed at pH 10 with casein.
[b] Assayed at pH 8 with histone.

proteins including alkaline proteinase were passed through. The neutral proteinase adsorbed was eluted with 0.1 *M* sodium acetate-HCl/0.2 *M* NaCl/2 *M* urea (pH 1.5). At this step, neutral proteinase was purified 920-fold with a yield of 12%.

Purified samples of alkaline and neutral proteinases were analyzed by SDS-PAGE and isoelectric electrophoresis. It was found that these two enzyme samples were homogenous, having molecular weights of 18 and 27 k for alkaline and neutral proteinases, respectively, and isoelectric points (pI) of about 11 for both.

The properties of alkaline and neutral proteinases are summarized in Tables 2 and 3. As mentioned above, these two enzymes had different specificities for histone and casein as substrates. The pH-activity curves of the purified enzymes resembled those with the enzymes bound to nuclei or chromatin. The optimum pH of the purified alkaline proteinase was 9.5, slightly lower than that of the bound one. The curves with histone and casein were practically the same for each of the purified enzymes; the activity with histone was about 18% of that with casein for alkaline proteinase, and the activity with casein was about 32% of that with histone for neutral proteinase. Alkaline proteinase hydrolyzed casein and nonhistone proteins similarly. With synthetic substrates, both enzymes cleaved the ester bonds with aromatic amino acids faster than that with arginine. Both alkaline and neutral proteinase activities were inhibited by chymostatin, a specific inhibitor for chymotrypsin, but hardly by leupeptin, antipain, or pepstatin. They were inhibited also by DFP and soybean trypsin inhibitor. Relatively high concentrations of phenylmethylsulfonyl fluoride (PMSF) and *p*-chloromercuribenzoate inhibited the neutral proteinase activity to a higher extent than the alkaline proteinase activity. The substrate-specificities and the sensitivities to the inhibitors indicate that alkaline and neutral proteinases were chymotrypsin-like, serine proteinases, although they showed distinctly different properties from those of a typical chymotrypsin.

It seems most likely that the neutral proteinase highly purified by us was the same as the ones reported by Carter et al.[21] and Furlan and Jericijo.[13] The difference between their values and ours of molecular weight (24 to 25 and 27 k, respectively) may be within an experimental error. However, the alkaline proteinase purified by us differs from the one reported by Kurecki and Toczko,[17] who found the latter enzyme to prefer histone over casein as the substrate, although their value of molecular weight (15.4 k) and ours (18 k) are similar to each other.

Table 3
EFFECT OF VARIOUS INHIBITORS ON ALKALINE AND NEUTRAL PROTEINASE ACTIVITIES IN RHODAMINE SARCOMA CHROMATIN[6]

		Relative activity (%)	
Inhibitor		**Alkaline proteinase[a]**	**Neutral proteinase[b]**
No addition		(100)	(100)
DFP[c]	1 m*M*	43	14
	5 m*M*	15	0
PMSF[d]	1 m*M*	80	40
	5 m*M*	70	9
STI[e]	100 μg/mℓ	44	41
	200 μg/mℓ	31	38
PCMB[f]	1 m*M*	87	69
	5 m*M*	63	16
EDTA	1 m*M*	84	94
	5 m*M*	84	84
Chymostatin	1 μg/mℓ	32	20
	10 μg/mℓ	6	9
Leupeptin	1 μg/mℓ	100	91
	10 μg/mℓ	100	91
Antipain	1 μg/mℓ	100	94
	10 μg/mℓ	100	81
Pepstatin	1 μg/mℓ	88	63
	10 μg/mℓ	84	60

[a] Assayed at pH 10 with casein.
[b] Assayed at pH 8 with histone.
[c] Diisopropyl fluorophosphate.
[d] Phenylmethylsulfonylfluoride.
[e] Soybean trypsin inhibitor.
[f] *p*-Chloromercuri benzoic acid.

IV. PROTEOLYSIS OF NUCLEAR PROTEINS AND ITS REGULATION

It is generally accepted that in vivo in the nuclei, nonhistone proteins have high turnover rates, while histones are highly conserved. However, in fractions of nuclei and chromatin, not only nonhistone proteins, but also histones are rapidly hydrolyzed by endogenous proteinases.[14,16,20] This suggests that in vivo the neutral proteinase activity is strictly depressed by a certain mechanism.

Using nuclear fractions from rhodamine sarcoma, we studied the in vitro hydrolysis of the proteins under four different conditions (Figure 3). The alkaline and neutral proteinase activities at pH 7.5 were about 41 and 220% of those at pH 10, respectively, which were practically the same as the activities of the enzymes bound to nuclei or chromatin. When nuclear fractions isolated from rhodamine sarcoma were incubated in 0.1 *M* Tris-HCl buffer (pH 7.5), the hydrolysis rates of the histones were in the order H1 ≧ H3 > H4 > H2B > H2A. When they were incubated in the buffer supplemented with 2 *M* NaCl and 5 *M* urea where the chromosomal proteins could be mostly dissociated from the DNA, the fast hydrolysis of the histones and the nonhistone proteins occurred without apparent specificities. When the nuclear fractions were incubated in 0.1 *M* sodium borate buffer (pH 10) with and without 2 *M* NaCl and 5 *M* urea, the hydrolysis rates of the histones and nonhistone proteins

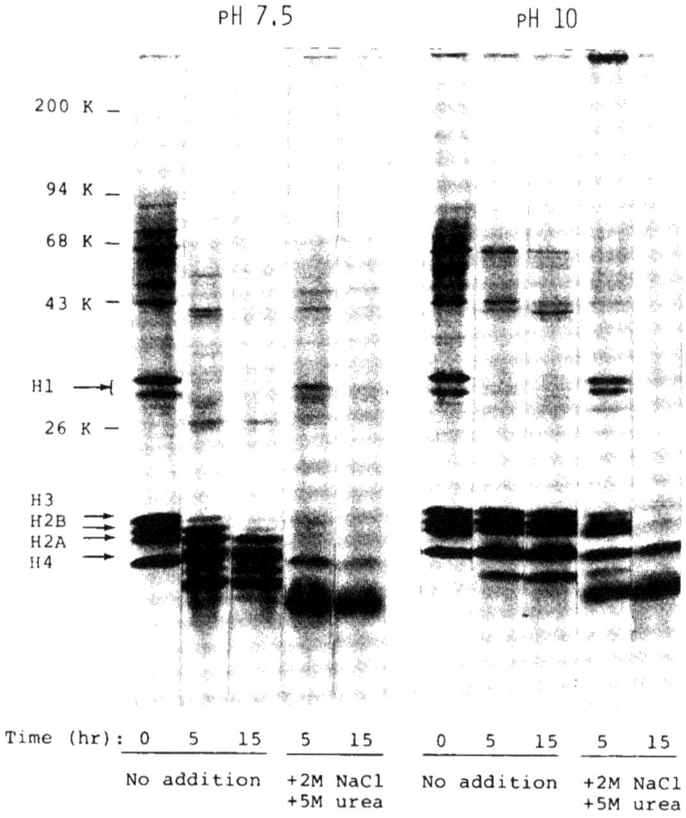

FIGURE 3. Hydrolysis of proteins in nuclear fraction from rhodamine sarcoma under four different conditions. Nuclear fraction from rhodamine sarcoma was incubated at 37°C for the indicated lengths of time in the presence and absence of 2 *M* NaCl and 5 *M* urea at pH 7.5 and at pH 10.0, and then analyzed by SDS-PAGE.

were lower and higher than those at pH 7.5, respectively. The relative hydrolysis rates of the histones were H1 $\gg$ H3 = H2B = H2A = H4 in the absence NaCl and urea and H2B $\geqq$ H1 = H3 = H2A > H4 in the presence of NaCl and urea. The major nonhistone proteins with molecular weights of 64 and 43 k were relatively resistant to proteolysis in the absence but not in the presence of NaCl and urea. The marked differences in the degradation of nuclear proteins with and without 2 *M* NaCl and 5 *M* urea indicate the importance of the localization of the two enzymes for their proteolytic activities. At pH 7 to 8, which is physiological, the alkaline proteinase activity was about 41% of that at pH 10, and its activity with histone was about 18% of that with casein, as mentioned above. On the other hand, the activity of neutral proteinase with casein was about 32% of that with histone. This suggests that in vivo in the nuclei, the histones are mostly hydrolyzed by neutral proteinase, whereas the nonhistone proteins are hydrolyzed similarly by both alkaline and neutral proteinases, provided that these two proteinases are present at similar activities to each other.

We have recently studied the distribution of alkaline and neutral proteinases in nuclear fractions from rhodamine sarcoma and livers of rats (Table 4). The nuclear fractions were divided into the two fractions, nucleoli and the remaining extranucleoli. The specific activity (activity per milligram of DNA) of alkaline proteinase was significantly higher with nucleoli than with extranucleoli, whereas that of neutral proteinase was appreciably higher with extranucleoli than with nucleoli. On the other hand, when chromatin fractions were digested

Table 4
SUBNUCLEAR DISTRIBUTION OF ALKALINE AND NEUTRAL PROTEINASE ACTIVITIES IN RHODAMINE SARCOMA AND NORMAL RAT LIVER

	Specific activity (mU/mg DNA)	
Fraction	Alkaline proteinase[a]	Neutral proteinase[b]
Rhodamine sarcoma		
Whole nuclei	238	59
Nucleoli	474	47
Extranucleoli	147	95
Normal rat liver		
Whole nuclei	53	33
Nucleoli	153	33
Extranucleoli	33	47

[a] Assayed at pH 10 with casein.
[b] Assayed at pH 8 with histone.

with micrococcal nuclease and then divided into fractions of eu- and heterochromatin, the neutral proteinase activity was found with both fractions to a similar extent. The fact that alkaline and neutral proteinases are firmly bound to chromatin suggests the possibility that the in vivo activities of alkaline and neutral proteinases are influenced by the movement of the DNA strings carrying these two enzymes during the cell cycle for proliferation.

Watson and Moudrianakis[31] digested chromatin fractions from calf thymus into their fragments with micrococcal nuclease, separated the resultant fragments, and detected the neutral proteinase activity in fractions of the monomer nucleosomes, but not in fractions of the spacer DNA; the tetramer of histones H3 and H4 was required for the binding of the proteinase to DNA. On the other hand, Lipinska and Klyszejko-Stefanowicz[25] reported that neutral proteinase was selectively extracted together with histone H2B from chromatin fractions of calf thymus and rat liver, suggesting that the proteinase was bound to the nucleosomal structure through its affinity to the histone.

Marushige et al.[32] reported that the acetylation of rat spermatid chromatin by acetic anhydride markedly increased the susceptibility of histones to a chromatin-bound proteinase. Ramponi et al.[23] also reported that the degradation of histones by neutral proteinase was stimulated when the proteins were acetylated, and depressed when they were carbamylated. Inagaki et al.[33] found that poly(ADP-ribose) inhibited the neutral proteinase activity in peritoneal macrophage chromatin. However, it is unknown whether the inhibitor reacted with the enzyme or with the substrate, although Okayama et al.[34] found that poly(ADP-ribose) can be covalently bound to histones H1 and H2B and other nuclear proteins in vivo. These results suggest the possibility that the chemical modification of nuclear proteins regulates the chromatin-bound proteinase activities.

V. CHANGES OF ALKALINE AND NEUTRAL PROTEINASE ACTIVITIES IN LIVER BY TUMOR-BEARING RATS

Miyazaki et al.[10] found that in rats, the kinds and amounts of nonhistone proteins present in nuclei significantly differed from one kind of tissue to another (tissue specificity). This specificity is probably due to differences in the gene expressions for the synthesis of the respective proteins and their turnover rates. As described above, the alkaline and neutral proteinase activities significantly vary for the kinds of tissues, suggesting that they may influence the turnover rates of nonhistone proteins.

Table 5
EFFECT OF TUMOR GROWTH ON ALKALINE AND NEUTRAL PROTEINASE ACTIVITIES IN LIVER NUCLEI OF HOST RATS

T/B[a]	Specific activity (mU/mg DNA)	
(%)	Alkaline proteinase[b]	Neutral proteinase[c]
0	181 (100%)	83 (100%)
4	227 (125%)	77 (93%)
14	261 (144%)	86 (104%)
24	283 (156%)	81 (98%)

[a] Ratio of tumor to body weight.
[b] Assayed at pH 10 with casein.
[c] Assayed at pH 8 with histone.

The growth of tumors in animals causes qualitative and quantitative changes of many enzymes in the liver.[35-38] We found that the growth of rhodamine sarcoma on the back of rats depressed the total content and the phosphorylation of nonhistone proteins, but it hardly influenced those of histones.[10,39] The decrease of seven kinds of nonhistone proteins with molecular weights of 30 to 52 k accounted for most of the decrease of the total content. On the other hand, the alkaline proteinase activity increased, but the neutral proteinase activity did not (Table 5). This suggests the possibility that the increase of the alkaline proteinase activity could cause the decrease of the total content and the phosphorylation of nonhistone proteins. Among five normal tissues, the higher the alkaline proteinase activity, the lower the total content of nonhistone proteins (see Table 1 and Reference 10). For example, the thymus is the highest in proteinase activity and the lowest in total content of nonhistone proteins. In contrast, the brain is the lowest in proteinase activity and the highest in protein content. However, this is not the case in cancerous tissues; rhodamine sarcoma is higher than the five normal tissues in both alkaline proteinase activity and total content of nonhistone proteins.

VI. PHYSIOLOGICAL FUNCTIONS OF ALKALINE AND NEUTRAL PROTEINASES

In nuclei, alkaline proteinase is bound to nucleoli, which have the gene for the ribosomal RNA (rRNA). Cells require the synthesis of ribosomes for their growth. Liau and Perry[40] reported that the amount of the proteins bound to immature ribosomes (ribonucleoproteins) gradually decreased during the maturation of ribosomes. Giese and Wunderlich[41] found that several proteinase inhibitors including PMSF and parahydroxymercuribenzoate depressed the transport of the ribonucleoproteins from the nuclei to the cytoplasm. Therefore, it seems likely that alkaline proteinase functions in the synthesis of ribosomes. Tsurugi and Ogata[42] found that ribosomal proteins newly synthesized in the regenerating rat liver were transported into the nuclei, where they were rapidly hydrolyzed by a thiol proteinase when the rRNA synthesis had been inhibited by the injection of actinomycin D. They reported that the thiol proteinase present in the nuclei had a molecular weight of about 40 k, and hydrolyzed ribosomal proteins as well as casein and histone in vitro with an optimum pH of 5.5.[43] On the other hand, Brown et al.[44] found that the cell lysate, if treated with a low concentration of trypsin, stimulated in vitro DNA synthesis in nuclei of chinese hamster fibroblasts at the G1 phase but not at the S phase, suggesting that this stimulation resulted from the direct action of the proteinase on DNA polymerase. There is a possibility that the action of alkaline and neutral proteinases stimulates DNA synthesis of proliferating cells. Chae et al.[45] reported that there was no strict relationship between the degradation rate of H1 histone in chromatin by neutral proteinase and the growth rates of four different hepatomas and normal rat liver. To the contrary, Bartley and Chalkley[14] and Lipinska and Klyszejko-Stefanowicz[25] reported that the activity of a chromatin-bound neutral proteinase was higher in tissues with high rates of cell turnover than those with low rates. Procaccini et al.[46] found a significant increase of histone-hydrolyzing activity in nuclear and cytosol fractions from the regenerating limbs of adult newt. In addition, Garrard and Bonner[47] reported that the turnover rate of histones increased in the regenerating rat liver.

As described above, the serine proteinases present in nuclei are alkaline and neutral proteinases, which are bound to chromatin. There is a possibility that these two proteinases may function in the turnover of nuclear proteins. In fact, histones and nonhistone proteins in isolated nuclei or chromatin are rapidly degraded by the proteinases, although in vivo the turnover rates of histones are negligible or extremely slow under most of the physiological conditions.[48,49] Marushige et al.[32] suggested that during spermatogenesis in trout testis, a chromatin-bound proteinase hydrolyzed the histones bound to DNA, converting the nucleohistone into nucleoprotamine. Zirkin et al.[50] found that rabbit sperm nuclei contained an acrosin-like proteinase, suggesting that it removed nuclear protamines from DNA in matured spermatozoa during fertilization. As mentioned above, the increase of the turnover rates of histones was also found in the regenerating rat liver after partial hepatectomy.[47]

Earlier, Nakahara and Fukuoka[51] extracted a substance from various cancer tissues which was capable of depressing the catalase activity in the livers when injected into normal animals (catalase-depressing substance). They called it "toxohormone". Miyazaki et al.[9] found that the catalase-depressing substance was a nonhistone protein bound to chromatin, which had pI from 4 to 7. When fractions of chromatin or nuclei from rhodamine sarcoma are injected into normal rats, catalase and some other enzymes in the livers are influenced in the same manner as in the livers of tumor-bearing rats.[35-37,52] Among the influenced enzymes, the spleen-type isoenzyme of pyruvate kinase, which is abundant in the spleen, the kidney, and various tumors, but little in the livers of normal rats, increases significantly in the livers of rats injected with the tumor chromatin as well as in those of tumor-bearing rats. Miyazaki et al.[9] compared the contents of the catalase-depressing substance in the nuclei from various normal and cancerous tissues of rats, and found that the higher the content of the substance, the lower the catalase activity. The catalase-depressing substance was purified from rhodamine sarcoma up to a homogeneous purity, which had a molecular weight 60 k and pI of 5.0 (60-k protein). Nakamura et al.[53] found that in vivo, the substance depressed the synthesis of catalase, but it did not stimulate the degradation of the enzyme. These findings suggest that 60-k protein controls the extent of the expression of the catalase gene. Recently, Hagiwara et al.[8] found that when fractions of nuclei or chromatin from rhodamine sarcoma were incubated at pH 7 to 8, 64-k protein (pI 7.5), which was the major nonhistone protein in the tumor, was hydrolyzed into 60-k protein (pI 5.0) and then further hydrolyzed. The conversion of 64-k protein into 60-k protein was not observed during incubation at pH 9 to 10, suggesting that this limited proteolysis was catalyzed by neutral proteinase rather than by alkaline proteinase. Such a limited proteolysis has not been observed during hydrolysis of purified 64-k protein (pI 7.5) by purified alkaline or neutral proteinase at the neutral and alkaline pH ranges, suggesting the possibility that the binding of 64-k protein to chromatin was required for the limited proteolysis. Matuo et al.[54] found that the proteolytic fragments of 60-k protein formed in vivo, if their molecular weights were higher than 7 k, still had the in vivo liver catalase-depressing activity. This suggests that in tumor-bearing animals, the active fragments of 60-k protein produced in the tumor nuclei may be transported to the liver, so that they can depress the expression of the catalase gene in the nuclei.

At the present time, it is still premature to present any conclusive opinion on the functions of chromatin-bound alkaline and neutral proteinases. Further studies are required to clarify the physiological roles of these two proteinases.

REFERENCES

1. **Stein, G. S., Stein, J. L., and Thomson, J. A.,** Chromosomal proteins in transformed and neoplastic cells: a review, *Cancer Res.*, 38, 1181, 1978.
2. **Kleinsmith, L. J., Stein, J., and Stein, G.,** Direct evidence for a functional relationship between nonhistone chromosomal protein phosphorylation and gene transcription, in *Chromosomal Proteins and Their Roles in The Regulation of Gene Expression*, Stein, G. and Kleinsmith, L. J., Eds., Academic Press, New York, 1975, 59.
3. **Kranias, E. G., Schweppe, J. S., and Jungmann, R. A.,** Phosphorylative and functional modifications of nucleoplasmic RNA polymerase II by homologous adenosine 3′:5′ monophosphate-dependent protein kinase from calf thymus and by heterologous phosphatase, *J. Biol. Chem.*, 252, 6750, 1977.
4. **Sekimizu, K., Kubo, Y., Segawa, K., and Natori, S.,** Difference in phosphorylation of two factors stimulating RNA polymerase II of Ehrlich ascites tumor cells, *Biochemistry*, 20, 2286, 1981.
5. **Phillips, D. M. P. and Johns, E. W.,** A study of the proteinase content and the chromatography of thymus histones, *Biochem. J.*, 72, 538, 1959.
6. **Hagiwara, H., Miyazaki, K., Matuo, Y., Yamashita, J., and Horio, T.,** Novel protease bound with chromatins in normal and tumorous tissues of rats, *Biochem. Biophys. Res. Commun.*, 94, 988, 1980.
7. **Hagiwara, H., Miyazaki, K., Matuo, Y., Yamashita, J., and Horio, T.,** Purification and characterization of alkaline protease and neutral protease from chromatin of rats, *Biochim. Biophys. Acta*, 660, 73, 1981.
8. **Hagiwara, H., Miyazaki, K., Matuo, Y., Yamashita, J., and Horio, T.,** Properties and functions of two different kinds of proteinses bound with chromatins, (in Japanese), *Tampakushitsu Kakusan Koso*, 25, 434, 1980.
9. **Miyazaki, K., Nagao, Y., Matumoto, K., and Nishikawa, K., and Horio, T.,** Nuclear proteins capable of depressing in vivo liver catalase, *Gann*, 64, 449, 1973.
10. **Miyazaki, K., Hagiwara, H., Nagao, Y., Matuo, Y., and Horio, T.,** Tissue-specific distribution of non-histone proteins in nuclei of various tissues of rats and its change with growth of rhodamine sarcoma, *J. Biochem.*, 84, 135, 1978.
11. **Sarkar, N. K. and Dounce, A. L.,** Relationship of protease action and hydrodynamic shearing to cleavage of thymus deoxyribonucleoprotein, *Arch. Biochem. Biophys.*, 92, 321, 1961.
12. **Furlan, M. and Jericijo, M.,** Protein catabolism in thymus nuclei. I. Hydrolysis of nucleoproteins by proteases present in calf-thymus nuclei, *Biochim. Biophys. Acta*, 147, 135, 1967.
13. **Furlan, M. and Jericijo, M.,** Protein catabolism in thymus nuclei. II. Binding of histone-splitting nuclear proteases to deoxyribonucleic acid, *Biochim. Biophys. Acta*, 147, 145, 1967.
14. **Bartley, J. and Chalkley, R.,** Further studies of a thymus nucleohistone-associated protease, *J. Biol. Chem.*, 245, 4286, 1970.
15. **Garrels, J. I., Elgin, S. C. R., and Bonner, J.,** A histone protease of rat liver chromatin, *Biochem. Biophys. Res. Commun.*, 46, 545, 1972.
16. **Chong, M. T., Garrard, W. T., and Bonner, J.,** Purification and properties of a neutral protease from rat liver chromatin, *Biochemistry*, 13, 5128, 1974.
17. **Kurecki, T. and Toczko, K.,** Purification and partial characterization of protease from calf thymus chromatin, *Acta Biochim. Pol.*, 21, 225, 1974.
18. **Kurecki, T., Kowalska-Loth, B., Toczko, K., and Chmielewska, I.,** Evidence that neutral protease from calf thymus chromatin is a serine type enzyme, *FEBS Lett.*, 53, 313, 1975.
19. **Brandt, W. F., Bohm, L., and Von Holt, C.,** Proteolytic degradation of histones and site of cleavage in histone F2a1 and F3, *FEBS Lett.*, 51, 88, 1975.
20. **Carter, D. B. and Chae, C.-B.,** Chromatin-bound protease: degradation of chromosomal proteins under chromatin dissociation conditions, *Biochemistry*, 15, 180, 1976.
21. **Carter, D. B., Efird, P. H., and Chae, C.-B.,** Chromatin-bound protease: [^{3}H]diisopropyl fluorophosphate labeling patterns of chromatin, *Biochemistry*, 15, 2603, 1976.

22. **Eickbush, T. H., Watson, D. K., and Moudrianakis, E. N.,** A chromatin-bound proteolytic activity with unique specificity for histone H2A, *Cell*, 9, 785, 1976.
23. **Ramponi, G., Nassi, P., Liguri, G., Cappugi, G., and Grisolia, S.,** Purification and properties of a histone-specific protease from rat liver chromatin. Effect on acylated histones, *FEBS Lett.*, 90, 228, 1978.
24. **Suzuki, Y. and Murachi, T.,** A chromatin-bound neutral protease and its inhibitor in rat peritoneal macrophages, *J. Biochem.*, 84, 977, 1978.
25. **Lipinska, A. and Klyszejko-Stefanowicz, L.,** The activity of chromatin-bound protease extracted selectively with histone H2B from calf thymus and rat liver, *Int. J. Biochem.*, 11, 299, 1980.
26. **Tsurugi, K. and Ogata, K.,** Studies on the serin proteases associated with rat liver chromatin, *J. Biochem.*, 92, 1369, 1982.
27. **Heinrich, P. C., Raydt, G., Puschendorf, B., and Jusic, M.,** Subcellular distribution of histone-degrading enzyme activities from rat liver, *Eur. J. Biochem.*, 62, 37, 1976.
28. **Destree, O. H. J., D'Adelhart-Toorop, H. A., and Charles, A.,** Cytoplasmic origin of the so-called nuclear neutral histone protease, *Biochim. Biophys. Acta*, 378, 450, 1975.
29. **Paik, W. K. and Lee, H. W.,** Enzymatic hydrolysis of histones in rat kidney microsomes, *Biochem. Biophys. Res. Commun.*, 38, 333, 1970.
30. **Lindsey, G., Thompson, P., and Von Holt, C.,** Lysosomal origin of chicken erythrocyte nuclear protease, *FEBS Lett.*, 135, 81, 1981.
31. **Watson, D. K. and Moudrianakis, E. N.,** Histone-dependent reconstitution and nucleosomal localization of a nonhistone chromosomal protein: the H2A-specific protease, *Biochemistry*, 21, 248, 1982.
32. **Marushige, K., Marushige, Y., and Wong, T. K.,** Complete displacement of somatic histones during transformation of spermatid chromatin: a model experiment, *Biochemistry*, 15, 2047, 1976.
33. **Inagaki, T., Miura, K., and Murachi, T.,** Identification of a protease inhibitor from rat peritoneal macrophages as poly(ADP-ribose), *J. Biol. Chem.*, 255, 7746, 1980.
34. **Okayama, H., Ueda, K., and Hayaishi, O.,** Purification of ADP-ribosylated nuclear proteins by covalent chromatography on dihydroxyboryl polyacrylamide beads and their characterization, *Proc. Natl. Acad. Sci. U.S.A.*, 75, 1111, 1978.
35. **Matuo, Y., Nishikawa, K., Horio, T., and Okunuki, K.,** Effect of growth of rhodamine sarcoma in rat on some liver enzyme activities, *Gann*, 59, 299, 1968.
36. **Nakamura, T., Matuo, Y., Nishikawa, K., Horio, T., and Okunuki, K.,** Effect of growth of rhodamine sarcoma in rat on isozymes of liver enzymes, with reference to the seesaw change of two pI-isozymes of glucose-6-phosphate dehydrogenase, *Gann*, 61, 177, 1970.
37. **Muroya, N., Nagao, Y., Miyazaki, K., Nishikawa, K., and Horio, T.,** Pyruvate kinase isozymes in various tissues of rat, and increase of spleen-type pyruvate kinase in liver by injecting chromatins from spleen and tumor, *J. Biochem.*, 79, 203, 1976.
38. **Herzfeld, A. and Greengard, O.,** The effect of lymphoma and other neoplasms on hepatic and plasma enzymes of the host rat, *Cancer Res.*, 37, 231, 1977.
39. **Nagao, Y., Arakaki, K., Nakamura, T., Hagiwara, H., Miyazaki, K., Matuo, Y., Yamashita, J., and Horio, T.,** Effect of partial hepatectomy and tumor-bearing on phosphorylation of nuclear protein in rats, *Gann*, 73, 549, 1982.
40. **Liau, M. C. and Perry, R. P.,** Ribosome precursor particles in nucleoli, *J. Cell Biol.*, 42, 272, 1969.
41. **Giese, G. and Wunderlich, F.,** Reduction of the *in vitro* nuclear export of ribosomal precursor particles by protease inhibitors, *Cell Biol. Int. Rep.*, 5, 1039, 1981.
42. **Tsurugi, K. and Ogata, K.,** Degradation of newly synthesized ribosomal proteins and histones in regenerating rat liver with and without treatment with a low dose of actinomycin D, *Eur. J. Biochem.*, 101, 205, 1979.
43. **Tsurugi, K. and Ogata, K.,** Presence of a thiol protease in regenerating rat-liver nuclei: partial purification and some properties, *Eur. J. Biochem.*, 109, 9, 1980.
44. **Brown, R. L., Clark, R. W., Chiu, J.-F., and Stubblefield, E.,** Protease activation of G1 nuclei isolated from chinese hamster fibroblasts, *Exp. Cell Res.*, 104, 207, 1977.
45. **Chae, C.-B., Smith, M. C., and Morris, H. P.,** Lack of relationship between activity of chromatin-bound protease and cell growth rates, *Biochem. J.*, 146, 281, 1975.
46. **Procaccini, D. J., Procaccini, R. L., and Pease, J.,** Histone hydrolase activity in regenerating limbs of the adult newt, *Diemictylus viridescens*, *Oncology*, 29, 265, 1974.
47. **Garrard, W. T. and Bonner, J.,** Changes in chromatin proteins during liver regeneration, *J. Biol. Chem.*, 249, 5570, 1974.
48. **Hancock, R.,** Conservation of histones in chromatin during growth and mitosis *in vitro*, *J. Mol. Biol.*, 40, 457, 1969.
49. **Dice, J. F. and Schimke, R. T.,** Turnover of chromosomal proteins from rat liver, *Arch. Biochem. Biophys.*, 158, 97, 1973.

50. **Zirkin, B. R., Chang, T. S. K., and Heaps, J.,** Involvement of an acrosin-like proteinase in the sulfhydryl-induced degradation of rabbit sperm nuclear protamine, *J. Cell Biol.*, 85, 116, 1980.
51. **Nakahara, W. and Fukuoka, F.,** A characteristic toxic substance produced by cancer tissues, *Gann*, 40, 45, 1949.
52. **Kannan, Y., Nishikawa, K., Matuo, Y., and Horio, T.,** Intracellular distribution of *in vivo* liver catalase-depressing substance in rhodamine sarcoma, *Gann*, 63, 201, 1972.
53. **Nakamura, A., Miyazaki, K., Nagao, Y., and Horio, T.,** Effect of injection of nuclear fraction from rhodamine sarcoma on turnover of liver catalase, *Gann*, 66, 75, 1975.
54. **Matuo, Y., Nishikawa, K., and Horio, T.,** Molecular size of *in vivo* liver catalase-depressing substance from rhodamine sarcoma of rat, *Gann*, 63, 95, 1972.

INDEX

A

B

C

D

E

F

G

H

I

K

L

M

N

O

P